Readings in
Psychological
Development
through Life

Readings in Psychological Development through Life

Edited by

Don C. Charles
Iowa State University

William R. Looft
Pennsylvania State University

HOLT, RINEHART AND WINSTON, INC.
New York   Chicago   San Francisco   Atlanta
Dallas   Montreal   Toronto

Preface

Why another book of readings? Such volumes have proliferated in recent years to an almost unbelievable number. The reason for this heavy production is, of course, instructor demand. If teaching a course by text and lecture alone was ever defensible, it is not today. Psychology is too complex and research approaches are too various for students to be exposed only to the generalizations found in textbooks. But many new and smaller schools lack adequate library facilities, and in the larger schools, undergraduate classes numbering in the hundreds render any library facilities inadequate. Hence the demand for books of readings: This is as true in developmental psychology as in any other area.

The production of this particular book of readings depends in part on the above conditions, but more on the editors' view of a specific, unmet need. The psychology of human development often is concerned only with infancy, childhood, or adolescence. But today, to an increasing degree, developmental psychologists are truly concerning themselves with developmental change, rather than with the static characteristics of a single stage of life. In addition, a variety of factors—not the least of which is the aging of subjects in several longitudinal studies—work to extend the range of the psychologist's interest over the whole lifespan. It is the purpose of this book of readings, therefore, to present theory and research with major concern for change, and with an emphasis on the usually neglected mature and older years.

Since this collection of papers is less narrowly focused than many, some range of difficulty will be found in the contents. Many, perhaps most, of the papers will be meaningful to the undergraduate with at least an introductory course in psychology behind him. A few selections will be challenging to advanced undergraduate or graduate students. We hope thus to stimulate and inform the relatively unsophisticated reader, and yet provide some thoughtful material for the more mature student (or professor). We hope therefore that the book, like its subject matter, will be useful over a wide range of individual scholarly development.

While these readings originally were planned to accompany a specific text (*Psychological Development through Life* by Don C. Charles, New York, Holt, Rinehart and Winston, Inc.), they are not keyed to specific chapters but rather to broad areas of psychological concern, and thus are usable in any developmental course. Five major areas of study are explored. First is an introduction to the field of developmental psychology, presenting papers concerned with theories and with methods. Second is the physiological basis of behavior and development, an often neglected aspect of developmental study. Third is the currently popular area of perceptual and cognitive development, including learning and language study. Fourth is the category of ability and achievement, emphasizing adult competence and functioning. Last comes the broad topic of social behavior and adjustment, again emphasizing the lives of adults in our society.

October 1972
Ames, Iowa   D. C. C.
University Park, Pennsylvania   W. R. L.

Contents

Readings in Psychological Development through Life

Part One

Introduction: The Field of Developmental Psychology

Developmental psychology may be broadly defined as the study of behavioral changes associated with age changes in human beings. But such a general statement does not tell us much about what valuable information we can really expect to learn through study of the discipline. To delineate any discipline, or division of a discipline, several possible approaches may be undertaken. One is to do a historical study, revealing how and into what form the field has evolved. Another approach is simply to define it by exposition. Still another approach is to examine the problems attacked and methods attempted in seeking solutions. The articles included in this section touch on all three approaches, but deal mostly with problems and methods.

Interest in human development antedates this century, but as a scientific discipline its history is essentially a twentieth-century phenomenon. Since no review of appropriate brevity is available for these readings, the gross outlines of historical influence on contemporary developmental psychology will be sketched here. The reason for our concern with history is, of course, that contemporary developmental psychology, like the persons it studies, is what it is because of what it has been. A discipline's strengths and weaknesses reflect the influences that formed it.

Late nineteenth-century and early twentieth-century studies of infants were descriptive, often-romanticized, "baby-biographies." Some reached a useful level of descriptive precision around the turn of the century. Then came G. Stanley Hall's "genetic psychology" movement, a bizarre collection of notions summed up in the catch-phrase, "ontogeny recapitulates phylogeny." The central idea, somewhat oversimplified, was that after conception each individual replicated first the biological evolution of man, and then the social evolution through his prenatal period and early developmental years. The lack of scientific support for the concept became evident before World War I, but the erroneous approach did stimulate scientific study of developing children. Clinical work with children, especially retarded and troubled youngsters, served to focus attention on the unique individual. Thus from the beginning, description and a helping attitude were central to psychological concern with development. The mental measurement movement, peaking in the 1920s and 1930s, served to quantify description. Educational psychology, whose growth paralleled child psychology, helped to maintain the focus on school-age children, as did the establishment of numerous child-study clinics in universities after World War I.

Concern with adult development and decline interested few researchers until after World War II; there were exceptions, of course, but except for college sophomores fortuitously available as subjects, most "developmental" psychology was child—or occasionally adolescent—centered. Postwar concern was stimulated primarily by the problems of the aged, rather than by any intrinsic interest in the developmental characteristics of mature human beings.

Developmental psychology by the 1950s was a discipline concerned primarily with school-age children and their problems (and thus was usually identified as "child psychology"). Its virtues lay in a concern for the individual and his problems, while its deficiencies were largely those of lack of scientific rigor. Descriptive research of earlier decades provided the necessary base of data on which to build, and the individual, sometimes clinical, focus provided a rich source of hypotheses to stimulate research. For the past two decades, for a variety of reasons, both the quantity and quality of research has increased. The quantity of work reflects in part the general vigor and growth of psychology, as well as a rise in interest in human development per se. Improved quality reflects concern of researchers with a greater variety of laboratory and field investigative techniques than were used in the past, as well as greater skill in their implementation. Developmental psychology, like other disciplines, has profited from computer and other improved data evaluation techniques. While most research has continued to focus on specific age groups, that is, six-year-olds, twenty-year-olds, sixty-five-year-olds, there is an increasing tendency for developmental psychologists to be truly developmental, that is, to look at *change* rather than at the static characteristics of a given age group, and to extend the interest in change to cover the entire lifespan. The lifespan emphasis seems to have risen out of data accumulated by certain psychologists who continued working into maturity or older age with subjects they originally studied as children. The continuities and discontinuities of change in certain aspects of the development of *persons* through their lives, rather than *populations* of subjects of different ages, provided both intriguing evidence and intriguing new problems to consider. This is not to say of course that *only* longitudinal or other studies designed to examine change are useful in understanding development. Age-specific study, whatever the intent of the researcher, may still provide insight into developmental phenomena or laws.

Concern with change, rather than with static description, calls for theory or model to make the data meaningful, to provide some structure to fit it into. Thus recent years have seen increasing concern for theory and less satisfaction with empirical evidence by itself.

We do not, of course, have any grand, all-encompassing schemes, or "schools" like those existing earlier in the history of psychology. Psychology in general and lifespan developmental psychology in particular have proved too complex and multifaceted to yield to even an elaborate theory construction. Of the major extant theories, psychoanalysis has concerned itself almost exclusively with emotional life, behaviorist formulations with learning, Piaget-Geneva School with cognition, and none has attended much to life beyond adolescence. Thus developmental psychologists, like others, tend to work with models or theories limited to that aspect of behavior under consideration; indeed, in many cases, the researcher views his task primarily as an empirical one.

Lifespan psychology is in a youthful stage of its own development, and the psychologists devoted to its study are consequently highly concerned with formulating research methods and eliciting meaning from data accumulated in the past. Part One, then, reflects the concern of certain psychologists for developmental meaning. Nancy Bayley and Mary Cover Jones, of the Institute of Human Development, University of California, Berkeley, have both been involved since the earliest days in what are perhaps the most comprehensive of all longitudinal studies, continuing at that institution for over four decades. Bayley discusses some problems and approaches in lifespan research, while Jones reviews the beginnings and evolution of the three Institute longitudinal studies

The late John E. Anderson brought the Institute of Child Development at the University of Minnesota to a position of eminence through his long tenure as director there. He was also an early leader in the psychological study of aging. Here he presents a model for aging, with applicability to all of life. Jerome Kagan was involved in longitudinal research at the Fels Institute, Yellow Springs, Ohio, before moving on to Harvard. In his piece he concerns himself with the problems that the absolutism of an earlier day has caused, and he suggests some remedies. Sheldon H. White, of the Laboratory of Human Development at Harvard University, presents an overview of the contributions to our thinking about the nature-versus-nurture issue as presented by a number of major figures in American psychology—G. S. Hall, James, McDougall, Watson, and Thorndike. In the last article of Part One, Sidney Bijou, of the University of Illinois, discusses the importance of study of the continuous processes of change and calls for deemphasis of the stage approach.

One

The Lifespan as a Frame of Reference in Psychological Research

Nancy Bayley

The study of change in persons over time is recognized as an important aspect of psychological research and theory, concerning both children (development) and the elderly (decline). There is less heed paid, however, to the evidence that the processes of change over the lifespan are continuous even during the relatively stable adult years. Such evidence indicates that any segment of the lifespan should be considered appropriate for the investigation of psychological change in relation to age. Furthermore, these investigations may appropriately be directed toward practically all major fields of psychology: experimental, learning, personality, emotions, social, clinical, educational, and very obviously in such areas as intelligence and motivation.

Certain methods and procedures of study, it is true, are age-specific; and to this extent the particular tools and procedures to be applied in studying changes over time must differ with the age of the subjects. Yet the tools that have been devised for the study of the young, the middle-aged or the old can often be adapted for other ages, or they can be used in ways that facilitate the observation of changes with age. What I should like to propose, therefore, is that investigators in the various fields of psychology seriously undertake to include in their research designs, and more particularly in their theoretical constructs, one frame of reference that is concerned with the processes of change with age.

THE CONCEPT OF MATURITY

Actually, many who study either children or the aging are interested in the entire lifespan. Questions are repeatedly raised about what a given developmental trend will lead to; or what may have preceded a given

From *Vita Humana*, 1963, 6, 125–139. Published with permission of author and publisher, S. Karger, Basel/New York.

Adapted from the Presidential Address, Division 20 of the American Psychological Association, August 29, 1958.

adult behavior pattern. There are also questions of the rates at which children approach mature status in a given function, and whether changes in senescence parallel, or are of the same character as, changes in childhood. For those whose interests are primarily with adults, much of the recent emphasis in research is on the nature of changes from mature to old, that is, the process of aging. This kind of research implies a concept of the mature condition as a point of reference against which the processes of growth and of aging can be juxtaposed for comparison.

If this is so, then we should inquire into the nature of such a reference point, to consider just what is meant by the term "mature" as it applies to behavioral processes and functions. This consideration leads to the further questions of when, under what conditions, and in what ways a person is to be considered mature. Maturity as a general concept applied to human adults is neither a specific point in time nor a static condition that extends over a span of years, but is rather a complex series of ever-changing processes. There may, however, be long periods of relative stability in a given process or function once that function has reached its full development.

For many behavioral functions it is difficult to know just when they become mature. It is difficult because there are individual differences at all ages in the degree, or quality attained in most kinds of behavior. For such characteristics we cannot tell whether a person has reached his own most mature functioning in a given structure or function except in relation to his own performances at other periods of his life. For example, with few exceptions the epiphyses in the long bones of the hand and arm close toward the end of the second decade of life, and for certain purposes, this condition may be called "mature." When we measure intelligence, however, there is no point or stage of adult functioning which is reached by all persons, no level to which one can point and say that now the person's intelligence is mature. If a point is picked arbitrarily, there will be many who never reach it, and many who will go far beyond.

Furthermore, there is evidence that different abilities and qualities of the personality become mature at different times in the life of the individual. Each characteristic may be thought of as having its own schedule. Some functions mature early in life, others late. For example, the well-known studies of Lehman (1953) show differing ages at greatest intellectual achievement for mathematicians, for poets, for historians, and for philosophers. Some abilities, including those that draw heavily upon vocabulary and upon accumulated knowledge, maintain a high degree of efficiency for a long time. Others tend to deteriorate rapidly after reaching an early peak. This latter course of rapid growth and decline tends to be true of certain kinds of physical or athletic abilities that require a combination of strength, speed, and skill. For some behaviors (that is, such personality factors as emotional independence) it is doubtful whether the term "mature" can be applied, even though reference is often made to persons' maturity or immaturity in respect to them. To the extent that maturational changes do occur in these respects, they may be very subject to environmental conditions that determine when, indeed whether, they ever mature.

This train of thought leads to the necessity, for a comprehensive approach to the study of behavior, that we consider the whole life process as a frame of reference. Any behavior that is being studied, if it is to be adequately interpreted, must be seen in reference to the age or ages of the individuals under study and to their probable status in the developmental cycle. For example, in studying learning, not only the fact that the subject is animal or human, but also the fact that he is six months, six years, twenty-six years, or sixty years old will determine the conditions of the experiment, the kinds of behaviors to be studied, and the selection of suitable rewards or punishments. In general, change is most rapid and most obvious at the two ends of the lifespan, in infancy-childhood and in old-age senility. However, even though it tends to be forgotten, and even though the various behavioral functions of the young adult are often treated as stable, change is continuous, right through the "mature" adult period. The processes of maturation of growth and subsequent decline never cease, whatever the structure or function being considered.

AGE AS A VARIABLE

For many practical reasons, it is useful and convenient in psychological experiments to study a function in its current condition, and to put aside for the time being any consideration of change over time. The nature of the function itself is the object of study. An experiment is set up to investigate learning, or perception, or aggression, or fear, or maternal behavior, or peer-group interactions; and the experiment is carried out on some selected population, usually all of whom are about the same age. This is obviously a procedure to be encouraged. It is important and necessary, and it provides the bulk of the experimentation in psychology. Psychological processes are so complex that they must at certain stages in knowledge be broken down into small, meaningful segments that can be studied in relative isolation.

A serious limitation of much of this type of research, however, is that the investigator often forgets to take into account the context from which he has abstracted. He continues to study various aspects of a specific bit of behavior, using essentially the same population of subjects for each new experiment. He then presents his findings as isolated items of information; or (often by implication only) he generalizes his findings and assumes that what he has learned, for instance, about group leadership in twenty-year-old males, or about shock avoidance learning in a given strain of albino rats, will apply equally to humans of all ages.

The inadequacy of this form of reasoning seems obvious when it is pointed out. Nevertheless, it tends to persist in practice. What is needed, perhaps, is a broader interpretation of the concept of comparative psychology. This term is usually applied to the study of animals as it throws light upon human behavior. To this end our animal experiments and observations are compared with pertinent aspects of human behavior. Investigators must be careful in making comparative interpretations to take account of the differences as well as the similarities between the animal under study and the human. Consistent with this reasoning, it should be kept in mind that if we study, for example, certain social responses of college sophomore men, then the results of such a study may apply to other male college sophomores, but not necessarily to populations who are older or younger, or females, or differently educated, or from a different cultural or ethnic group.

The relevant characteristics of a population to be studied for a given psychological function are legion. It should go without saying that if these relevant characteristics are not controlled or accounted for, the experimenter's conclusions may go far astray. If, then, the investigator is constantly alert to seek out and to identify these relevant factors, he will continue to find new and often unexpected ones that must be taken into account.

The earlier studies on the effects of environment on mental growth, for example Bayley and Jones (1937), were concerned with obvious variables such as years of schooling, and types of occupation. They took no cognizance, however, of some of the more subtle variations such as emotional climate in the home, or the impact of different cultural value systems upon intellectual motivation. Such factors as the latter are now regularly considered to be environmental variables with which one must reckon. The monograph of Sontag, Baker, and Nelson (1958) is a good example of a study in which parental attitudes and behaviors were included in investigating the causes of change in I.Q.

We are only beginning to be aware of the possible effects on various kinds of mental functioning of *age differences* in such things as motivation, the quality of the subject's education, and of his accumulated life experiences. This is an area that is open for much exploration and one that has great potentialities. We have learned through experience with infants and young children to adapt experimental procedures to their interests and capacities. We find, for example, that children usually cannot be motivated for learning tasks with the same lures as are used for monkeys; and that the verbal incentives used to encourage school children to strive for success in intellectual tasks are inapplicable with six-month-olds. We have not yet considered very seriously the need to find out, or to utilize in our experimental designs, what we already know about pertinent aspects of older persons' motivations, preoccupations and physical limitations.

For example, in studying the nature of intellectual functions and in developing tests for measuring intelligence we have become aware of the need to adapt the specific test items and procedures to the different cultural and socioeconomic populations to be studied. In the same way, we need to make the test items and procedures age-appropriate, to make them more relevant to the motivations, the life situations, and the goals of persons at each period in the lifespan. In this respect we have done fairly well with infants, children, and youths, but we have been remiss with intelligence tests for older people. It is customary to apply to older populations those tests that were originally designed for young (usually college-age) adults. In doing so, we have overlooked important areas in motivation, preoccupations, and daily experiences that are most common in older people. Consequently our interpretations of the intellectual processes and capacities of older people may well be faulty.

Pressey has been concerned with this problem, and he and Demming (Pressey and Demming, 1957) have devised tests of intelligence that are especially designed for measuring the abilities of older persons. Their results show consistently higher scores for populations over thirty than are found with the conventional intelligence tests. As has been noted before, whether measured intelligence decreases, increases, or remains unchanged after age of twenty-five, is at least in part a function of the tests used. We should keep in mind the question whether the tests are equally "fair" (in terms of meaningfulness and motivating power) at all ages.

The foregoing is but one way in which psychological researchers need to gain sophistication about the conditions that may be important in determining behavior. Let us consider some others which could profitably be reviewed through a lifespan orientation.

When measuring a behavior out of context, as in an "unnatural" laboratory environment, we may inadvertently change the relevant aspects of the conditions for response so that the particular responses are not representative of the behaviors we started out to measure. This has been pointed out by animal ecologists, who have shown many differences between the behavior of the rat or the monkey, for example, when in its natural "wild" environments and when in its laboratory cages. It is often necessary repeatedly to check the results of laboratory experiments by returning to natural life situations for further observations of the behaviors under study. What is more, the relative importance of the laboratory-natural life differences may shift with the populations under study, and with the age at which an animal started his life in a cage.

The point to be made here is that we have tended to neglect the important environmental differences that are related to the age of the persons under study. Some of these differences are obvious; others are less so; and some may not become evident until revealed by further studies. Let us consider some of the factors, both external and internal, as they relate to the infant under one year, the preschooler, the school-age child, the adolescent, the young adult, the middle-aged, the elderly, and the senescent. Within the organism itself there are such age-related variables as sensory acuity, energy, motor coordination and agility, knowledge and experience (or apperceptive mass), awareness of and outlook toward life expectancies, and attitudes toward other people. Also, as Barker and his associates (1955) have shown, social interactions, and roles change markedly with age. Many of these conditions are related, as both cause and effect, to the attitudes toward and treatment of these individuals by those persons with whom they interact.

Cultural stereotypes attribute different characteristics to the different age groups, but an individual may not see himself as having these conventionally assigned attributes. Depending on the nature of the problem, any of these and other variables could, if not controlled or taken into account, influence the data of an experiment.

LONGITUDINAL STUDIES

Nothing has thus far been said about longitudinal studies, in which the same subjects are remeasured or retested at successive intervals as they grow up or grow older. The points already mentioned could apply to either cross-sectional or longitudinal studies. However, some kinds of inquiry are better adapted to one or the other of these methods.

There are certain kinds of information, of particular relevance to the lifespan frame of reference, that can be obtained only by means of longitudinal studies. Fortunately, a longitudinal study need not span the entire life cycle of any group of persons. Of necessity, we deal with portions and segments, usually a few years, of our subjects' lives. The longer the time period over which the researcher can reassess the same population, the more we can learn about individual patterns of growth and change. By studying the same persons over time, we can begin to see sequences of etiological complexes. We become alerted to dynamic interactions between the organism and his environment. When a person is studied only once, as he appears in relation to the rest of the sample, there is a tendency to make static interpretations and evaluations. Subject X, for instance, on the basis of a single set of observations and tests, is defined as being anxious, as lacking in energy, as having an average I.Q., as having a slow reaction time. Observed repeatedly over time, this same subject will be seen to change in many of these respects. We find we must look for the reasons for the change both in X and in his environment. Why, for example, was he anxious at one time, but less so at another time? We become able to note such things as the effects of long-continued emotional climates as they operate to decrease or to enhance intellectual functions.

Just such a study as this was made recently by Sontag and his associates (1958) on records from the Fels Institute population. By observing growth changes over time, we learn something of the processes of maturation of inherent characteristics, and we are able to study the effects of specific kinds of experiences on the course of development. Another study relevant to this point is that of Macfarlane, Allen, and Honzik (1954) in which the incidence of problems reported in a longitudinal sample was found to change over time with the age and relative maturity of the children. A series of studies by Bayley and Schaefer, so far published only in part (Schaefer and Bayley, 1960; Bayley and Schaefer, 1960a, 1960b) indicate changes in both maternal behaviors and child behaviors over time.

As already mentioned, most studies of this kind have been made for those parts of the lifespan in which change is most rapid. We now have a considerable accumulation of data on development of the child, and we are rapidly getting data on some aspects of the retrogressive processes in the elderly. Examples of the latter are the studies of Kallmann and his associates (1951, 1961) of aged twins.

The processes of change and of adaptations during the middle years are much less adequately documented. However, several populations who were studied as children are now being studied as young adults. Perhaps the pioneer study in this respect is Terman and Oden's series of reappraisals of gifted children as adults (1947, 1959). With many of these subjects being studied again at an average age of about fifty-five, the study has covered a span of almost forty-five years in the lives of these subjects. In several studies, such as those of Lowell Kelly (1955) and of Owens (1953), adult populations have been remeasured after long intervals. Other follow-ups of adult populations first studied as children are based on research programs that were started somewhat more recently, and most of the subjects, therefore, are at present in their thirties. These include the three growth studies at Berkeley, California (for example, Tuddenham, 1959), the Minnesota adult studies of populations studied as school children (Anderson and Harris, 1959) and, to some degree, the continued observations of subjects at the Fels Institute and at the University of Colorado. In most of these studies there are large accumulations of records, including both mental and physical data.

From these various populations, we should soon be getting much valuable information about relationships and growth trends between childhood and adulthood for both mental and physical characteristics. The very fact that these are longitudinal studies gives assurance that the research programs will be oriented toward processes of change with age. There will be an effort made to discover relationships between adult mental processes (both intellectual and emotional) and all available childhood records that might be relevant, concerning both the children themselves and their childhood environments.

Considerable thought and effort should be put into the planning of adult follow-up programs. For example, there is the importance, in assessing intelligence, of maintaining continuity by use of the same instrument at different ages. The fact that the test is objectively the same (for example, the Terman Group Test, or the Stanford-Binet, or even the Wechsler Scales) does not insure that the tests tap the same, or the most relevant, intellectual functions at all ages. In order to study age changes in intellectual capacity, it will be necessary to devise and standardize (on a cross-sectional sample) a series of tests that are designed to secure high motivation and to be valid measures of ability for several different adult age-groups. If we can devise and standardize such a series of age-appropriate scales, which for convenience may be referred to as tests B, C, D, and E, then in longitudinal studies it might be possible to evaluate changes in intellectual functions by giving these tests in a succession of partial repetitions. That is, if, in the original study, some standard test, A, had been given at age sixteen and earlier, the first of the adult follow-ups might include test A and the new test B. Subsequent rounds of testing could follow some such pattern as: tests B and C, A and C, B and D, D and E, and C and E. Such a procedure would keep the testing sessions from becoming too long at any one time, and would have the further advantage of dealing with test material that would always be appropriate for the subjects.

There may be less of a problem in adapting, for age-appropriateness, such instruments as projective tests and other measures of emotional and personality variables. However, it is pertinent to investigate this aspect of any test one plans to use. Again it may be pointed out that this kind of age-adjustment in personality tests has been considered more often in making adaptations for young children than for older adults.

RELATIONS BETWEEN GROWTH AND DECLINE

Because the beginning and end of the life cycle are both periods of rapid change, there is a tendency to compare the two periods and to look for similarities in them. Also, because the deterioration in abilities makes the older person more dependent and in many ways more "childlike," there is the temptation to consider these deteriorative processes as direct reversals of developmental processes. There are, however, many differences. The young nervous system and the entire anatomy of the child are very different from those of the aged individual. The reasons for less than mature or optimal functioning are very different in the young and the old. The processes of change have different rates, and they appear to be controlled by very different, perhaps quite uncorrelated, conditions.

An area of research that has been little explored, but that could prove to be very fruitful, is that of comparisons between infancy and senescence —or, to expand the range, the total period of growth and of decline. We may inquire into the nature of individual differences in rates of growth versus rates of decline, and ask what are the determinants of these opposing processes. In what ways do they seem to be interdependent? Do those persons who mature early in a function show earlier decline, possibly a shorter lifespan, a shorter span of mature function, or a more rapid decline once the senescent process has started? It is possible that the rates of change in the three stages (growth, maturity, and decline), are quite unrelated, or that there are differently correlated rates in different processes.

Across species there is clear evidence that animals with shorter lifespans have relatively rapid growth and early decline. Whether or not this is true within a given species is another question, although it seems plausible that this might be so. We can find in humans some meager evidence that seems to support such a hypothesis. For example, in several longitudinal studies of early behavior development (Bayley, 1933; Furfey and Muehlenbein, 1932), there is a slight negative correlation between mental test scores earned in the first six months of life and scores earned after three years. A study of Bantu infants (Geber and Dean, 1957) reports great precocity, by our test standards, in infants under six months; but this precocity diminishes with age, and disappears by the time the children are two or three years old. There is some evidence that children who by school-age score highest in intelligence often start slowly, and are relatively retarded in the first six or eight months of life. At the other end of the lifespan, a number of investigators (Jones, 1955) have found relatively greater decrement in scores earned by those persons in the lower educational and occupational levels. None of these studies has been carried out with sufficient controls to give clear-cut answers to the question of possible relations between rates of change and levels of ability. Some full lifespan studies may be necessary before we can know whether those slow-starting babies are the same persons who continue to be intellectually alert in old age. Other conditions, such as adult life experiences, could be the important determinants of rates of senescent decline.

CONSTANCIES OF PERSONALITY

Another recurring concern of those who study the same persons over time is to identify those basic, underlying constancies of personality that characterize the individual, and to chart the course of their developmental changes. Many behaviors, as well as physical characteristics, change with age. Being age-specific, a behavior that is appropriate at one stage of development is often very inappropriate at another. A two-month-old is expected to communicate his wants by crying; a four-year-old, by gestures and words. An age-reversal of these behaviors would certainly be remarkable. The manner of expressing one's emotions changes with age, and in order to evaluate degrees of "emotionality" at different ages, we must work out a means of relating these emotional behaviors to scales of intensity that include the age-appropriateness of each behavior pattern.

In many respects there appears to be no continuity in traits or reaction-tendencies, and no possibility of predicting from observations made at one age what can be expected in the same person a few years later. Yet in spite of the fact that our measures often indicate little or no consistency, we intuitively feel, and believe that we recognize, constancies of personality and behavioral tendencies. The problem is to be able to devise ways of documenting and scoring these characteristics so that they may be compared, evaluated, and used to predict later patterns of behavior.

This is an important and baffling problem, for example, in studies of the etiology of schizophrenia and other mental disorders. Studies of run-of-the-mill behavior problems in childhood (for example, Macfarlane, Allen, and Honzik, 1954) have shown that for the most part problems tend to be age-limited. Certain problem behaviors occur at certain ages, and not at other ages. One set of problems is "out-grown," and replaced by another set. The occurrence of a problem in a child at one age does not usually signify that this child will continue to have problems, even in a different guise, at his next stage of development.

However, some basic reaction-tendencies, or deviations in patterns of sensitivity, seem to make some people more and others less vulnerable to environmental hazards. If these individual differences in sensitivity or vulnerability persist, perhaps they can be measured and studied in relation to their age-related manifestations, and to tendencies toward given patterns of pathology. The framework within which to look for these underlying constancies may be available in Freudian theory, or in some variant conception of dynamic developmental processes; or it may be found in such physical variables as hormonal or biochemical balance, or in inherent genetic characteristics. Schaefer and Bayley (1963), for example, found for the Berkeley Growth Study children that the most consistent behaviors observed were activity and speed of action. These behaviors in the first two years were correlated with positive task-oriented behaviors in boys through age twelve, but for girls the *r*'s were negative with task-oriented behaviors in childhood, though positive with adolescent extraverted, aggressive behaviors. These correlational trends point to the further probability of sex differences in the processes of change or stability over time.

There is thus the hope that we may find persistent underlying attitudes and reaction-tendencies that determine indentifiable sequences of age-related or maturity-related behavioral patterns. In the past, the efforts to identify such tendencies and patterns have been mostly by means of retrospective studies of the mentally ill or the emotionally disturbed. This approach has been found inadequate because of both the atypical selection of subjects, and the selective biases in the subjects' memories. Researchers in this field are now turning more often to longitudinal studies (which are prospective rather than retrospective), or to the records of extant studies, to check their hypotheses.

In the analysis of personality factors and of mental pathologies the age-specific behavior tendencies may well be studied profitably by cross-sectional methods. But the durability and persistence of the basic personalities can be learned only by the more tedious longitudinal process.

In this paper I have touched upon a variety of ways in which it seems to me that psychological research will benefit when it is planned, carried out, and interpreted within the frame of reference of the lifespan and of the continuous processes of change that characterize all behavior. This frame of reference will affect our understanding of the concept of maturity; it will emphasize the relevance of changing functions, experiences and motivations for any particular behaviors under consideration; and it will help direct attention to ways of improving various testing instruments.

The longitudinal studies lend themselves to investigation of many important questions that can only be answered definitely by observations of the same persons over time. We should continue to explore ways in which to utilize more fully the data from the longitudinal studies that are now in progress, as well as to institute new ones. One relatively unexplored field in longitudinal research is the comparison of the processes of development with those of decline. Another important but difficult problem is to identify underlying persistent and predictable qualities of personality, to chart their course over the lifespan, and to study their causes and their responsiveness to re-education. And, of course, the last word has by no means been said about age-trends in intellectual capacities. Better tools can give us much better information about the nature of intelligence.

Let us hope that there will be others, in addition to those whose research is specifically oriented toward processes of change in the young and the old, who will work within this frame of reference in carrying on their researches. More adequate knowledge of the processes of behavioral change with age should serve to improve both the science and the practice of psychology.

SUMMARY

Psychological theory and research will benefit in many ways if the research is planned, carried out, and interpreted within the frame of reference of the lifespan and the continuous processes of change that characterize all behavior. This frame of reference will affect our understanding of the concept of maturity; it will emphasize the relevance of changing functions, experiences, and motivations to any particular behaviors under consideration; and it will help direct attention to ways of improving various testing instruments. It is of relevance both to cross-sectional and longitudinal studies. Not only are cross-sectional studies needed for age-specific normative data. All behavioral research, including studies of restricted and atypical populations and behaviors, should be oriented in regard to the subjects' age or stage of development as well as their other characteristics. Longitudinal studies lend themselves to investigation of many important questions that can only be answered definitely by observations of the same subjects over time. Questions to be explored by this method include comparisons of the processes of development with those of decline, the identification of underlying persistent and predictable qualities of personality, and individual age-trends in intellectual capacities.

REFERENCES

Anderson, J. E., and D. B. Harris. *A survey of children's adjustment over time: a report to the people of Nobles County.* Minneapolis: University of Minnesota, 1959.

Barker, R. G., and H. F. Wright. *Midwest and its children.* Evanston: Row, Peterson, 1955.

Bayley, Nancy. "Mental growth during the first three years. A developmental study of sixty-one children by repeated tests." *Genet. Psychol Monogr. 14:* No. 1, 1933.

Bayley, Nancy, and H. E. Jones. "Environmental correlates of mental and motor development. a cumulative study from infancy to six years." *Child Develpm. 8:* 329–341, 1937.

Bayley, Nancy, and E. S. Schaefer. "Relationships between socioeconomic variables and the behavior of mothers toward young children." *J. Genet. Psychol.* 96: 61–77, 1960.

Bayley, Nancy, and E. S. Schaefer. "Maternal behavior and personality development: data from the Berkeley Growth Study." *Psychiatric Res. Rep.* 13: 155–173, 1960.

Demming, J. A., and S. L. Pressey. "Tests 'indigenous' to the adult and older years." *J. Counsel. Psychol.* 4: 144–148, 1957.

Furfey, P. H., and J. Muehlenbein. "The validity of infant intelligence tests." *J. Genet. Psychol.* 40: 219–223, 1932.

Geber, M., and R. F. A. Dean, "The state of development of newborn African children." *Lancet i*: 1216, 1957.

Jones, H. E. *Age changes in mental ability. Old age in the modern world.* London: Livingstone, 1955, pp. 267–274.

Kallmann, F. J. "Genetic factors in aging: comparative and longitudinal observations on a senescent twin population." In: Hoch, P. H., and J. Zubin (eds.), *Psychopathology of aging.* New York: Grune & Stratton, 1961.

Kallmann, F. J., Lissy Feingold, and Eva Bondy. "Comparative adaptational, social, and psychometric data on the life histories of senescent twins." *Amer. J. Hum. Genet.* 3: 65–73, 1951.

Kelley, E. L. "Consistency of the adult personality." *Amer. Psychologist* 10: 659–681, 1955.

Lehman, H. C. *Age and achievement.* Princeton: Princeton University Press, 1953.

Macfarlane, Jean W., Lucile Allen, and Marjorie P. Honzik. *A developmental study of the behavior problems of normal children between twenty-one months and fourteen years.* University of California Publications in Child Development, 2; p. 222. Berkeley: University of California Press, 1954.

Owens, William A., Jr. "Age and mental abilities: a longitudinal study." *Genet. Psychol. Monogr.* 48: 3–54, 1953.

Schaefer, E. S., and Nancy Bayley. "Maternal behavior, child behavior and their intercorrelations from infancy through adolescence." *Monogr. Soc. Res. Child Develpm.* 28: 1963.

Sontag, L. W., C. T. Baker, and V. L. Nelson. "Mental growth and personality development: A longitudinal study." *Monogr. Soc. Res. Child Develpm.* 23: 1958.

Terman, L. M., and M. H. Oden. *The gifted child grows up: Twenty-five years' follow-up of a superior group, genetic studies of genius.* Stanford, Calif.: Stanford University Press, 1947.

Terman, L. M., and M. H. Oden. *The gifted group of midlife: Thirty-five years' follow-up of the superior child, genetic studies of genius.* Stanford, Calif.: Stanford University Press, 1959.

Tuddenham, R. D. "Constancy of personality ratings over two decades." *Genetic Psychology Monographs* 60: 3–29, 1959.

Two

A Report on Three Growth Studies at the University of California

Mary Cover Jones

The belief that "the child is father of the man" has considerable common-sense justification, but until recently we have had very little systematic evidence on the persistence or change of personality traits over extended periods of time. Long through-time studies of the same individuals are just beginning to report findings in which development in the growing years can be related to adult status.

Longitudinal studies are being used to trace relationships of earlier to later and later to earlier characteristics and to determine which combinations of behavior at which ages have the greatest predictive powers to what other age periods and behavior patterns.

From the Institute of Human Development, where growth studies have been in process for over thirty years, data are becoming available on more than 300 persons who are in their thirties and forties.

These individuals have been enrolled in one of three longitudinal projects: the Berkeley Growth Study, the Guidance Study, and the Oakland Growth Study. Each of these studies has had a principal staff member over the long period of data collection, an important contribution to the success of the projects. Nancy Bayley has been identified with the Berkeley Growth Study, Jean Macfarlane with the Guidance Study, the late Harold Jones with the Oakland Growth Study (formerly the Adolescent Growth Study), and Herbert Stolz with the biological growth program for all three studies. It is largely due to the personal relationship between permanent staff and study members that, in spite of some inevitable attrition, selective factors, such as socioeconomic status, have operated at a minimum (Bayley, 1966b; Macfarlane, 1966; Haan, 1962; Elder[1]).

From *The Gerontologist*, 1967, 7, 49–54. Published with the permission of the author and *The Gerontologist*, the official publication of the Gerontological Society for the professional practitioner in gerontology.

Adapted from a presentation for a Symposium on Longitudinal Studies at the Seventh International Congress of Gerontology, Vienna, Austria, June 27, 1966.

THE BERKELEY GROWTH STUDY

The Berkeley Growth Study began in 1928 with a relatively homogeneous sample of 61 infants born in Berkeley. With some attrition, cases were added to expand the total sample to 74. According to Bayley:

> The plan of the Berkeley Growth Study was to test and measure the mental, motor and physical development and the health of a sample selected as full term healthy newborns. The scheduled testing periods, at each of which most of the subjects were seen, to date have occurred at fifty-six ages, the first at three to four days of age, the most recent at thirty-six years. . . . Fifty-four (or 73%) of this sample comprise the current adult (36 year) sample. (Bayley, 1966b)

THE GUIDANCE STUDY

The Guidance Study has followed an initial sample of 248 individuals, representative of those born in Berkeley from January 1928 to July 1929, from the time they were twenty-one months of age through eighteen years. A follow-up study 12 years later of 169 individuals at the age of thirty represented 68 percent of the original sample. Macfarlane (1963) describes the objectives and procedures of the study as follows:

> (1) to delineate physical, mental and personality growth and development in a normal group and to ascertain the variations among and within individuals at different developmental periods over a long time span; (2) to see the relationships of these findings (a) to the biological facts (constitutional makeup, sex, health and rate of maturity), (b) to environmental facts—physical, socio-economic, intellectual and social, including interpersonal relationships to family members, playmates, classmates, teachers and important others, with their varying personalities and impacts upon individual children; (3) to throw light upon critical combinations of facts whereby some individuals are able to realize their full potentials while others fall far short of such realization; some individuals develop mature and sturdy personalities whereas others rigidly or neurotically cling to immature or ineffectual patterns; some individuals under stress gather new strengths and others give up the struggle or disintegrate; and (4) to assess how much confidence we should have in the predictive usefulness for the short haul or over the long developmental period of our tools of appraisal of personality characteristics, mental ability, etc., of the child.
>
> With the above objectives we had to repeat systematically over the years a wide range of measurements covering biological, environmental, familial and behavioral aspects of the growing organism. Direct measures such as developmental x-rays, body build measures, mental tests and projective tests were used; also interviews furnished materials as seen by parents, teachers, brothers, sisters, the child himself and the

professional interviewers (clinical psychologists and physicians). Classmates' appraisal was secured by sociometrics and cumulative achievement from yearly school records.

THE OAKLAND GROWTH STUDY

In 1932, approximately 200 members of the fifth and sixth grades of five Oakland elementary schools serving adjacent areas were enrolled in the Adolescent Growth Study (now termed the Oakland Growth Study). The program of physical, physiological, and psychological measurements was designed to study the complex ways in which developmental changes and behavioral potentials and tendencies are interrelated (Jones, 1958b). Clausen (1964), in describing the plans for the data analysis of the Oakland Growth Study, writes as follows:

> In viewing the adolescent subject, one tends to see his development as the resultant of the interplay between genetic potentiality, general patternings of relationship and expectation surrounding him, and salient experiences of self and others, all within a social-cultural matrix that provides general definitions of individual experience. Stature, physical attractiveness, rate of maturing, intellectual potential, and temperament interact with family structure, parental personalities, and the subcultural and ecological correlates of position within the social structure. As individuals mature, the constraints of their original social matrix become less compelling. They not only respond to environmental pressures and potentialities, they select them. They commit themselves to lines of activity and to other persons, as well as to conceptions of themselves. As a consequence, one expects not only changing personal attributes and changing saliences of attributes and values but also a changed relationship between the way the person sees himself and the ways that he is seen by others.

The representative group participated in an intensive schedule from ten and a half years of age until graduation from high school with several follow-up projects in adulthood. Since most of these study members knew each other as classmates in the same junior and senior high schools, this acquaintance provided a unique situation for the observation and recording of social interactions. The children were from varied social and economic backgrounds, although, in accordance with the town's basic structure at that time, they were all white, American born, urban, and predominantly middle class. The representativeness of the follow-up sample has been assessed by comparing the sample at the beginning of the study, at the end of high school and at age thirty-eight on a number of variables, including socioeconomic status, intelligence, and adjustment. At their present age, in the mid-forties, these more than 100 individuals are the oldest and thus those nearest the age to provide candidates for gerontological research.

THE GROWTH OF INTELLIGENCE

Bayley (1966a) and Eichorn (1963a) have used the Berkeley Growth Study principally to follow intensively the course of physical and mental growth over the years from infancy to the middle thirties. Hearteningly, Bayley reports that motivation and drive and ample time rather than small variations in intelligence are the important determiners for much learning in adults. The intellectual potential for continued learning seems unimpaired through thirty-six years at least (the age of her sample).

The most consistent and marked increase in scores from eighteen to thirty-six years are found for both sexes in the highly verbal tests. Other subtests, requiring analysis and synthesis, level off after twenty-six years and the performance subtests may show a slight decline. The women tend to show a drop in scores, especially on some of the performance tests and on arithmetic after twenty-six years. On the basis of these findings and those from other studies, Bayley predicts that increase in general verbal capacity may well be maintained through fifty years or longer.

The children of the original sample are now being studied under a program administered by Eichorn. Analyses include comparison of children's scores on a number of measures with those of their parents (a) as adults and (b) at the child's age. Data from both the Berkeley Growth Study and the Guidance Study report that parent-child resemblances are likely to be greater in childhood and adolescence than in infancy (Eichorn, 1963a; Bayley, 1954; Honzik, 1957, 1963).

Bayley, using the Berkeley Growth Study, and Honzik the Guidance Study data, have examined consistency in mental abilities and have reported on the environmental correlates of mental growth and on cross-sex findings. Bayley and Schaefer (1964) suggest that girls' intelligence scores are to a greater extent genetically determined, while boys' are more influenced by the emotional climate of their early environment, especially maternal behaviors. Honzik (1967) indicates the relevance of parental compatability and the father's affectional relationship to the daughter's mental growth and the influence of the mother's behavior patterns and attitudes for the cognitive development of boys and girls.

STYLES OF BEHAVIOR

Honzik (1964), in summarizing the findings of the Guidance Study, which has emphasized the clinical approach, reports that the most consistent personality dimensions are styles of behavior. For example, certain characteristic response tendencies, such as level of activity and degree of expressiveness or responsiveness, have emerged as being stable and good predictors of later behavior. Similarly Schaefer and Bayley (1963), using ratings of Berkeley Growth Study members made on observed, overt behavior in a testing situation, find the most stable dimensions to be active, extroverted versus inactive, introverted behaviors. Additionally, Tuddenham (1959), in a study to be discussed later, reports that for the Oakland Growth Study, variables closely related to an extroversive-introversive dimension are among the most stable. Although these "central orientations" tend to be consistent over time for individuals, they may be revealed in different behavior for the two sexes and for different developmental periods.

Bronson (1966) reports for the Guidance Study that the expressive-outgoing versus reserved-withdrawn pattern relates to sociability and expressiveness in adulthood for both sexes but that the style of interaction is different. The behavioral similarity is more marked for boys (the expressive outgoing boy becomes the gregarious expressive man). The preadolescent years are more predictive of adult behavior than earlier periods. Women who are outgoing and expressive show domineering attitudes as well. In early childhood they tend to be defiant and irresponsible but also with self-doubts. In late adolescence the pattern that is related to the adult syndrome suggests a righteous and self-satisfied involvement in interpersonal relations.

Bronson suggests that although there is consistency in underlying personality syndromes, their expression may be altered by environmental expectations and stresses.

Analysis for both the Guidance Study and the Oakland Growth Study data have contributed information concerning the potential of various periods in childhood for prediction of adult personality. Block and Haan[2] found the junior high school years to be more predictive of adult status than the intervening high school period.

In a study using the Guidance Study sample, Livson and Peskin (1967) found ratings of behavior made during the preadolescent and early adolescent period (ages eleven, twelve, thirteen) to be more predictive of adult mental health than were data from the early childhood or the later adolescent years. The authors suggest, as one possible explanation, that the preadolescent and early adolescent period is most "trustworthy" in the sense that behavior is more direct, and hence more reliably assessed at this time.

Stout (1965) has studied the adaptiveness to the maternal role of the women of the Guidance Study in relation to the stresses and strengths in the mother's family during childhood. The adaptive mothers were more likely to be first born, in a lower class family with a foreign-born parent. The marriage was intact even though it may not have been a happy one. The mother-daughter relationship was likely to be strained, although the mother was a more stable person than the father.

The unadaptive mother is likely to have been an over-indulged and over-protected, last born child in a socially mobile family.

The families of partially adaptive mothers are more likely to be middle class; the parental marriage, if intact, is happy; the parental relationship warm. In the event of parental friction, a separation is likely. Stout discusses her results in terms of contribution of various family factors to the development of cognitive and affective predispositions toward the parental role and their facilitation or interference with parental identifications that are maturing, ambivalent, or inadequate. For example, the foreign-born family of the adaptive parent may have presented a more coherent family orientation in which attitudes of social tolerance and self-acceptance are learned. This mother, more used to discord and dissent in the parental home, may be better prepared to cope in her own maternal role.

ADULT CHARACTERISTICS RELATED TO ADOLESCENT PHYSICAL DEVELOPMENT

Reports on the relationship between physical characteristics, especially the rate of maturing in adolescence, and behavior have come principally from the Oakland Growth Study. These more than a dozen publications have been reviewed by Eichorn (1963b). Briefly summarized, for boys, they show that those who are accelerated in physical development are advantaged socially in the peer culture (Jones and Bayley, 1950) and reveal more favorable self-concepts (Mussen and Jones, 1957).

The physical differences noted in adolescence were largely eliminated at maturity. There were few differences in educational level, marital status, or family size between the two maturity groups. However, a number of significant differences remained in measured aspects of personality (Jones, 1965). The early-maturing boys, now adults, are on the average more responsible, cooperative, sociable, conforming, and self-controlled. It has been suggested that these boys took on the pattern of the adult culture at an early age and therefore were able to conform more readily to the valued mores of adult society. This seems to have led to a somewhat rigid as well as conforming pattern of adaptation.

Although adult measures for men showed a continuation of the impulsivity and assertiveness associated with late-maturing in adolescence, there is also indication of greater insightfulness and ability to cope with new situations. It may be that in the course of having to adapt to difficult status problems in adolescence, the late maturers have gained some insights which permit greater flexibility. Peskin (1967), using the Guidance Study cases, interprets somewhat similar findings in psychoanalytic terms.

SOCIAL BEHAVIOR IN ADOLESCENCE AND IN ADULTHOOD

In Ames' observations (1957) of the Oakland Growth Study sample, which focuses on the relationship between earlier and later social characteristics, it was found, rather surprisingly, that, for men, adult social participation in both informal activities and organized social activities was correlated more closely with rate of physical maturing (early maturers were more social) than with social behavior in adolescence.

Girls who acquired social prestige in high school were, as adults, more active in clubs and structured organizations. However, women's active participation in community organizations was more highly correlated with their husbands' occupational status than with their own high school social participation. These results suggest that determinants of sociality, as measured for this sample, lie more in persistent personal traits in the case of men and more in social class and related institutional factors in the case of women (Jones, 1960).

ROLE IDENTIFICATION IN ADOLESCENCE AS RELATED TO ADULT CHARACTERISTICS

Mussen (1961), using the Oakland Growth Study sample, is now able to report that boys who were classified in adolescence as more masculine or less masculine (an attribute associated with the strength of their identification with their fathers) are found, in general, to maintain their relative positions in adulthood. Those who were low in masculinity in their youth are found as adults to have fewer masculine interests and less masculine attitudes. However, Mussen also points out that the linkage between high masculinity and good adjustment in adolescence does not necessarily continue into adulthood. Societal expectations favor strong masculinity responses in adolescent boys, but this requirement may not be maintained for adult behavior.

PERSONALITY CHARACTERISTICS OVER THE YEARS

Some of the personality variables that had been measured by observational and drive ratings in adolescence were used to assess the Oakland Growth Study individuals in their early thirties. The adult judgments were made on the basis of interview situations. Tuddenham (1959) found the most stable variables for both men and women to be those which connoted expressiveness and expansive spontaneity versus inhibition—a syndrome closely allied to the expressive-outgoing and reserved-withdrawn dimension discussed earlier in this paper.

Variables equally well agreed about by judges at both periods of assessment but unstable in the sense that they did not show consistency from adolescence to adulthood were those related to physical attractiveness, self-confidence, and among men only, leadership, and popularity. This later shifting of social characteristics over the years appears also in Ames' study of the continuity of social roles, mentioned earlier.

Tuddenham's conclusion from this study is one that needs to be emphasized in all studies reporting findings for groups. Although there is a clearly significant measure of temporal stability in personality across the developmental span from early adolescence to adulthood, in this sample, the relationships are too low to permit individual predictions. In individual cases, marked shifts in adjustment have occurred, both in a favorable and an unfavorable direction, which could hardly have been predicted from anything that was known earlier.

The adolescent drive ratings used by Tuddenham to compare adult with adolescent personality characteristics were also used by McKee and Turner (1961) as the basis of comparison with self-report inventory scores obtained by the study members at the time of the follow-up study when they were in their early thirties.

Some of the relationships which were most positive over time were between the drive rating in adolescence for aggression and the adult self-report score on dominance, between the drive rating in adolescence on recognition and the adult self-report scores on dominance and self-acceptance, and between the drive rating for escape and a negative correlation with the self-report score for self-control. Agreement between the adolescent and adult measures (indicating continuity in characteristics) was found for both sexes, but the relationships were higher for women than for men. The authors interpret this sex difference as reflecting the fact that adolescent girls in the American culture are more socially mature than are adolescent American boys. Therefore their behavior in the developmental period might be expected to reflect adult behavior to a greater extent.

Skolnick (1966) compared the Thematic Apperception Test (TAT) responses of this sample at age seventeen with those at age thirty-seven. She scored these for motives of achievement, aggression, affiliation, and power. Males tended to remain stable in power and aggression, females in achievement and affiliation. She sees some suggestion of a pro-social and "anti" social dimension in these findings.

Stewart (1962) classified a group of these adults in their early thirties with regard to criteria of adjustment—a group with psychosomatic symptoms, those with behavior maladjustment, and a symptom-free group—and compared these classifications with earlier measures of personality. Those with poorer adjustment in the thirties were found to have had childhood records indicating family maladjustment, feelings of inferiority, generalized tensions, physical symptoms, less interest in the opposite sex, and poorer relationships with their classmates.

Jones (1960), relating these adult adjustment classifications to galvanic skin reflex (GSR) data from adolescence, found a significantly better prognosis for the high (internally) reactives with regard to adjustment in middle maturity.

Recently, California Q set ratings (Block, 1961) have been used with each of the three samples in adulthood, and for the Guidance Study and Oakland Growth Study samples also for the junior and senior high school periods. A wide variety of material (self-reports, projective tests, interviews, and ratings) has been assessed in an effort to summarize the voluminous data of the respective studies.

For the Oakland Growth Study sample, Block (1966) reports,

> Over-simplifying, non-changers are individuals who at preadolescence enjoyed well-formed character structures and moved onward to deeper but essentially similar maturity or they are individuals who at preadolescence were guarded, brittle personalities who even as adults are closed off from much of adult life. Changers typically were diffuse and unformed during their early years but as adults have moved toward either a rich, savoring, coping approach to life or a dysphoric psychopathology. Efforts are proceeding to understand how these different paths of personality development were determined.

A study of the drinking patterns (Jones, 1968) of the Oakland Growth Study sample, using the California Q set for the junior high school and senior high school years and for the adult interviews at age thirty-eight, shows marked consistency in ratings over these periods, especially for the problem drinkers, who are impulsive, unpredictable, and rebellious, and the abstainers, who are bland, overcontrolled, and ethical. The junior high school period, again, seems to predict somewhat better to adult behavior than does the senior high school period.

Jones (1958a) has summarized the problems of research in the area of personality continuity and change as follows:

The problem here may lie partly in the fact that over a long period behavior consistency, when it occurs, may be countered by changes in the environment. The adaptive significance of a given behavior pattern can thus be interpreted only with reference to changing demands in the life situation.

Those who have spent their life careers in the study of human development are impressed not so much with the consistency or the change in personality over time as with the sturdiness of the human organism. Jones and Jones (1957) have summarized this with the statement:

Within a normal range of subjects and of situations, the degree of consistency is striking evidence of the organism's ability to maintain patterns and to resist fundamental changes.

And Macfarlane (1952) reacts to her long years of experience with the Guidance Study Sample with these words:

The urge for good mental health is so strong in most of us that in spite of ups and downs and periods of distress, we either work out solutions or develop enough defenses so that we not only survive but also get real satisfactions out of our lives. . . .

NOTES

[1] Elder in a careful study of sample attrition for the Oakland Growth Study reports a slightly disproportionate loss of working-class men from the follow-up samples. G. Elder, *Children of the Depression*, in preparation.

[2] Personal Communication, 1962. See also J. Block, and N. Haan. *Ways of personality development: Continuity and change from adolescence to adulthood*, 1971.

REFERENCES

Ames, R. "Physical maturing among boys as related to adult social behavior." *Calif. J. Educ. Res.* 8: 69–75, 1957.

Bayley, N. "Some increasing parent-child similarities during the growth of children." *J. Educ. Psychol.* 45: 1–21, 1954.

Bayley, N. "Learning in adulthood. The role of intelligence." In: Klausmeier, H. J., and C. W. Harris. (eds.) *Analysis of conceptual learning.* New York: Academic Press, 1966, pp. 120–136. (a)

Bayley, N. *Methodological problems in longitudinal research.* Read at the Sixth International Congress of Child Psychiatry, Symposium on Problems of Research and Methodology, Edinburgh, Scotland, July 24–29, 1966. (b)

Bayley, N., and E. S. Schaefer. "Correlations of maternal and child behaviors with the development of mental abilities—Data from the Berkeley Growth Study." *Monogr. Soc. Res. Child Develpm.* 29 (serial no. 97), 3–80, 1964.

Block, J. *The Q sort method in personality assessment and psychiatric research.* Springfield, Ill.: Charles C Thomas, 1961.

Block, J. *Implications of continuity and change from preadolescence to adulthood.* Proceedings of the XVIII International Congress of Psychology, Symposium on Longitudinal Studies in Child Development, pp. 167–168 (Abstract), Moscow, August 1966.

Bronson, W. C. "Central Orientations: A study of behavior organization from childhood to adolescence." *Child Develpm.* 37: 125–155, 1966.

Clausen, J. A. "Personality measurement in the Oakland Growth Study." In: Birren, J. E. (ed.), *Relations of development and aging.* Springfield, Ill.: Charles C Thomas, 1964, pp. 165–175.

Eichorn, D. H. "Two-generation similarities in height during the first five years." In: Jersild, A. T.; *Psychology of adolescence* (2d ed.). New York: Macmillan, 1963, p. 61. (a)

Eichorn, D. H. "Biological Correlates of Behavior." In: Stevenson, H. W. (ed.), *Yearb. Nat. Soc. Stud. Educ., Part 1 Child Psychology,* 1963, pp. 4–61. (b)

Haan, N. *Some comparisons of various Oakland Growth Study subsamples on selected variables.* Dittoed. Berkeley, California: Institute of Human Development, 1962.

Honzik, M. P. "Developmental studies of parent-child resemblance in intelligence." *Child Develpm.* 28: 215–228, 1957.

Honzik, M. P. "A sex difference in the age of onset of the parent-child resemblance in intelligence." *J. Educ. Psychol.* 54: 231–237, 1963.

Honzik, M. P. "Personality consistency and change—some comments on papers by Bayley, Macfarlane, Moss and Kagan, and Murphy." *Vita Humana* 7: 139–143, 1964.

Honzik, M. P. "Environmental correlates of mental growth: Prediction from the family setting at 21 months." *Child Develpm.* 38: 337–364, 1967.

Jones, H. E. "Consistency and change in early maturity." *Vita Humana* 1: 43–51, 1958. (a)

Jones, H. E. "The Oakland Growth Study—Fourth decade." *Newsltr. Geront. Soc.* 5: No. 1, pp. 3, 10, 1958. (b)

Jones, H. E. "The longitudinal method in the study of personality." In: Iscoe, I., and H. W. Stevenson (eds.), *Personality development in children.* University of Texas Press, 1960, pp. 3–27.

Jones, H. E., and M. C. Jones. *Adolescence.* Berkeley, University Extension, University of California, 1957.

Jones, M. C. "The later careers of boys who were early- or late-maturing." *Child Develpm.* 28: 113–128, 1957.

Jones, M. C. "The psychological correlates of somatic development." *Child Develpm.* 36: 899–911, 1965.

Jones, M. C. "Personality Antecedents of Drinking Patterns in Adult Males." *J. Consult. Psychol.* 32: 2–12, 1968.

Jones, M. C., and N. Bayley. "Physical maturing among boys as related to behavior." *J. Educ. Psychol.* 41: 129–148, 1950.

Livson, N., and H. Peskin. "The prediction of adult psychological health." *J. Abnorm. Psychol.* 72: 509–518, 1967.

Macfarlane, J. W. *Research findings from a twenty-year study of growth from birth to maturity.* Mimeographed. Berkeley, California: Institute of Human Development, 1952.

Macfarlane, J. W. "From infancy to adulthood." *Childh. Educ.* 39: 336–342, 1963.

Macfarlane, J. W. Unpublished material. Institute of Human Development, Berkeley, 1966.

McKee, J. P., and W. S. Turner. "The relations of 'drive' ratings in adolescence to CPI and EPPS scores in adulthood." *Vita Humana* 4: 1–14, 1961.

Mussen, P. H. "Some antecedents and consequents of masculine sex typing in adolescent boys." *Psychol. Monogr.* 75: 1961, no. 2 (Whole no. 506).

Mussen, P. H., and M. C. Jones. "Self-conceptions, motivations and interpersonal attitudes of late- and early-maturing boys." *Child Developm.* 28: 243–256, 1957.

Peskin, H. "Pubertal onset and ego functioning: A psychoanalytic approach." *J. Abnorm. Psychol.* 72: 1–15, 1967.

Schaefer, E. S., and N. Bayley. "Maternal behavior, child behavior, and their intercorrelations from infancy through adolescence." *Monogr. Soc. Res. Child Developm.* 28, no. 3 (serial no. 87), 79–96, 1963.

Skolnick, A. "Stability and interrelationships of thematic test imagery over twenty years." *Child Developm.* 37: 389–396, 1966.

Stewart, L. H. "Social and emotional adjustment during adolescence as related to the development of psychosomatic illness in adulthood." *Genet. Psychol. Monogr.* 65: 175–215, 1962.

Stout, A. M. "Adaptiveness to the maternal role at age 30, and stresses and strengths in the mother's family during childhood." Dittoed. Berkeley, California: Institute of Human Development, 1965.

Tuddenham, R. D. "Constancy of personality ratings over two decades." *Genet. Psychol. Monogr.* 60: 3–29, 1959.

Three

A Developmental Model for Aging

John E. Anderson

Aging is part of the life cycle which begins with conception and ends with death. In this cycle a complex organism moves forward in time and goes through a series of transformations with which we as scientists are concerned. Minute by minute transactions are carried on with the environment, of which some result in cumulations which affect the future and some do not. By virtue of the genetic potential the limits of development can be described with some accuracy. Recently my concern has been with a model for development from birth to maturity (Anderson, 1957). I now propose to examine the aging process in terms of some of the concepts which emerged.

At the outset may I direct attention to common processes and away from variation. Kuhlen several years ago (1952) pointed out the extent of the overlapping between age levels and the masking of trends by individual variation. He pointed out the need of studies of intra-individual variability and of inter-individual variability. Because of sampling errors in cross-section studies, we must look to future longitudinal studies, which control some of this variance, for answers to many puzzling questions. Let us turn to classical models of aging of which I list four: (a) one factor models, (b) machine models, (c) factor of safety or stress models, and (d) reversal of developmental models.

ONE FACTOR MODELS

Many one factor models have been proposed. Whether the basic mechanism is nervous, hormonal, metabolic, circulatory or psychological, a wide variety of complex results are traced to a single factor (Anderson, 1956). Such extreme reductionism seems to be based upon generalizations

From *Vita Humana*, 1958, *1*, 5–18. Reprinted with permission of the publisher, S. Karger, Basel/New York.

Report to the Fourth International Gerontological Congress, Merano, 1957.

from particular pathologies which characterize small proportions of aging persons. Although some extreme pathological conditions make it impossible for the whole to function, it is clear that for most people, homeostatic mechanisms function above a lower limit beyond which single factor theories offer no solution. They are based on conditions which though necessary for function do not determine it when these limits are met.

We have not been very successful in correlating aging with changes in tissue or with the elementary components of the organism. For example, some deterioration in sensory functioning correlates with changes in the eye and with loss of fibers in the ear. Deposits of waste products in tissues tend to choke function. When, however, we ask whether surviving brain cells deteriorate or not, no very clear answer is found and controversy arises over the facts. Some students point out that the effects of aging are more apparent on the surface of the organism and in total behavior than they are on the inside of the organism or within parts or subsystems. This raises the question whether or not we will ever be able to subsume aging under a single physical or chemical formula.

MACHINE MODEL

This classical model based upon our experience with machines, assumes aging to be a summation of progressive inadequacies. Man's machines wear out by deterioration of parts and the failure of components. Breakdown from either cause is likely to be sudden.

What determines behavior when a part becomes ineffective or is destroyed? It has long been assumed that the resulting disturbance is directly related to the missing part. This view assumes a one-to-one correspondence between packets of behavior and particular nervous and physiological structures. But a more acceptable hypothesis holds that the organism adjusts with what is left behind and that behavior is the outcome of what is left, not of what is taken away. In many cases there is no substantial change. Minor destruction may produce little effect and major disturbance a disproportionate effect. But minor defects at critical points may produce major disturbances. Hence functional efficiency must be evaluated directly some time after loss. Because the machine lacks this capacity to react with what is left behind, we must reject the machine model.

But the machine also lacks the ability of the organism to repair itself over time and to adjust by substituting one component for another. An automobile cannot itself repair a defective generator, nor can it meet a defect in the generator by increasing the flow of gasoline, that is by an adjustment in a different modality. This adjustive and regenerative capacity over time is one of the most striking characteristics of the living system. Within limits, living systems build themselves up while machines wear out in converting energy into action.

FACTOR OF SAFETY OR STRESS MODELS

This variant of the machine model is based on the principle that the organism progressively loses its capacity to endure stress and to meet emergencies. In addition to a narrowing of the range of adaptation the time needed to recover normal functioning is increased. A young system is extraordinarily able to withstand physiological and psychological stress; the older system is less able to do so. Only in a rare emergency is the younger system called upon to put forth all its strength, speed, and variable behavior. The factor of safety built into every bodily structure and homeostatic mechanism in order to insure survival under stress is tremendous. Not only are many organs double but within each organ there is a great excess of cells above those normally used. Hence function is maintained. In addition there is some capacity to substitute across modalities and thus preserve function.

With this model the question of how the organism learns to go near his breaking point without exceeding it becomes important. The experience of pain, discomfort and fatigue when nearing capacity and of discomfort during recovery from over-exertion may create signals which warn and thus constrict functioning within safe limits. In the DeSilva (1938) experiment older persons gradually lowered their driving speeds because they were uncomfortable at speeds which younger persons took as a matter of course.

For this model we must ask whether the facts upon which it is based even if correct, are adequate for explanatory purposes. The model is essentially negative, in that it considers not what the person does but what he might do. Moreover, because it is based upon rare rather than common situations aging becomes an accident of the situation rather than a positive internal process.

MODEL WHICH REVERSES DEVELOPMENT

This model assumes that behavior is built up in orderly fashion by integrating smaller units into larger and more complex patterns which thereafter function at a higher level. In downgrading, these complex structures disappear in reverse order with the most complex going first and the simplest last. A second childhood offers a neat and simple paradigm.

Unfortunately, the results of studies do not reveal such an orderly reversal. Some very complex functions remain stable long after simple functions are disturbed. Moreover, the studies of regression (Cameron, 1938) in which psychotics and children are compared in similar situations show that the so-called child-like responses only superficially resemble children's performances. Moreover, studies of development in animals and humans show irreversible processes with lasting changes in internal and external relations which make a return to the former state impossible. Presumably the same principles hold for true decrements.

In development, much early time goes, as Piaget (1954) points out, to inter-sensory checking and repetitive and exploratory movements. From these later complex perceptual and motor skills evolve. The simple perceptions and skills are over-learned to a degree seldom attained by later patterns. When deterioration occurs these remain. But here we need much research concerned with the order and level of change throughout many patterns studied simultaneously rather than studies of single or isolated patterns. From these models we may turn to a comparison with a model of development.

NATURE OF THE DEVELOPING ORGANISM

The person is a complex manifold moving through time. As a system it is basically open and with high input and high outgo. While much of its internal and external action is transactional it carries some experience forward and thus modifies itself. In one sense it is a physico-chemical system converting food into energy. In another sense it is a sensory-neuromuscular mechanism engaged in channelling information into the controls which direct this energy into a wide variety of specific and integrated actions. A fundamental characteristic of the living whole is the irreversibility of its relations as it moves from one state to another. This cumulation and progressive modification differentiate the living system from man-made machines. During the early part of the life cycle growth and cumulation are at a maximum. The mature organism is concerned with maintenance and cumulation, not with growth.

We may now check the aging or third portion of the life cycle against these descriptions. Consider first the openness of the system. The older system is less open than the younger because so much selection has gone forward. The infant is multipotential; as he grows he purchases efficiency at the cost of versatility: He continuously faces choice points which present alternatives. Whatever the choice, subsequent development is more limited. The longer the system operates the greater the number of choices that have been made and the more stable is the pattern. The mechanization, the ritualization, the constancies that result enable him to master situations and in a sense free him for reacting to other situations. But these organizations bulk larger and larger in the total person. While they constitute a reservoir of behavior for meeting life effectively, at the same time they constrain and limit. Hence radical transformations of the environment which produce marked effects in children are less likely to do so in older persons.

In this developmental process two mechanisms operate. One is growth, maturation, unfolding. This age-bound process is marked by differential increases with time in almost every function and process. There seems to be no comparable process in terms of the range and variety of effects in decline. The other mechanism is learning. A living organism placed in situations in which stimulation is repeated, builds new patterns of action, which are thenceforth relatively stable. As far as I can see some learning capacity is retained throughout life with the result that at every stage some capacity to cumulate experience remains. In general, the younger system has more capacity to build complex behavior out of simple behavior. It has been suggested that in complex integrations the older person has difficulty in programming because the controls which are automatic in the younger person have to be replaced by conscious controls.

Perhaps the most important adjustive mechanism in the human being is found in the symbolic devices through which he both controls and communicates. This process appears at the end of infancy and mediates and modifies all behavior. Some aspects of the symbolic process increase in later maturity and old age (vocabulary, information, and so forth) while other aspects decrease (conceptualization). However, no comprehensive studies of the entire area in relation to aging that compare with those at younger ages have been undertaken.

In summary then, checking against the model for development we can say that growth has disappeared, that openness, mechanization and activation are probably reduced by virtue of the selective and channelling process that living in time entails, that possibilities of learning and adjusting remain in substantial form, that cumulation continues, that the possibility of complex integrations may be limited by constraints that grow out of the mass of ordered behavior and by breakdown in programming and that symbolization remains as an active mechanism subject, however, to limitations that resemble those for other complex integrations.

THE NATURE OF TASKS

In order to understand the relation between aging and the demands made by the environment some attention must go to the nature of tasks. These demand an integration of traits based on what is to be done rather than upon internal characteristics. This view differs from the traditional view of psychologists who emphasize individual capacities and slice the changing organism in terms of inherent traits. But a job is a system set by society in the broad sense, which once set selects persons in terms of their capacity to develop the required organization from their own combination of abilities. Thus, the English boy learns cricket and the American boy baseball, activities which bear some resemblance to one another and which demand similar patterns of ability (speed, strength, fine coordination, gross coordination, and so forth). But the rules of the games differ. Most human activities in a social setting are constrained in the same sense. For example, sitting on a chair which is a man-made instrument peculiar to some cultures and not to others, demands particular coordinations which are quickly acquired by the young child, if chairs are available. But in some societies children squat rather than sit. In a complex skill such as riding a bicycle the combination of leg, arm, trunk, neck, head, and eye movements is set by the nature of the bicycle, a man-made object. It involves a wide variety of simultaneous and successive perceptions, actions, and coordinations. A similar analysis could be made of running a typewriter, throwing a spear, handling a tennis racket, painting a picture, and writing a letter. In many fields much effort has gone to working out the sequences in detail in order that teachers and coaches may aid persons integrating the components demanded by the activities in question.

In thinking about tasks we must not fall into the error of thinking in terms of expert performers. The ordinary person reaches a level of competence which enables him to meet the practical demands of life and to earn a living in competition with others. In this process learning is usually carried well beyond the psychologist's ordinary criteria. But time is, however, relatively unimportant if competence can be acquired since once competence is attained, the individual carries the skill or organization with him as a permanent asset to be called upon when needed. One other difference exists between the expert and the ordinary person. The expert is put under high stress in competition and called upon to show maximum performance. The ordinary man, however, meets continuing day-to-day demands at an ordinary level. His job seldom calls upon all his resources. When age changes take place, demands can be met by readjusting, by compensation, by shifting emphasis to other phases of the process, long after failure or inadequacy has occurred in particular parts. For example, the worker may compensate by more careful visual attention as kinesthetic and touch controls become less effective or use special supports or mechanical devices.

It is but a step from this view to the view that different individuals with their differing traits which vary in content and in degree, build up unique ways of meeting demands. Thus, tasks are highly selective integrations or systems which establish their own rules as they develop. Because of this interplay and variation it is difficult to resolve performances into firm constituents or elements. What emerges is a concept of functional efficiency which is itself a system in relation to demands which are themselves systems. Components may vary substantially from age level to age level.

STABILITY

From our analysis it is clear that we deal with the relations between a very complex evolving system which we know as the person and an environment which calls upon this system to adjust. Adjustment involves the organization of systems of behavior which mediate between the person and the environmental demands and which have stability in themselves. Thus there enters into our conceptual system the idea of stability by which we refer to the capacity of a system or subsystem to function as a whole and to carry through to some sort of adjustment in spite of defects. The basic concept is that the properties cannot be predicted from knowledge of the parts alone, but only from a knowledge of their inter-relations and how they function as a whole. A living system may even have a considerable capacity to integrate pathology. More than 400 years ago Leonardo defined an arch thus: "An arch is nothing other than a strength caused by two weaknesses; for the arch in buildings is made up of two segments of a circle and each of these segments being in itself very weak desires to fall and as the one withstands the downfall of the other, the two weaknesses are converted into a single strength."

In *Design for a Brain*, Ashby (1954) develops at some length the characteristics of a complex system and shows that as the number of parts increases, the system becomes less dependent upon any particular part of any particular subsystem. A very complex system thus becomes *ultra-stable* or able to adjust under a very wide set of circumstances. With ultra-stability, homeostasis is modified in the direction of less proportionate deviation to particular stresses.

With aging there is an enormous increase in the complexity of the system as a whole, viewed in terms of the number, variety and level of reaction mechanisms and the number of factors which affect present behavior. To meet the stimulation of the immediately present the older persons have available a lifetime of experience and involvement. Hence, theoretically he should be more stable, that is, less affected by experiences which come to him at random. Second, because stimulation and stress are proportionately less disturbing, response may be constricted.

We may now consider the additional factors above those operating in development that have to be considered in constructing a model for aging. There seems to me to be two that must be given serious consideration: The first is concerned with the ordering of experience in the face of the cumulation of effect within the system, a sort of psychological choking which arises because of the time limitations that inhere in a system which possesses the power to accumulate. The second is concerned with the progressive loss of speed in the functioning of the organism. Both of these are general, not specific, factors which affect a wide range of functions.

PROGRESS FROM RANDOM TO ORDERED BEHAVIOR

When Doctor Watson first met Sherlock Holmes he mentioned the fact that the earth went around the sun. Holmes said that he had not known it, and that he would promptly forget it since he could not clutter up his mind with unrelated facts, however interesting they might be. In one sense, development is a process of locating the relations between the apparently isolated facts of experience and ordering them into an integrated and meaningful whole. In the early years of life stimulation from many sources impinges in a somewhat random (to the child) manner. For the child the introduction of new randomized stimulation does not disturb or confuse because his world is essentially disorganized in the sense that he has still to develop the constancies. The random stimulation, however, should be confusing to the older person since it lies outside of the established patterns.

With aging more and more stimulation can be brought within the person's ken and assimilated. But life keeps pouring in and at some point in terms of the sheer mass of past experience the material to be cognized and organized approaches the system's capacity to organize. Many are not as effective in ignoring as Sherlock Holmes. Thus a breakdown comes in the assimilative mechanism. Perhaps old people know too much—the world has poured in on them for a long span of years—the amount of information exceeds the capacity of the channels, and functioning decreases. This is similar to the proaction phenomenon described by Underwood (1957). He found that naive persons reproduced large proportions of lists of nonsense syllables after an interval, while the experienced nonsense syllable learner reproduced small proportions of such lists because of the number of interfering bonds that were carried forward. Retroactive inhibition would function in a similar way.

For logical material, however, the result would be the opposite, that is, the more experienced the person the greater the capacity to organize in meaningful ways. But how far can this go? The illness and death of friends and relatives, accidents to loved ones, the complexities of the community, the nation, and the world become more obvious and the "ifs" of one sort and another occupy more time. There is some evidence that older people are more concerned with religious and philosophical interests; they seem to be attempting to put meaning into a world of stimulation the random characteristic of which has gone beyond their ability to organize in meaningful terms. In their struggle with life the mass of material to be integrated exceeds the capacity of the channels.

This view may also offer some hint of an answer to the question of why decline in mental function is differential with respect to level of ability or attainment. While the facts are not completely clear, the literature suggests that older people of higher levels of intelligence, of higher educational levels and who participate more actively in learning, hold up longer. This may mean more effective handling of the random stimulation coming in from the environment because of greater capacity to organize and integrate experience and thus decrease its randomness of nonsensical character. This may be a kind of ultra-stability that preserves functioning.

SPEED

The second general factor which bulks large in the literature of aging concerns the decline in the speed of functioning in aging which has appeared in many studies of reaction time, serial action, and motility, and in studies of motor and intellectual performance. While many factors affect speed (nerve conduction, refractory phase, loss of inhibition, programming), we may for the moment think of it as a generalized factor and inquire into its effects in terms of the input-outgo energy transformation aspect of the model.

Every total process, whether single and simple or multiple and complex, is a summation of a series of timed reactions. At the minimum there is the time for the incoming stimulation, then time for the transmission of the nervous impulse, then time of decision or choice within the system, and finally the time of response. In addition, the range of response mechanisms, and the difficulty of decision or choice between alternatives affect action.

While we may recognize the fact that in various crafts and arts the quality of the product is independent of time (for example, it makes little difference whether a good poem is written in one week or a year; whether an invention is complete in one month or two years, provided each are made and put into the social stream) from the standpoint of our model when we think of the input-outgo function, it is clear that a reduction in speed or rate of flow of information through the channels of the system and of decision will affect every function and process in some degree and serial or successive functions in marked degree. Its effects will obviously be greater on those activities for which society demands completion within fixed time limits as in production lines or timed tests.

But is the reduction of speed a mere reduction in quantity or does it affect the quality of response as well? One could take the position that response, except in those situations in which speed is essential as in quick reaction time to a signal, would not be affected, that it would merely take longer for qualitatively the same level of response to appear. But it might equally be held that delays in timing change the character as well as the speed of response. Lehman (1953) in his studies of outstanding performers in a great many fields found that the qualitatively better performances came generally between thirty and thirty-five years. When quantity rather than quality measures were used some fields showed that with age the quantity of production remained high and the quality fell. Here is an area for research.

Decision time in relation to social demands may be a more important factor than transmission time. The person is a great device for storing memories and experiences and holding them available for use at appropriate times. Man now can build amazing storage in computers and so make information available promptly and efficiently. But the difference between man and the machine is that man's performance is sharply limited by the speed of his communication system which is very slow compared to what can be built into a computer. Effective reaction involves some selection from among the many alternatives presented in incoming streams of information. The machine sorts out possibilities very quickly, man does so slowly and has only a limited time available. Possibly, given infinite time, the human could accomplish what the machine does quickly, but the human does not have the time, as it lives only twenty-four hours a day and must select. No matter what his potentiality he has both biological limits and a system of constraints imposed by the pressures put upon him. Hence, timing and speed become important dimensions of action and slight reductions in transmission, cumulate into substantial retardations.

Speed of functioning may not only be a matter of transmission time, it may also involve the phenomenon earlier described as the outcome of the cumulation of experience. As experience accumulates, as alternatives increase, as the past flows in to affect the present, decision may become progressively more difficult and a choking of the mechanism be added to the delays in transmission time. In other words, life has a way of becoming too much for the organism in terms of its capacity to integrate and organize experience.

SUMMARY

1. The various models proposed to explain the psychological aspects of aging, namely the one factor model, the machine model, the factor of safety model, and the reversal of development model, when examined, appear to be deficient in one respect or another.

2. If we view the human as a very complex manifold moving forward in time and made up of interacting systems which also interact with the environment, we can check aging against growth and development. When this is done, the older manifold is less open because for many years it has been making binary choices which progressively limit functioning and produce highly selected response patterns. Stability and rigidity result. Although the older manifold retains the capacity to learn and to adjust, this capacity is exercised less often because of the reliance upon the repertoire of existing patterns. Nevertheless, the aging manifold continues to cumulate patterns. The symbolic process also remains at a high level and continues to facilitate all processes.

3. A general factor that comes into the manifold with aging affects many psychological processes. It is like the proaction phenomenon which affects learning of nonsense syllables and reveals itself as inadequate memory. The mass of experience which must be ordered and integrated within a system increases with its age. In the young child stimulation is essentially random and the capacity for organization is high. With development, ordering occurs which makes possible the handling of many diverse stimuli. In the older person because of cumulation, the sheer mass of material that has to be tied together approaches the limits of organizing capacity. The fact that decline in mental function tends to be preserved in persons of high ability and attainment suggests that the greater capacity to integrate a mass of diverse stimuli is preservative of function. But the end result of the proaction factor is kind of choking which is akin to the accumulation of debris within the physical and physiological system that is thought by some to lead to breakdown and death.

4. A second factor which comes into the manifold with aging is the decline in speed, particularly in motility or serial speed. It first appears about the age of thirty-five and thereafter increases with age. This decline affects complex functions because they are a series of integrated time relations in which a disturbance of rate throughout, or a differential disturbance of part rates affects adversely the programming of the whole response pattern.

5. There is some possibility that the proaction and interference factor affects speed and timing since choice points come to be surrounded with more alternatives. If factors such as loss of speed and proaction are operative, the resulting changes are nonreversible. However, a continua-

tion of participation and learning and more particularly of the organizing activities and of the integration which results from them, should delay downward changes.

REFERENCES

Anderson, J. E. "Dynamics of Development: System in Process." In: Harris, D. B. *The Concept of Development*, pp. 25–46. Minneapolis: University of Minnesota Press, 1957.

Anderson, J. E. *Psychological Aspects of Aging*, 323 pp. Washington: American Psychological Association, 1956.

Ashby, W. R. *Design for a Brain*, 260 pp. New York: Wiley, 1954.

Cameron, N. "Reasoning, Regression and Communication in Schizophrenics." *Psychol. Monogr. 50:* 54, 1938.

DeSilva, H. R. "Age and Highway Accidents." *Sci. Mon. 47:* 536–546, 1938.

Kuhlen, R. *Individual Differences in Aging.* Abstract in Division 20, A.P.A. Newsletter *1:* 1–2, 1952.

Lehman, H. C. *Age and Achievement.* Princeton: Princeton University Press, 1953.

Piaget, J. *The Construction of Reality in the Child*, 386 pp. New York: New York Basic Books, 1954.

Underwood, B. J. "Interference and Forgetting." *Psychol. Rev. 64:* 49–60, 1957.

Four

On the Need for Relativism

Jerome Kagan

The psychology of the first half of this century was absolutistic, outer directed, and intolerant of ambiguity. When a college student carries this unholy trio of traits he is called authoritarian, and such has been the temperament of the behavioral sciences. But the era of authoritarian psychology may be nearing its dotage, and the decades ahead may nurture a discipline that is relativistic, oriented to internal processes, and accepting of the idea that behavior is necessarily ambiguous.

Like her elder sisters, psychology began her dialogue with nature using a vocabulary of absolutes. Stimulus, response, rejection, affection, emotion, reward, and punishment were labels for classes of phenomena that were believed to have a fixed reality. We believed we could write a definition of these constructs that would fix them permanently and allow us to know them unequivocally at any time in any place.

Less than seventy-five years ago biology began to drift from the constraints of an absolute view of events and processes when she acknowledged that the fate of a small slice of ectodermal tissue depended on whether it was placed near the area of the eye or the toe. Acceptance of the simple notion that whether an object moves or not depends on where you are standing is a little over a half century old in a science that has five centuries of formalization. With physics as the referent in time, one might expect a relativistic attitude to influence psychology by the latter part of the twenty-third century. But philosophical upheavals in one science catalyze change in other disciplines and one can see signs of budding relativism in the intellectual foundations of the social sciences.

The basic theme of this paper turns on the need for more relativistic

Reprinted from *The American Psychologist*, 1967, Vol. 22, pp. 131–142. Copyright 1967 by the American Psychological Association, and reproduced by permission of publisher and author.

Preparation of this paper was supported in part by research Grant MH–8792 from the National Institute of Mental Health, United States Public Health Service. This paper is an abridged version of a lecture presented at the Educational Testing Service, Princeton, New Jersey, January 1966.

definitions of selected theoretical constructs. "Relativistic" refers to a definition in which context and the state of the individual are part of the defining statement. Relativism does not preclude the development of operational definitions, but makes that task more difficult. Nineteenth-century physics viewed mass as an absolute value; twentieth-century physics made the definition of mass relative to the speed of light. Similarly, some of psychology's popular constructs have to be defined in relation to the state and belief structure of the organism, rather than in terms of an invariant set of external events. Closely related to this need is the suggestion that some of the energy devoted to a search for absolute, stimulus characteristics of reinforcement be redirected to a search for the determinants of attention in the individual.

It is neither possible nor wise to assign responsibility to one person or event for major changes in conceptual posture, but Helson's recent book on adaptation-level theory (Helson, 1964), Schachter's (Schachter and Singer, 1962) hypothesis concerning the cognitive basis of affects, and Hernández-Peón's demonstration of the neurophysiological bases of selective attention (Hernández-Peón, Scherrer, and Jouvet, 1956) are contemporary stimulants for a relativistic view of psychological phenomena.

Three messages are implicit in the work of these men.

1. If a stimulus is to be regarded as an event to which a subject responds or is likely to respond then it is impossible to describe a stimulus without describing simultaneously the expectancy, and preparation of the organism for that stimulus. Effective stimuli must be distinct from the person's original adaptation level. Contrast and distinctiveness, which are relative, are part and parcel of the definition of a stimulus.

2. The failure of one individual to respond to an event that is an effective stimulus for a second individual is not always the result of central selection after all the information is in, but can be due to various forms of peripheral inhibition. Some stimuli within inches of the face do not ever reach the interpretive cortex and, therefore, do not exist psychologically.

3. Man reacts less to the objective quality of external stimuli than he does to categorizations of those stimuli.

These new generalizations strip the phrase "physical stimulus" of much of its power and certainty, and transfer the scepter of control—in man, at least—to cognitive interpretations. *Contrast, cognitively interpreted, becomes an important key to understanding the incentives for human behavior.* Since contrast depends so intimately on context and expectancy, it must be defined relativistically.

The issue of relativism can be discussed in many contexts. Many existing constructs are already defined in terms of contextual relations.

The concept of authority only has meaning if there are fiefs to rule. The role of father has no meaning without a child. The concept of noun, verb, or adjective is defined by context—by the relation of the word to other constituents. We shall consider in some detail the ways in which a relativistic orientation touches two other issues in psychology: the learning of self-descriptive statements (the hoary idea of the self-concept), and, even more fundamentally, some of the mechanisms that define the learning process.

THE CONCEPT OF THE SELF

The development and establishment of a self-concept is often framed in absolute terms. The classic form of the statement assumes that direct social reinforcements and identification models have fixed, invariant effects on the child. Praise and love from valued caretakers are assumed to lead the child to develop positive self-evaluations; whereas, criticism and rejection presumably cause self-derogatory beliefs. The presumed cause-effect sequences imply that there is a something—a definable set of behaviors—that can be labeled social rejection, and that the essence of these rejecting acts leads to invariant changes in the self-concept of the child. Let us examine the concept of rejection under higher magnification.

The concept of rejection—peer or parental—has been biased toward an absolute definition. Witness the enormous degree of commonality in conceptualization of this concept by investigators who have studied a mother's behavior with her child (Baldwin, Kalhorn, and Breese, 1945; Becker, 1964; Kagan and Moss, 1962; Schaefer, 1959; Schaefer and Bayley, 1963; Sears, Maccoby, and Levin, 1957). These investigators typically decide that harsh physical punishment and absence of social contact or physical affection are the essential indexes of an attitude called maternal rejection. It would be close to impossible for an American rater to categorize a mother as high on both harsh beating of her child and on a loving attitude. A conventionally trained psychologist observing a mother who did not talk to her child for 5 hours would probably view the mother as rejecting. This may be a high form of provincialism. Alfred Baldwin [personal communication] reports that in the rural areas of northern Norway, where homes are 5 to 10 miles apart, and the population constant for generations, one often sees maternal behaviors which an American observer would regard as pathognomonically rejecting in an American mother. The Norwegian mother sees her four year old sitting in the doorway blocking the passage to the next room. She does not ask him to move, but bends down, silently picks him up and moves him away before she passes into the next room. Our middle-class observer would be tempted to view this indifference as a sign of dislike. However, most mothers in this Arctic outpost behave this way and the children do not behave the way rejected children should by our current theoretical propositions.

An uneducated Negro mother from North Carolina typically slaps her four-year-old across the face when he does not come to the table on time. The intensity of the mother's act tempts our observer to conclude that the mother hates, or at best, does not like her child. However, during a half-hour conversation the mother says she loves her child and wants to guarantee that he does not grow up to be a bad boy or a delinquent. And she believes firmly that physical punishment is the most effective way to socialize him. Now her behavior seems to be issued in the service of affection rather than hate. Determination of whether a parent is rejecting or not cannot be answered by focusing primarily on the behaviors of the parents. Rejection is not a fixed, invariant quality of behavior qua behavior. Like pleasure, pain, or beauty, rejection is in the mind of the rejectee. It is a belief held by the child; not an action by a parent.

We must acknowledge, first, a discontinuity in the meaning of an acceptance-rejection dimension before drawing further implications. We must distinguish between the child prior to thirty or thirty-six months of age, before he symbolically evaluates the actions of others, and the child thereafter.

We require, first, a concept to deal with the child's belief of his value in the eyes of others. The child of four or five years is conceptually mature enough to have recognized that certain resources parents possess are difficult for the child to obtain. He views these resources as sacrifices and interprets their receipt as signs that the parents value him. The child constructs a tote board of the differential value of parental gifts—be they psychological or material. The value of the gift depends on its scarcity. A $10.00 toy from a busy executive father is not a valued resource; the same toy from a father out of work is much valued. The value depends on the child's personal weightings. This position would lead to solopsism were it not for the fact that most parents are essentially narcissistic and do not readily give the child long periods of uninterrupted companionship. Thus, most children place high premium on this act. Similarly, parents are generally reluctant to proffer unusually expensive gifts to children, and this act acquires value for most youngsters. Finally, the child learns from the public media that physical affection means positive evaluation and he is persuaded to assign premium worth to this set of acts. There is, therefore, some uniformity across children in a culture in the evaluation of parental acts. But the anchor point lies within the child, not with the particular parental behaviors.

This definition of acceptance or rejection is not appropriate during the opening years. The one-year-old does not place differential symbolic worth on varied parental acts, and their psychological significance derives from the overt responses they elicit and strengthen. A heavy dose of vocalization and smiling to an infant is traditionally regarded as indicative of maternal affection and acceptance. This bias exists because we have accepted the myth that "affection" is the essential nutrient that produces socially adjusted children, adolescents, and adults. The bias maintains itself because we observe a positive association between degree of parental smiling and laughing to the infant and prosocial behavior in the child during the early years. The responses of smiling, laughing, and approaching people are learned in the opening months of life on the basis of standard conditioning principles. This conclusion is supported by the work of Rheingold and Gewirtz (1959) and Brackbill (1958). However, phenotypically similar behaviors in a ten- or twenty-year-old may have a different set of antecedents. The argument that different definitions of rejection-acceptance must be written for the pre- and postsymbolic child gains persuasive power from the fact that there are no data indicating that degree of prosocial behavior in the child is stable from six months to sixteen years. Indeed, the longitudinal material from the Fels Research Institute study of behavior stability (Kagan and Moss, 1962) showed no evidence of any relation between joy or anxiety in the presence of adults during the first 2–3 years of life and phenotypically similar behaviors at six, twelve, or twenty-four years of age. The child behaviors that are presumed, by theory, to be the consequences of low or high parental rejection do not show stability from infancy through adolescence. This may be because the childhood responses, though phenotypically similar to the adult acts, may be acquired and maintained through different experiences at different periods.

It seems reasonable to suggest, therefore, that different theoretical words are necessary for the following three classes of phenomena: (*a*) an attitude on the part of the parent, (*b*) the quality and frequency of acts of parental care and social stimulation directed toward the infant, and (*c*) a child's assessment of his value in the eyes of another. All three classes are currently viewed as of the same cloth. The latter meaning of "rejection" (that is, a belief held by a child) is obviously relativistic for it grows out of different experiences in different children.

SELF-DESCRIPTIVE LABELS

Let us probe further into the ideas surrounding the learning of self-evaluation statements, beyond the belief, "I am not valued." The notion of a self-concept has a long and spotted history and although it has masqueraded by many names in different theoretical costumes, its intrinsic meaning has changed only a little. A child presumably learns self-descriptive statements whose contents touch the salient attributes of the culture. The mechanisms classically invoked to explain how these attributes are learned have stressed the invariant effects of direct social reinforcement and identification. The girl who is told she is attractive, annoying, or inventive, comes to believe these appellations and to apply these qualifiers to herself. We have assumed that the laws governing the learning of self-descriptive labels resemble the learning of other verbal habits with frequency and contiguity of events being the shapers of the habit. Identification as a source of self-labels involves a different mechanism, but retains an absolutistic frame of reference. The child assumes that he shares attributes with particular models. If the model is viewed as subject to violent rages, the child concludes that he, too, shares this tendency.

Theory and data persuade us to retain some faith in these propositions. But relativistic factors also seem to sculpt the acquisition of self-descriptive labels, for the child evaluates himself on many psychological dimensions by inferring his rank order from a delineated reference group. The ten-year-old does not have absolute measuring rods to help him decide how bright, handsome, or likable he is. He naturally and spontaneously uses his immediate peer group as the reference for these evaluations. An immediate corollary of this statement is that the child's evaluation is dependent upon the size and psychological quality of the reference group, and cannot be defined absolutely. Specifically, the larger the peer group, the less likely a child will conclude he is high in the rank order, the less likely he will decide he is unusually smart, handsome, or capable of leadership. Consider two boys with I.Q.s of 130 and similar intellectual profiles. One lives in a small town, the other in a large city. It is likely that the former child will be the most competent in his peer group while the latter is likely to regard himself as fifth or sixth best. This difference in perceived rank order has obvious action consequences since we acknowledge that expectancies govern behavior. In sum, aspects of the self-descriptive process appear to develop in relativistic soil.

LEARNING AND ATTENTION

A second issue that touches relativistic definitions deals with a shift from external definitions of reinforcement—that is, reward or pleasure—to definitions that are based more directly on internal processes involving the concept of attention. Failure to understand the nature of learning is one of the major intellectual frustrations for many psychologists. The query, "What is learning?" has the same profound ring as the question, "What is a gene?" had a decade ago. Our biological colleagues have recently had a major insight while psychology is still searching. The murky question, "What is learning?" usually reduces to an attempt to discover the laws relating stimuli, pain, and pleasure, on the one hand, with habit acquisition and performance, on the other. Pain, pleasure, and reinforcement, are usually defined in terms of events that are external to the organism and have an invariant flavor. Miller (1951) suggested that reinforcement was isomorphic with stimulus reduction; Leuba (1955) argued for an optimal level of stimulation, but both implied that there was a level that could be specified and measured. We should like to argue first that sources of pleasure and, therefore of reinforcement, are often relative, and second, that the essence of learning is more dependent on attentional involvement by the learner than on specific qualities of particular external events.

The joint ideas that man is a pleasure seeker and that one can designate specific forms of stimulation as sources of pleasure are central postulates in every man's theory of behavior. Yet we find confusion when we seek a definition of pleasure. The fact that man begins life with a small core set of capacities for experience that he wishes to repeat cannot be disputed. This is a pragmatic view of pleasure and we can add a dash of phenomenology to bolster the intuitive validity of this point of view. A sweet taste and a light touch in selected places are usually pleasant. Recently, we have added an important new source of pleasure. It is better to say we have rediscovered a source of pleasure, for Herbert Spencer was a nineteenth-century progenitor of the idea that *change in stimulation* is a source of pleasure for rats, cats, monkeys, or men. But, change is short-lived, quickly digested, and transformed to monotony. Popping up in front of an infant and saying peek-a-boo is pleasant for a three-month-old infant for about fifteen minutes, for a ten-month-old infant for three minutes and for a thirty-month-old child, a few seconds. This pleasant experience, like most events that elicit their repetition a few times before dying, is usually conceptualized as a change in stimulation. The source of the pleasure is sought in the environment. Why should change in external stimulation be pleasant? The understanding of pleasure and reinforcement in man is difficult enough without having to worry about infrahuman considerations. Let us restrict the argument to the human. The human is a cognitive creature who is attempting to put structure or create schema for incoming stimulation. A schema is a representation of an external pattern; much as an artist's illustration is a representation of an event. A schema for a visual pattern is a partial and somewhat distorted version of what the photograph would be. Consider the usefulness of the following hypothesis:

The creation of a schema for an event is one major source of pleasure. When one can predict an event perfectly, the schema is formed. As long as prediction is not perfect the schema is not yet formed. The peek-a-boo game works for 15 minutes with a twelve-week-old for it takes him that long to be able to predict the event—the "peek-a-boo." Charlesworth (1965) has demonstrated the reinforcing value of "uncertainty" in an experiment in which the peek-a-boo face appeared either in the same locus every trial, alternated between two loci, or appeared randomly in one of two loci. The children persisted in searching for the face for a much longer time under the random condition than under the other two conditions. The random presentation was reinforcing for a longer period of time, not because it possessed a more optimum level of external stimulation than the other reinforcement schedules, but because it took longer for the child to create a schema for the random presentation and the process of creating a schema is a source of pleasure.

Consider another sign of pleasure beside persistence in issuing a particular response. Display of a smile or laugh is a good index of pleasure. Indeed, Tomkins' (1962) scheme for affect demands that pleasure be experienced if these responses appear. Consider two studies that bear on the relation between pleasure and the creation of schema. In our laboratory during the last two years, we have seen the same infants at four, eight, and thirteen months of age and have shown them a variety of visual patterns representative of human faces and human forms. In one episode, the four-month-old infants are shown achromatic slides of a photograph of a regular male face, a schematic outline of a male face, and two disarranged, disordered faces. The frequency of occurrence of smiling to the photograph of the regular face is over *twice* the frequency observed to the regular schematic face—although looking time is identical—and over *four times* the frequency shown to the disordered faces. In another, more realistic episode, the four-month-old infants see a regular, flesh-colored sculptured face in three dimensions and a distorted version of that face in which the eyes, nose, and mouth are rearranged. At four months of age the occurrence of smiling to the regular face is over three times the frequency displayed to the distorted version, but looking time is identical. There are two interpretations of this difference (Kagan, Henker, Hen-Tov, Levine, and Lewis, 1966). One explanation argues that the mother's face has become a secondary reward; the regular face stands for pleasure because it has been associated with care and affection from the mother. As a result, it elicits more smiles. An alternative interpretation is that the smile response has become conditioned to the human face via reciprocal contact between mother and infant. A third interpretation, not necessarily exclusive of these, is that the smile can be elicited when the infant matches stimulus to schema—when he has an "aha" reaction; when he makes a cognitive discovery. The four-month-old infant is cognitively close to establishing a relatively firm schema of a human face. When a regular representation of a face is presented to him there is a short period during which the stimulus is assimilated to the schema and then after several seconds, a smile may occur. The smile is released following the perceptual recognition of the face, and reflects the assimilation of the stimulus to the infant's schema—a small, but significant act of creation. This hypothesis is supported by the fact that the typical latency between the onset of looking at the regular face (in the four-month-old) and the onset of smiling is about three to five seconds. The smile usually does not occur immediately but only after the infant has studied the stimulus. If one sees this phenomenon live, it is difficult to avoid the conclusion that the smile is released following an act of perceptual recognition.

Additional data on these and other children at eight months of age support this idea. At eight months, frequency of smiling to both the regular and distorted faces is *reduced dramatically*, indicating that smiling does not covary with the reward value of the face. The face presumably has acquired more reward value by eight months than it had at four months. However, the face is now a much firmer schema and recognition of it is immediate. There is no effortful act of recognition necessary for most infants. As a result, smiling is less likely to occur. Although smiling is much less frequent at eight than four months to all faces, the frequency of smiling to the distorted face now *equals* the frequency displayed to the regular face. We interpret this to mean that the distorted face is sufficiently similar to the child's schema of a regular face that it can be recognized as such.

The pattern of occurrence of cardiac deceleration to the regular and distorted three-dimensional faces furnishes the strongest support for this argument. A cardiac deceleration of about eight to ten beats often accompanies attention to selected incoming visual stimuli in adults, school-age children, and infants. Moreover, the deceleration tends to be maximal when the stimuli are not overly familiar or completely novel, but are of intermediate familiarity. One hypothesis maintains that a large deceleration is most likely to occur when an act of perceptual recognition occurs, when the organism has a cognitive surprise. Let us assume that there is one trial for which this type of reaction occurs with maximal magnitude. If one examines the one stimulus presentation (out of a total of 16 trials) that produces the largest cardiac deceleration, a lawful change occurs between four and eight months of age. At four months of age more of the infants showed their largest deceleration to the regular face (45 percent of the group: $n = 52$) than to the scrambled (34 percent), no eyes (11 percent), or blank faces (10 percent). At eight months, the majority of the infants ($n = 52$) showed their largest deceleration to the scrambled face (50 percent to scrambled versus 21 percent to regular face). This difference is interpreted to mean that the scrambled face now assumes a similar position on the assimilation continuum that the regular face did sixteen weeks earlier.

At thirteen months of age these infants are shown six three-dimensional representations of a male human form and a free form matched for area, coloration, and texture with the human form. The stimuli include a faithful representation of a regular man, that same man with his head placed between his legs, the same man with all limbs and head collaged in an unusual and scrambled pattern, the man's body with a mule's head, and the mule's head on the man's body, the man's body with three identical heads, and a free form. The distribution of smiles to these stimuli is leptokurtic, with over 70 percent of all the smiles occurring to the animal head on the human body and the three-headed man, forms that were moderate transformations of the regular man, and stimuli that required active assimilation. The free form and the scrambled man rarely elicited smiles from these infants. These stimuli are too difficult to assimilate to the schema of a human form possessed by a thirteen-month-old infant. It is interesting to note that the regular human form sometimes elicited the verbal response "daddy" or a hand waving from the child. These instrumental social reactions typically did not occur to the transformations. The occurrence of cardiac deceleration to these patterns agrees with this hypothesis. At thirteen months of age, the man with his head between his legs, the man with the animal head, or the three-headed man, each elicited the largest cardiac decelerations more frequently than the regular man, the scrambled man, or the free form ($p < .05$ for each comparison). Thus, large cardiac decelerations and smiles were most likely to occur to stimuli that seemed to require tiny, quiet cognitive discoveries—miniaturized versions of Archimedes' "Eureka."

It appears that the act of matching stimulus to schema when the match is close but not yet perfect is a dynamic event. Stimuli that deviate a critical amount from the child's schema for a pattern are capable of eliciting an active process of recognition, and this process behaves as if it were a source of pleasure. Stimuli that are easily assimilable or too difficult to assimilate do not elicit these reactions.

A recent study by Edward Zigler [unpublished paper; personal communication] adds important support to the notion that the smile indicates the pleasure of an assimilation. Children in Grades 2, 3, 4, and 5 looked at cartoons that required little or no reading. The children were asked to explain the cartoon while an observer coded the spontaneous occurrence of laughing and smiling while the children were studying the cartoons. It should come as no surprise that verbal comprehension of the cartoons increased in a linear fashion with age. But laughing and smiling increased through Grade 4 and then declined markedly among the fifth-grade children. The fifth graders understood the cartoons too well. There was no gap between stimulus and schema and no smiling. Sixteen-week-old infants and eight-year-old children smile spontaneously at events that seem to have one thing in common—the event is a partial match to an existing schema and an active process of recognitory assimilation must occur.

The fact that a moderate amount of mismatch between event and schema is one source of pleasure demands the conclusion that it is not always possible to say that a specific event will always be a source of pleasure. The organism's state and structure must be in the equation. This conclusion parallels the current interest in complexity and information uncertainty. The psychologist with an information-theory prejudice classifies a stimulus as uncertain and often assumes that he does not have to be too concerned with the attributes of the viewer. This error of the absolute resembles the nineteenth-century error in physics and biology. This is not a titillating or pedantic, philosophical issue. Psychology rests on a motive-reinforcement foundation which regards pleasure and pain as pivotal ideas in the grand theory. These constructs have tended to generate absolute definitions. We have been obsessed with finding a fixed and invariant characterization of pleasure, pain, and reinforcement. Melzack and Wall (1965) point out that although the empirical data do not support the notion of a fixed place in the brain that mediates pain, many scientists resist shedding this comfortable idea. Olds' (1958, 1962) discovery of brain reinforcing areas has generated excitement because many of us want to believe that pleasure has a fixed and absolute locus. The suspicious element in this discovery of pleasure spots is that there is no habituation of responses maintained by electrical stimulation to hypothalamic or septal nuclei, and minimal resistance to extinction of habits acquired via this event. Yet, every source of pleasure known to phenomenal man does satiate—for awhile or forever—and habits that lead to pleasant events do persist for awhile after the pleasure is gone. These observations are troubling and additional inquiry is necessary if we are to decide whether these cells are indeed the bed where pleasure lies.

We are convinced that contiguity alone does not always lead to learning. Something must ordinarily be added to contiguity in order to produce a new bond. Psychology has chosen to call this extra added mysterious something reinforcement, much like eighteenth-century chemists chose to label their unknown substance phlogiston. If one examines the variety of external events that go by the name of reinforcement it soon becomes clear that this word is infamously inexact. A shock to an animal's paw is a reinforcement, a verbal chastisement is a reinforcement, an examiner's smile is a reinforcement, a pellet of food is a reinforcement, and a sigh indicating tension reduction after watching a killer caught in a Hitchcock movie is a reinforcement. These events have little, if any, phenotypic similarity. What then, do they have in common? For if they have nothing in common it is misleading to call them by the same name. Learning theorists have acknowledged their failure to supply an independent a priori definition of reinforcement and the definition they use is purely pragmatic. A reinforcement is anything that helps learning. And so, we ask: What has to be added to contiguity in order to obtain learning? A good candidate for the missing ingredient is the phrase "attentional involvement." Let us consider again the events called reinforcements· a shock, food, a smile, each of these acts to attract the attention of the organism to some agent or object. They capture the organism's attention and maybe that is why they facilitate learning. Consider the idea that what makes an event reinforcing is the fact that it (a) elicits the organism's attention to the feedback from the response he has just made and to the mosaic of stimuli in the learning situation and (b) acts as an incentive for a subsequent response. The latter quality is what ties the word "reinforcement" to the concepts of motivation and need, but much learning occurs without the obvious presence of motives or needs. Ask any satiated adult to attend carefully and remember the bond syzgy-aardvark. It is likely that learning will occur in one trial. It is not unreasonable to argue that a critical component of events that have historically been called reinforcement is their ability to attract the organism's attention. They have been distinctive cues in a context; they have been events discrepant from the individual's adaptation level. If attention is acknowledged as critical in new mental acquisitions it is appropriate to ask if attention is also bedded in relativistic soil. The answer appears to be "Yes." The dramatic experiments of Hernández-Peón and his colleagues (1956) are persuasive in indicating that attention investment may not be distributed to many channels at once. One has to know the state of the organism. Knowledge of the organism's distribution of attention in a learning situation may clarify many controversial theoretical polemics that range from imprinting in chickens to emotion in college undergraduates. For example, comparative psychologists quarrel about which set of external conditions allow imprinting to occur with maximal effect. Some say the decoy should move; others argue that the young chick should move; still others urge that the decoy be brightly colored (for example, Bateson, 1964a, 1964b; Hess, 1959; Klopfer, 1965; Thompson and Dubanoski, 1964). The quarrel centers around the use of phenotypically different observable conditions. Perhaps all these suggestions are valid. Moving the decoy, or active following by the infant chick, or a distinctively colored decoy all maximize the organism's level of attention to the decoy. The genotypic event may remain the same across all of these manipulations.

A similar interpretation can be imposed on Held's (1965) recent hypothesis concerning the development of space and pattern perception.

Held controlled the visual experience of pairs of kittens. The only exposure to light was limited to a few hours a day when one kitten was placed in a gondola and moved around by an active, free kitten in an arena whose walls were painted in vertical stripes. After 30 hours of such experience each kitten was tested. The free kitten showed appropriate visual reactions. It blinked when an object approached; it put up its paws to avoid collision when carried near to a surface; it avoided the deep side of a visual cliff. The passive passenger kitten did not show these normal reactions. Why? Held, focusing on the obvious external variable of activity versus no activity, concludes that the sensory feedback accompanying movement is necessary to develop visual-motor control. This conclusion rests on the assumption that the passive kitten sitting in the gondola was attending to the stripes on the wall as intently as the free walking kitten. This assumption may be gratuitous. If the passive kitten were staring blankly—as many human infants do—then one would not expect these animals to develop normal perceptual structures. This interpretation may not be better, but it has a different flavor than the one suggested by Held.

A final example of the central role of attention is seen in Aronfreed's (1964, 1965) recent work on the learning of self-critical comments. Aronfreed states that the learning of a self-critical comment proceeds best if the child is first punished and then hears a social agent speak the self-critical statement. He interprets this result in drive reduction language. However, suppose one asks which sequence is most likely to maximize a child's attention to the adult's comment—Punish first and then speak to the child? Or speak first and then punish? The former sequence should be more effective. The punishment is a violation of what the child expects from a strange adult and recruits the child's attention to the adult. The child is primed to listen to the self-critical commendation and thus more likely to learn it.

DISTINCTIVENESS OF CUES

The above examples suggest that the organism's distribution of attention is a critical process that should guide our search for the bases of many diverse phenomena. One of the critical bases for recruitment of attention pivots on the idea of distinctiveness of the signal. Jakobson and Halle (1956) argue that the chronology of acquisition of phonemes proceeds according to a principle of distinctive elements. Distinctive elements capture the child's attention and give direction to the order of learning.

The importance of *relative distinctiveness of cues* finds an interesting illustration in the concept of affect. The concept of emotion has lived through three distinct eras in modern times. The pre-Jamesian assumed the sequence was: stimulus event–cognition–visceral response. James interchanged events two and three and said that the visceral afferent feedback occurred before the cognition. But Cannon quieted Jamesian ideas until Schachter's ingenious studies and catching explanations suggested that the individual experiences a puzzling set of visceral afferent sensations and integrates them cognitively. The language integration of visceral feelings, cognition, and context is an affect. This imaginative suggestion may be maximally valid for Western adults but perhaps minimally appropriate for children because of a developmental change in the relative distinctiveness of visceral cues.

Let us share a small set of assumptions before we proceed with the argument. Aside from pain and its surrogates, the major psychological elicitors of unpleasant visceral afferent sensations are violations of expectancies (uncertainty); anticipation of receiving or losing a desired goal; anticipation of seeing or losing a nurturant person; blocking of goal attainment; and anticipation of harm to the integrity of the body. Each of these event situations becomes conditioned to visceral afferent feedback early in life. These events—or conditioned stimuli—are salient and maximally distinctive for children and affect words are attached to the events, not primarily to the visceral afferent sensations. Thus, the six-year-old says he is mad because mother did not let him watch television; he says he is sad because the cat died; he says he is happy because he just received a prized toy. Affect words are labels for a set of external events. With development, individuals—with the help of social institutions—learn to protect themselves against most of the unpleasant sources of visceral afferent feedback—against the apocalyptic horsemen of uncertainty, loss of nurturance, goal blocking, and bodily harm. Moreover, they erect defenses against recognizing these events. They defend against recognition that they are confused, rejected, unable to attain a goal, or afraid. Thus, when events occur that are, in fact, representations of these situations, the events are not salient or distinctive and are not labeled. However, the conditioned visceral afferent sensations do occur, as they always have in the past. In the adult, the visceral afferent sensations become more distinctive or salient; whereas, for the child, the external events were salient and distinctive. The adult provides us with the situation Schachter and his colleagues have described. The adult often has visceral afferent sensations but cannot decide why he has them or what they mean. So he scans and searches the immediate past and context and decides that he is happy, sad, alienated, uncommitted, or in love. The essence of this argument is that for the child the external event is more distinctive than the visceral afferent sensations and the affect word is applied to external events. In the adult, the visceral afferent sensations are relatively more distinctive and the affect words are more often applied to them.

The personality differences ascribed to children in different ordinal positions are the result, in part, of differences in relative distinctiveness of social agents. For the firstborn, the adult is the distinctive stimulus to whom to attend; for the second born the older sibling has distinctive value and competes for the attention of the younger child. Only children lie alone for long periods of uninterrupted play. A parent who enters the room and speaks to the infant is necessarily a distinctive stimulus. For a fifth-born whose four older siblings continually poke, fuss, and vocalize into the crib, the caretaking adult, is, of necessity, less distinctive and, as a result, less attention will be paid to the adult. The importance of distinctiveness with respect to adaptation level engages the heated controversy surrounding the role of stimulus enrichment with infants and young children from deprived milieux. The pouring on of visual, auditory, and tactile stimulation willy-nilly should be less effective than a single distinctive stimulus presented in a context of quiet so it will be discrepant from the infant's adaptation level. If one takes this hypothesis seriously, a palpable change in enrichment strategies is implied. The theme of this change involves a shifting from a concern with increasing absolute level of stimulation to focusing on distinctiveness of stimulation. Culturally disadvantaged children are not deprived of stimulation; they are deprived of distinctive stimulation.

The early learning of sex role standards and the dramatic concern of school children with sex differences and their own sex role identity becomes reasonable when one considers that the differences between the sexes are highly distinctive. Voice, size, posture, dress, and usual locus of behavior are distinctive attributes that focus the child's attention on them.

One of the reasons why the relation between tutor and learner is important is that some tutors elicit greater attention than others. They are more distinctive. Those of us who contend that learning will be facilitated if the child is identified with or wants to identify with a tutor believe that one of the bases for the facilitation is the greater attention that is directed at a model with whom the child wishes to identify. A recent experiment touches on this issue.

The hypothesis can be simply stated. An individual will attend more closely to an initial stranger with whom he feels he shares attributes than to a stranger with whom he feels he does not share attributes, other things equal. The former model is more distinctive, for a typical adult ordinarily feels he does not share basic personality traits with most of the strangers that he meets. The subjects in this study were 56 Radcliffe freshmen and sophomores preselected for the following pair of traits. One group, the academics, were rated by four judges—all roommates—as being intensely involved in studies much more than they were in dating, clubs, or social activities. The second group, the social types, were rated as being much more involved in dating and social activities than they were in courses or grades. No subject was admitted into the study unless all four judges agreed that she fit one of these groups.

Each subject was seen individually by a Radcliffe senior, and told that each was participating in a study of creativity. The subject was told that Radcliffe seniors had written poems and that two of the poets were selected by the Harvard faculty as being the best candidates. The faculty could not decide which girl was the more creative and the student was going to be asked to judge the creativity of each of two poems that the girls had written. The subjects were told that creativity is independent of I.Q. for bright people and they were told that since the faculty knew the personality traits of the girls, the student would be given that information also. The experimenter then described one of the poets as an academic grind and the other as a social activist. Each subject listened to two different girls recite two different poems on a tape. Order of presentation and voice of the reader were counterbalanced in an appropriate design. After the two poems were read the subject was asked for a verbatim recall of each poem, asked to judge its creativity, and finally, asked which girl she felt most similar to. Incidentally, over 95 percent of the subjects said they felt more similar to the model that they indeed matched in reality. Results supported the original hypothesis. Recall was best when a girl listened to a communicator with whom she shared personality traits. The academic subjects recalled more of the poem when it was read by the academic model than by the social model; whereas, the social subjects recalled more of the poem when it was read by the social model than the academic model. This study indicates that an individual will pay more attention to a model who possesses similar personality attributes, than to one who is not similar to the subject. Distinctiveness of tutor is enhanced by a perceived relation between learner and tutor.

Myths and superstitions are established around the kinds of experimental manipulations teachers or psychologists should perform in order to maximize the probability that learning will occur. When one focuses on the kind of manipulation—providing a model, giving a reinforcement, labeling the situation, punishing without delay—there is a strong push to establish superstitions about how behavioral change is produced. Recipes are written and adopted. If one believes, on the other hand, that a critical level of attention to incoming information is the essential variable, then one is free to mix up manipulations, to keep the recipe open, as long as one centers the subject's attention on the new material.

The most speculative prediction from this general argument is that behavioral therapy techniques will work for some symptoms—for about twenty years. A violation of an expectancy is a distinctive stimulus that attracts attention. The use of operant shaping techniques to alleviate phobias is a dramatic violation of an expectancy for both child and adult, and attention is magnetized and focussed on the therapeutic agent and his paraphernalia. As a result, learning is facilitated. But each day's use of this strategy may bring its demise closer. In time, a large segment of the populace will have adapted to this event; it will be a surprise no more and its attention getting and therapeutic value will be attenuated. Much of the power of psychoanalytic techniques began to wane when the therapist's secrets became public knowledge. If therapy is accomplished by teaching new responses, and if the learning of new responses is most likely to occur when attention to the teacher is maximal, it is safe to expect that we may need a new strategy of teaching patients new tricks by about 1984.

Let us weave the threads closer in an attempt at final closure. The psychology of the first half of this century was the product of a defensively sudden rupture from philosophy to natural science. The young discipline needed roots, and like a child, attached itself to an absolute description of nature, much as a five-year-old clings to an absolute conception of morality. We now approach the latency years and can afford to relax and learn something from developments in our sister sciences. The message implicit in the recent work in psychology, biology, and physics contains a directive to abandon absolutism in selected theoretical areas. Conceptual ideas for mental processes must be invented, and this task demands a relativistic orientation. Learning is one of the central problems in psychology and understanding of the mechanisms of learning requires elucidation and measurement of the concept of attention. Existing data indicate that attention is under the control of distinctive stimuli and distinctiveness depends intimately on adaptation level of subject and context, and cannot be designated in absolute terms.

These comments are not to be regarded as a plea to return to undisciplined philosophical introspection. Psychology does possess some beginning clues as to how it might begin to measure elusive, relative concepts like "attention." Automatic variables such as cardiac and respiratory rate appear to be useful indexes, and careful studies of subtle motor discharge patterns may provide initial operational bases for this construct.

Neurophysiologists have been conceptualizing their data in terms of attention distribution for several years, and they are uncovering some unusually provocative phenomena. For example, amplitude of evoked potentials from the association areas of the cortex are beginning to be regarded as a partial index of attention. Thompson and Shaw (1965) recorded evoked potentials from the association area of the cat's cortex—the middle suprasylvian gyrus—to a click, a light, or a shock to the forepaw. After they established base level response to each of these "standard" stimuli, the investigators presented these standard stimuli when the cat was active or when novel stimuli were introduced. The novel events were a rat in a bell jar, an air jet, or a growling sound. The results were unequivocal. Any one of these novel stimuli or activity by the cat produced reduced cortical evoked responses to the click, light, or shock. The authors suggest that the "amplitude of the evoked responses are inversely proportional to attention to a particular event (p. 338)." Psychology is beginning to develop promising strategies of measurement for the murky concept of attention and should begin to focus its theorizing and burgeoning measurement technology on variables having to do with the state of the organism, not just the quality of the external stimulus. The latter events can be currently objectified with greater elegance, but the former events seem to be of more significance. Mannheim once chastised the social sciences for seeming to be obsessed with studying what they could measure without error, rather than measuring what they thought to be important with the highest precision possible. It is threatening to abandon the security of the doctrine of absolutism of the stimulus event. Such a reorientation demands new measurement procedures, novel strategies of inquiry, and a greater tolerance for ambiguity. But let us direct our inquiry to where the pot of gold seems to shimmer and not fear to venture out from cozy laboratories where well-practiced habits have persuaded us to rationalize a faith in absolute monarchy.

REFERENCES

Aronfreed, J. "The origin of self criticism." *Psychological Review* 71: 193–218, 1964.

Aronfreed, J. "Internalized behavioral suppression and the timing of social punishment." *Journal of Personality and Social Psychology* 1: 3–16, 1965.

Baldwin, A. L., J. Kalhorn, and F. H. Breese. "Patterns of parent behavior." *Psychological Monographs* 58 (3, Whole No. 268), 1945.

Bateson, P. P. G. "Changes in chicks' responses to novel moving objects over the sensitive period for imprinting." *Animal Behavior* 12: 479–489, 1964. (a)

Bateson, P. P. G. "Relation between conspicuousness of stimuli and their effectiveness in the imprinting situation." *Journal of Comparative and Physiological Psychology* 58: 407–411, 1964. (b)

Becker, W. C. "Consequences of different kinds of parental discipline." In: Hoffman, M. L., and L. W. Hoffman (eds.), *Review of child development research.* Vol. 1. New York: Russell Sage Foundation, 1964, pp. 169–208.

Brackbill, Y. "Extinction of the smiling response in infants as a function of reinforcement schedule." *Child Development* 29: 115–124, 1958.

Charlesworth, W. R. *Persistence of orienting and attending behavior in young infants as a function of stimulus uncertainty.* Paper read at Society for Research in Child Development, Minneapolis, March 1965.

Held, R. "Plasticity in sensory motor systems." *Scientific American 213* (5): 84–94, 1965.

Helson, H. *Adaptation level theory: An experimental and systematic approach to behavior.* New York: Harper & Row, 1964.

Hernández-Peón, R., H. Scherrer, and M. Jouvet. "Modification of electrical activity in cochlear nucleus during attention in unanesthetized cats." *Science 123:* 331–332, 1956.

Hess, E. H. "Two conditions limiting critical age for imprinting." *Journal of Comparative and Physiological Psychology 52:* 515–518, 1959.

Jakobson, R., and M. Halle. *Fundamentals of language.* The Hague: Mouton, 1956.

Kagan, J., B. A. Henker, A. Hen-Tov, J. Levine, and M. Lewis. "Infants' differential reactions to familiar and distorted faces." *Child Development 37:* 519–532, 1966.

Kagan, J., and H. A. Moss. *Birth to maturity.* New York: Wiley, 1962.

Klopfer, P. H. "Imprinting: A reassessment." *Science 147:* 302–303, 1965.

Leuba, C. "Toward some integration of learning theories: The concept of optimal stimulation." *Psychological Reports 1:* 27–33, 1955.

Melzack, R., and P. D. Wall. "Pain mechanisms: A new theory." *Science 150:* 971–979, 1965.

Miller, N. E. "Learnable drives and rewards." In: Stevens, S. S. (ed.), *Handbook of experimental psychology.* New York: Wiley, 1951, pp. 435–472.

Olds, J. "Self-stimulation of the brain." *Science 127:* 315–324, 1958.

Olds, J. "Hypothalamic substrates of reward." *Physiological Review 42:* 554–604, 1962.

Rheingold, H., J. L. Gewirtz, and H. Ross. "Social conditioning of vocalizations in the infant." *Journal of Comparative and Physiological Psychology 52:* 68–73, 1959.

Schachter, S., and J. E. Singer, "Cognitive, social and physiological determinants of emotional states." *Psychological Review 69:* 379–399, 1962.

Schaefer, E. S. "A circumplex model for maternal behavior." *Journal of Abnormal and Social Psychology 59:* 226–235, 1959.

Schaefer, E. S., and N. Bayley. "Maternal behavior, child behavior and their intercorrelations from infancy through adolescence." *Monographs of the Society for Research in Child Development,* 28, No. 87, 1963.

Sears, R. R., E. E. Maccoby, and H. Levin. *Patterns of child rearing.* Evanston, Ill.: Row, Peterson, 1957.

Thompson, R. F., and J. A. Shaw. "Behavioral correlates of evoked activity recorded from association areas of the cerebral cortex." *Journal of Comparative and Physiological Psychology 60:* 329–339, 1965.

Thompson, W. R., and R. A. Dubanoski. "Imprinting and the law of effort." *Animal Behavior 12:* 213–218, 1964.

Tomkins, S. S. *Affect imagery consciousness.* Vol. 1. *The positive affects.* New York: Springer, 1962.

Five

The Learning-Maturation Controversy: Hall to Hull

Sheldon H. White

Within fifty years after the publication of Darwin's *The Origin of Species* in 1859, the theory of evolution had crystallized its influence upon developmental psychology not once but several times. As happens with a broad and powerful ideology, different people took different messages from evolutionism and occasionally those messages could come into conflict. Since we are interested in the controversy between those who held for the influence of learning versus those who held for the influence of maturation on child development, we are particularly interested in three veins of post-Darwinian activity.

The first was a theory of mental evolution advocated by G. Stanley Hall that was an ancestor, or better a collateral, of the contemporary psychoanalytic and comparative-developmental points of view. The second was a body of instinct theories, the most influential of which were those of James and McDougall. The instinct theories were to wane, but they promulgated attitudes about children's learning which we today recognize as notions of "readiness" and "critical periods," and schematizations of development which were vaguer versions of those being suggested today by ethologists. The third offshoot of Darwinism was a tradition of research on the adaptation of animals to their environment which was initiated by Thorndike and Pavlov, which survived and grew strong, and which had direct descent through Watson to the contemporary-learning theory point of view. This third tradition had no outspoken advocates in developmental psychology—perhaps Watson in the 1920s and some experimental child psychologists in the 1950s and 1960s—but it was an important polarizing force in American psychology and its indirect influence has undoubtedly been considerable.

From *Merrill-Palmer Quarterly*, 1968, *14*, pp. 187–196. Reprinted with permission of author and publisher.

Paper presented at the symposium on History of the Heredity-Environment Controversy in Child Development, biennial meeting of the Society for Research in Child Development, March 1967.

We are accustomed to think of the Child Development Movement of the 1930s as the historic organization of developmental psychology's activities, but actually that movement was a second coming, a second institutionalization. The first was the Child Study Movement in the 1890s. The most significant individual for this movement was Granville Stanley Hall. During the 1880s and 1890s Hall was influential on all fronts. In his massive questionnaire studies, he set forth a great deal of information about what was known and what was not known by children. In 1888, becoming President of Clark University, he was an incessant activist in attempts to introduce child study, genetic psychology, as a basis for psychology and pedagogy. In 1891, he founded the *Pedagogical Seminary* as a journal for studies of children—his journal continues today as the *Journal of Genetic Psychology*—and this journal was the principal vehicle for studies of children, except for the *American Journal of Psychology*, which Hall had earlier founded, at Johns Hopkins in 1884. He was the teacher of John Dewey at Hopkins, of Lewis Terman and Arnold Gesell at Clark.

Under Hall's influence there was an active and activist Child Study Movement going in America by the time of the 1900s and 1910s. Most of the basic kinds of research on children had been attempted—not only were there the well-known baby biographies and Hall's questionnaires, but there were studies of sensory capacities, measurements of physical growth, cross-sectional studies of motor ability, behavior abnormalities, studies of learning, and so forth. The thrust of this movement was clearly towards education, and the intent of this activity was to establish a better knowledge of children so that one might better know how to go about educating them. Thus, one of Hall's signal political triumphs came when, in 1893, he persuaded the National Education Association to set up a Department of Child Study.

In a general sense, all of this activity was directed at education, but there was not yet any considerable research activity on the process of learning, nor any body of information about it. Hall's two-volume magnum opus on adolescence (Hall, 1904), contains almost no material on the topic of learning as we would today define it. As late as 1918, there were almost no psychological textbooks available on learning—just Thorndike's text on educational psychology and an English translation of a German work by Meumann (Hunter, 1952).

Hall is remembered as "the Founder"; but the title that thrilled him was "the Darwin of the Mind." He was an enthusiast, an evangelist on the theoretical and practical fronts, and his evangelism was built around his understanding of Darwin. He regarded developmental psychology as the cornerstone of education and thus, when he had converted the National Education Association to child study, he said:

. . . the issues of this hour, which I think the future historian will not forget, lie in this single fact, that unto you is born this day a new Department of Child Study. I am not sure, lusty as this infant is, and visible as I believe it to be, that we shall be likely to overestimate the importance of this event, which gives, as I believe, a new scientific character to education . . . (Strickland and Burgess, 1965).

Hall had a large, fuzzy theory of child development which embraced enormous amounts of data, anecdotal material, and speculation in a rather loose clasp. It is difficult to convey the breadth and scope of Hall's views. In brief, Hall believed that the development of the mind of the child recapitulated the evolutionary development of mankind. He took seriously the writings of Haeckel, who took seriously the embryological findings of the day which suggested that the development of the embryo rehearses the evolutionary history of the species in compressed form. Add to this premise an active mind, which collected a wide body of information and theorizing in biology, psychology, pedagogy, philosophy, and history, personal anecdotes, which accepted this diversity of information without being fussy about soundness, and which arranged it all in a grand evolutionary scheme which envisioned Mind in a vast unbroken continuum from amoeba to man. Add an ability to write furiously, to turn out in the span of his professional career 14 books and over 400 articles. (There was an early group of articles—conventional articles, impeccably within the scientific tradition of the 1880s, on the temperature sense, on bilateral assymetries of reaction, on reaction time, on associations. Then there was a middle set of articles reporting on the famous large-scale questionnaire studies about children's information. And there was a final group, towards the end of his career, which appeared regularly in the psychological journals but which had lost the flavor of psychological papers, and which seem to be mostly occasional pieces, essays, reminiscences.)

He offers as a principle what is called a "general psychonomic law which assumes that we are influenced in our deeper more temperamental dispositions by the life-habits and codes of conduct of we know not what unnumbered hosts of ancestors, which like a cloud of witnesses are present throughout our lives, and that our souls are echo-chambers in which their whispers reverberate" (Hall, 1904, vol. 2, p. 61).

He speaks poetically about minds beneath our conscious mind, of "archeopsychisms" which float most of the time below consciousness, occasionally hovering on the periphery of awareness and he criticizes the introspectionists for their excessive concern for the strong and clear signals in consciousness.

He is, in a sense, anti-intellectual:

. . . the feeling-instincts of whatever name are the psychophores or bearers of mental heredity in us, some of which persist below the threshold of consciousness throughout our lives, while others are made over as instincts or are transformed to habits. . . . We have to deal with the archeology of mind, with zones or strata which precede consciousness as we know it, compared to which even it, and especially cultured intellect, is an upstart novelty. . . . Both the degree and the direction of development of intellect vary more with age, sex, environment, etc., and sharpen individuality, while the instinct-feelings in each person are broader, deeper, and more nearly comprehensive of the traits of the whole human race (Hall, 1904, vol. 2, p. 61).

Rousseau would leave prepubescent years to nature and to these primal hereditary impulsions and allow the fundamental traits of savagery their fling until twelve. Biological psychology finds many and cogent reasons to confirm this view if only a proper environment could be provided. The child revels in savagery, and if its tribal, predatory, hunting, fishing, fighting, roving, idle, playing proclivities could be indulged in the country and under conditions that now, alas!, seem hopelessly ideal, they could conceivably be so organized and directed as to be far more humanistic and liberal than all that the best modern school can provide (Hall, 1904, vol. 1, p. x).

I am quoting, as most people do, the rather more systematic and ideological passages of Hall's writings, and perhaps Hall was at his worst (or, some might say, at his best) in his systematic writings. There is no system to his system; it consisted in a visionary, near-poetic, near-philosophical, unification of psychology with biology. This vision was not groundless. There is plenty of good solid scholarship in Hall's two volumes on adolescence and those volumes should be, I would think, a good, sound source for the developmental research work of the time.

Hall was not followed—his synthesis was personal and I do not see how he could have been followed—and his views did not have any overt line of descent. We remember him casually today as the advocate of Recapitulationism; we associate to his name the slogan "ontogeny recapitulates phylogeny." After all, his ideas were not "testable" (the word seems puny as applied to the vast scope of his ideas); he proposed no program of research, no outline, no organization, no method. Part of his outlook on developmental psychology may have survived in the work of Arnold Gesell, his student. Writing an autobiographical article in 1952, Gesell still speaks feelingly of the genius of G. Stanley Hall and, like his mentor, he still sees wisdom in the unification of psychology and embryology:

Huxley was right when he insisted that the study of embryology subtends the entire life cycle. The higher as well as lower orders of behavior were built up by evolutionary processes, and they survive only through embryological (ontogenetic) processes, however much they bear the final impress of acculturation. Learning is essentially growth; and even creative behavior is dependent upon the same kind of neuronic growth which fashions the capacities of the archaic motor system, *in utero* and *ex utero*. The performances of genius belong to a hierarchical continuum, because there is only one physiology of development. There is but one embryology of behavior (Gesell, 1952, p. 139).

But Gesell never elaborated the evolutionary organization of mind that Hall espoused. That kind of elaboration has been transmitted to developmental psychology by other hands, through other routes of descent. Something very much like it was advocated by Herbert Spencer, a contemporary of Darwin. Spencer, like Hall, was too speculative—he was elegantly and acidly dismissed by Thomas Huxley, who said, "Spencer's idea of a tragedy is a deduction killed by a fact." (For Hall himself, the dismissal came at the hands of Edward L. Thorndike, who concluded that "Hall was essentially a literary man rather than a man of science.")

A more influential version of the ideas of Herbert Spencer and G. Stanley Hall is to be found in the writings of Hughlings Jackson, whose "dissolution" theory of cerebral pathology assumed that progressively higher brain centers had been laid down in the phylogenetic history of man, and that pathology led to an attack on higher centers and release of more archaic behavior patterns. Dissolution—a reversal of evolution (Alexander and Selesnick, 1966, p. 100). Hughlings Jackson's writings are known to have influenced Freud, in turn, and it is possible to see Freud's hierarchical mental layers as analogous to the cerebral layers of Jackson (Oldfield, 1961). But perhaps Hall would have found his most kindred spirit on the contemporary scene, appropriately enough, at Clark University. The late Heinz Werner's *Comparative Psychology of Mental Development* (Werner, 1948) is a tighter, more logical presentation of what appears to be the core of Hall's reasoning. Hall would probably have appreciated Myrtle McGraw's classic study of infants, *The Neuromuscular Maturation of the Human Infant* (McGraw, 1945).

Let us examine now a second descendant of evolution, the instinct theories. After Hall, a rather more orderly, tamer view of man's animal nature did not find favor in the 1910s and 1920s. It was common in textbook writing to treat man's instincts in parallel with his habits. The authoritative work of William James set the pattern:

> With the first impulses to *imitation*, those to significant *vocalization* are born. *Emulation* rapidly ensues, with *pugnacity* in its train. *Fear* of definite objects comes in early, *sympathy* much later, though on the instinct . . . of sympathy so much in human life depends. *Shyness* and *sociability*, *play*, *curiosity*, *acquisitiveness*, all begin very early in life. The *hunting instinct*, *modesty*, *love*, the *parental instinct*, etc., come later. By the age of 15 or 16 the whole array of human instincts is complete. It will be observed that *no other mammal, not even the monkey, shows so large a list*. In a perfectly founded develop-

ment every one of these instincts would start a habit toward certain objects and inhibit a habit toward others. Usually this is the case; but in the one-sided development of civilized life, it happens that the timely age goes by in a sort of starvation of objects, and the individual then grows up with gaps in his psychic constitution which future experiences can never fill (James, 1892, p. 407).

James's curiously definite assertion that man's instinctive life was fuller than that of animals was to receive a lot of controversy and rationalization in later writing. For example, Angell in 1918 reasoned that animals have more articulated and detailed instincts, while man may have a greater number of rudimentary instinctive impulses.

James explained himself a little elsewhere:

It is often said that man is distinguished from the lower animals by having a much smaller assortment of native instincts and impulses than they, but this is a great mistake. Man, of course, has not the marvellous egg-laying instincts which some articulates have; but, if we compare him with the mammalia, we are forced to confess that he is appealed to by a much larger array of objects than any other mammal, that his reactions on these objects are characteristic and determinate in a very high degree. The monkeys, and especially the anthropoids, are the only beings that approach him in their analytic curiosity and width of imitativeness (James, 1900, p. 43).

Otherwise James's views represent the instinctivism of the day. The notion of "instinct" is defined in general terms. There is a list of instincts; but the basis on which just this list is proposed and no other is not made clear, nor is it shown what tests these traits have passed to be considered instincts. Watson's charge seems justified—that this is a faculty psychology of instincts. It is suggested that the instincts form a substrate out of which habits are elaborated. Implications for behavior development are suggested which are today familiar to us under terminology like "readiness," "critical period," or "stages"—that is, that the ontogenesis of behavior may be such that there are times when it is too early and too late for learning to occur. As James put it:

In children we observe a ripening of impulses and interests in a certain determinate order. Creeping, walking, climbing, imitating vocal sounds, constructing, drawing, calculating, possess the child in succession; and in some children the possession, while it lasts, may be of a semi-frantic and exclusive sort. Later, the interest in any one of these things may wholly fade away. Of course, the proper pedagogic moment to work skill in, and to clench the useful habit, is when the native impulse is most acutely present. Crowd on the athletic opportunities, the mental arithmetic, the verse-learning, the drawing, the botany, or what not, the moment you have reason to think the hour is ripe. In this way you economize time and deepen skill; for many an infant prodigy, artistic or mathematical, has a flowering epoch of but a few months (James, 1900, p. 61).

Compare now, McDougall's famous list of instincts promulgated in edition after edition of his *Social Psychology* (McDougall, 1914). There is a list of seven primary instincts, each coupled with one of seven primary emotions (Flight–Fear, Repulsion–Disgust, Curiosity–Wonder, Pugnacity–Anger, Self-Abasement–Subjection, Self-Assertion–Elation, Parental–Tenderness). There are four instincts without definite accompanying emotion (Reproduction, Gregariousness, Acquisition, Construction), plus a category of "minor instincts, such as those that prompt to crawling and walking." Then there are non-specific innate tendencies, or pseudo-instincts, which have no specific behavioral form, no definite accompanying emotion, no implicit purpose (Sympathy, Suggestion, Imitation, Play, Temperament).

We remember McDougall today primarily as a foil to Watson. It is one of the little legends of psychology, which we solemnly perpetuate to each upcoming class of freshmen, that McDougall overemphasized heredity with long lists of instincts, that Watson overemphasized environment, and that modern psychology has finally come to its senses and has decided that behavior is a joint function of both heredity *and* environment. The story is fairy tale, of course, and it does serious injustice to both men. Both men were quite subtle enough to give place in their systems to heredity and environment. This much can be said for the story, though: Watson tried hard to live up to it.

A good many of Watson's famous dogmatisms about environment come from a series of three Powell lectures which he gave at Clark University in 1925 (Watson, 1926). It is instructive to read what comes between those famous quotes. In that extroverted way which is Watson's unique characteristic, he is visibly struggling with himself. On the one hand, he wants to say and he does say that the instinctivist theories which have gone before are all poppycock and that the limits of training have been seriously underestimated; on the other, he feels forced to recognize that there are reflexes and there are individual genetic differences in structure which produce individual and racial differences in ability and style. The discussion goes back and forth. What we emerge with is a qualified environmentalism, in which Watson argues that any innate elements in the infant's behavior are inextricably interwoven with learning very early in life.

Watson here is reacting against James, his instincts, his stream of consciousness. He proposes the activity stream (Watson, 1930, p. 138). There is an unlearned substrate—again, a list—and it is proposed that learning builds out from items on the list. (Love behavior–conditioned loves; rage behavior–conditioned rages; fear behavior–conditioned fears; sneezing; hiccoughing; feeding reactions–conditioned feeding responses; trunk and leg movements–crawling [conditioned]–walking [conditioned]; vocal responses–talking [conditioned]–thinking [silent talking]; circulation and respiration–conditioned circulation and respiration; grasping–reaching and manipulation, acts of skill, vocations; defecation and urination–conditioned elimination responses; crying and other duct gland activity–conditioned glandular activity; erection and other sex organ responses–conditioned sex organ responses; smiling and laughter–conditioned smiling and laughter; "defensive" movements–fighting, boxing, and so on [conditioned]; Babinski reflex; blinking.)

The environmentalism of Watson is at least a compromised environmentalism. It is interesting that succeeding behaviorists were to accomplish *de facto* what Watson could not quite establish *de jure*. In fact, if not in principle, succeeding behaviorists were to downplay the instinctive and the innate in human behavior. We have now to briefly consider the tradition of Thorndike and Watson and Hull.

Thorndike, as well as the other turn-of-the-century figures we have been considering, derived a good deal of impetus from Darwin. His demonstrations of the laborious nature of trial-and-error learning were undertaken in part to demonstrate certain exaggerations of animal faculties in the post-Darwinian period. (It must be remembered that while G. Stanley Hall and McDougall were trying to find the animal in man, certain biological followers of Darwin—notably, Romanes—were as strenuously speculating about the humanoid qualities of animals.) Indeed, the logic of Thorndike's model of behavioral adaptation is rather strikingly similar to some of Darwin's discussions of progressive species adaptation. Where Darwin is talking about the progressive adaptation of the species to the environment through natural selection of favorable characters for survival, Thorndike's connectionism proposes a natural selection of response tendencies through reinforcement and nonreinforcement.

Thorndike's methods, and Pavlov's, were the progenitors of most of the research from which Tolman, and Hull, and Skinner, were to lay down their systematic positions. There is little need to rehearse the behavior theories here; obviously they did not treat instinctive and innate elements except by cursory acknowledgement. For that matter, they left a need for considerable further systematization of learning processes. It seems evident that the behavior theorists placed emphasis on "theory" more than behavior, on the attempt to build a scientifically sound and logically adequate behavioristic psychology rather than on comprehensive, naturalistic scope. Thus, the contents of key books of the era of behavior theory belie the breadth of their titles, books like *Purposive Behavior in Animals and*

Men, Principles of Behavior, Principles of Psychology, The Science of Human Behavior. They were essays into system-building, system-building which could necessarily only be undertaken on a narrow and clear base of fairly homogeneous experimental methods.

Unfortunately, a kind of parsimony developed around the methods, variables, and schematizations of the theories, which has argued that a vague hypothesis outside the scope of the theories is less worthwhile, because it is less testable, than a more precise hypothesis from within. The individual who would talk of covert genes, instincts, and maturation has been at a disadvantage with another who talks of reinforcement history. The behavior theorists did place some burden of proof on anyone who would set forth explanations of behavior on an innate basis; the recent re-entry of behavioral genetics and ethnologists to the councils of developmental theory has not been without a little show of justifiably bruised pride.

Let me try, now, to summarize this discussion. I do not find much convincing evidence of a controversy between learning and maturation in the material we have been considering. James, Hall, McDougall, Watson, Thorndike, Hull, Skinner—all agree that behavior proceeds from unlearned, innate, reflexive, or instinctive components to elaboration based upon learning. Controversy comes in on the fringes of this agreement.

A certain kind of controversy has come in because some writers, like McDougall, have spoken mostly about learned elements of behavior, while others, like Watson or Skinner, have written mostly about the learned elements. One can, perhaps, take issue with a writer's emphases but the day seems to be over when psychologists form into schools and argue hard about emphases and phrasings.

There has in recent years been an emphasis on research into learning. It seems intrinsically easier to use research to demonstrate and study learning processes, than to study maturational processes. Following the employment of our armamentarium of methods to deal with learning, there has been reluctance about the admission of less testable maturational hypotheses. Perhaps what is controversial in this is an issue of method rather than behavior causation. Are testable hypotheses about learning necessarily more significant than vaguer, more intuitively reasonable, maturational hypotheses?

The fundamental problem about learning versus maturation is not the question about which is more important (whatever that means). It is the problem of going beyond typologies of either. It is a problem of amassing evidence, large masses of missing evidence, about what the child does or does not do with experience, of admitting correlative evidence about animal behavior, brain structure, physiological development, of proposing and slowly perfecting a general theory or system which articulates their interaction. We admit to stages today, most of us. Piaget has convinced us. Some years ago, Piaget (1952) said:

> I hope to be able some day to demonstrate relationships between mental structures and stages of nervous development, and thus to arrive at the general theory of structures to which my earlier studies constitute merely an introduction.

It is to be doubted that Piaget will have time to attain that hope. Perhaps some one else will. That might be just the sort of thing G. Stanley Hall was looking for.

REFERENCES

Alexander, F. G., and S. T. Selesnick. *The history of psychiatry.* New York: Harper & Row, 1966.

Angell, J. R. *Introduction to psychology.* New York: Henry Holt, 1918.

Boring, E. G. *A history of experimental psychology.* New York: Appleton-Century-Crofts, 1950.

Darwin, C. *The origin of species.* New York: Collier, 1909 (orig. 1859).

Gesell, A. In: Murchison, C. (ed.), *A history of psychology in autobiography.* Worcester, Mass.: Clark University, 1952.

Hall, G. S. *Adolescence: Its psychology and its relations to physiology, anthropology, sociology, sex, crime, religion and education.* (2 vols.) New York: Appleton, 1904.

Hunter, W. S. In: Murchison, C. (ed.), *A history of psychology in autobiography.* Worcester, Mass.: Clark University, 1952.

James, W. *Psychology.* New York: World, 1948 (orig. 1892).

James, W. *Talks to teachers on psychology.* London: Longmans, Green, 1939 (orig. 1900).

McDougall, W. *An introduction to social psychology.* (8th ed.) Boston: Luce, 1914.

McGraw, M. B. *The neuromuscular maturation of the human infant.* New York: Hafner, 1966 (orig. 1945).

Oldfield, R. C. "Changing views of behavior mechanisms." In: Crombie, A. C. (ed.), *Scientific change.* London: Heinemann, 1961.

Piaget, J. In: Murchison, C. (ed.), *A history of psychology in autobiography.* Worcester, Mass.: Clark University, 1952.

Strickland, C. E., and C. Burgess. *Health, growth, and heredity: G. Stanley Hall on natural education.* New York: Teachers College Press, 1965.

Watson, J. B. "What the nursery has to say about instincts. Experimental studies on the growth of emotions. Recent experiments on how we lose and change our emotional equipment." In: Murchison, C. (ed.), *Psychologies of 1925.* Worcester, Mass.: Clark University, 1926.

Watson, J. B. *Behaviorism.* Chicago: University of Chicago Press, 1930.

Werner, H. *Comparative psychology of mental development.* New York: Science Editions, 1961 (orig. 1948).

Six

Ages, Stages, and the Naturalization of Human Development

Sidney W. Bijou

Currently, developmental psychology is expanding at a greater rate than any other branch of psychology. It is being incorporated into small departments of psychology and expanded in large ones; it is attracting first-rate graduate students; it is increasing its ranks with well-trained, eager young men and women; and it is being generously subsidized. Such acceleration will probably continue; confidence in the potentialities of a scientific study of human development is at a high peak. For one thing, the public is convinced that research in human development will ameliorate, in a substantial way, many of the pressing problems of child rearing, early education, and behavioral remediation.

How long will this favorable situation continue? It is difficult to say. However, the answer in large measure will be contingent upon the nature of what is produced. On one hand, confidence can be expected to remain high, if after a reasonable period the productions clearly advance basic knowledge of the historical-developmental component of psychological events in the form of the concepts and principles they generate and if they establish new guidelines to applied problems in the form of demonstrated empirical relationships. On the other hand, confidence can be expected to wane, if the field continues to yield products which are peripheral to general psychological theory and offers practical solutions which turn out to be fads, gimmicks, and verbal prescriptions with only captivating face validities.

I am convinced that future productions will "live up" to the promise

Reprinted from the *American Psychologist*, 1968, Vol. 23, pp. 419–427. Copyright 1968 by the American Psychological Association, and reproduced by permission of author and copyright holder.

Address of the President, Division of Developmental Psychology, American Psychological Association, New York, September 1966. These reflections have grown out of the research on normal and deviant children supported by the National Institute of Mental Health, United States Public Health Service, Grant MH-12067. The author is deeply appreciative of this support.

made, and, in many instances, surpass them, but I am not convinced that they will be achieved with reasonable dispatch. The time of initial delivery could be so extended that our supporters, professional and lay alike, would surely become discouraged. I am not saying that I expect the large augmentation of manpower and facilities to produce immediate and spectacular advancements, for I am in agreement with Leonard Ross (1966) that developmental problems are intricate; good research strategy and design come slowly with experience; data-gathering procedures in long-range studies are particularly time-consuming; and interpretations which advance theory require deliberation, discussion, and successive reformulations. My misgivings are based on the possibility of prolonged delay because of preoccupation with formulations based on tenous assumptions and misconceptions. If this turns out to be the case, it will be a pity for all of us.

In the hope of forestalling this possibility or of accelerating its demise, I have selected for discussions five issues I believe could be the heart of much misdirected effort. I shall indicate the nature of these problems, their assumptions, and offer some reevaluations.

I realize it is folly to discuss any five issues in the time allotted. Yet I have chosen to do so because I believe that these five are equally important. It is my hope that by at least touching on all five issues, this presentation will stimulate further and broader discussion than if I had restricted myself to two or three.

Two of the issues are external—problems concerning general intersystematic relationships: (a) the relationship between general psychology and developmental psychology and (b) the relationship between the psychology of learning and the psychology of development.

The other three are internal—problems concerning relationships within developmental psychology: (a) the criteria of developmental stages, (b) the concept of the longitudinal method, and (c) the relationship between developmental psychology and applied child research.

RELATIONSHIP BETWEEN GENERAL PSYCHOLOGY AND DEVELOPMENTAL PSYCHOLOGY

In order to deal with the first general intersystematic issue—the relationship between general psychology and developmental psychology —I must discuss the philosophy of science inherent in a natural science approach (for example, Kantor, 1959; Skinner, 1963) and describe what I believe general psychology is, should be, or should become. From this approach psychology is perceived as a division of the scientific enterprise which specializes between a total functioning organism and observable environmental events. Psychological events involve stimulus and response functions which evolve from the genetic endowment of the individual, the current situation, and the interactional history. This point of view has much in common with others in postulating that psychological events have these three components. However, it differs sharply with other viewpoints in the way these components are observed, conceptualized, and interre lated in the form of empirical laws. With respect to interactional history, for example, past events are conceptualized as past events. They are not transformed into hypothetical constructs.

From this point of view, developmental psychology is related to general psychology as the specialized branch concerned with the historical aspects of psychological events. One might say that developmental psychology is the study of progressive changes in interactions between a biologically changing organism (maturing and aging) and sequential changes in environmental events through a series of life periods (Bijou and Baer, 1961). Such a description of the field assumes that events and concepts of psychological development stretch in a line of successive intervals and bears on the second major concern of the field—analysis of psychological events within developmental stages.

Let me compare briefly developmental psychology, so conceived, with some of the other subdivisions of psychology. Developmental psychology is not like physiological or social psychology. These branches concentrate on a class of stimulus variables in psychological events. It is not like psychophysics or experimental psychology, since they are tied to a method of investigation. Developmental psychology is most like comparative psychology, which concentrates on similarities and differences in interactions within and between species. In fact, many, including Werner (1948), have referred to developmental psychology as a comparative psychology of individual development and have pointed out the overlapping interests of both fields. For example, both branches are interested in the develop- ment of different species under similar circumstances, and in the same species under different circumstances (cultures).

I turn now to a brief consideration of developmental psychology in light of the characteristics which constitute a strong subsystem, a strong subsystem in the sense of one which will undoubtedly survive in the future evolution of general psychology.

A substantial subsystem is concerned with problems that affect the entire discipline. Because it specializes in problems which bear on all analyses, it can contribute to the postulates of general theory. In the case of developmental psychology, the postulates relate to the nature of stages and their subdivisions, and the interaction of past and current events in the determination of behavior change.

The other characteristic of an essential subsystem is that it centers on an investigative approach. In developmental psychology, it is, of course, the longitudinal method. As we all know, the field has been greatly concerned with the nature of the longitudinal method and its role in developmental research. Such issues are still very much alive. At the Eighteenth International Congress of Psychology, Zazzo (1966) chaired a symposium entitled, "Diversity, Reality, and Fiction of the Longitudinal Method," in which, as is apparent from the title, he and the members of the panel sought to evaluate the approach. In his introductory remarks Zazzo stated: "The longitudinal approach as a universal method is an illusion which originates from this archaic belief that observation must blend with the observed reality." The concept of the longitudinal method is, indeed, one of the critical issues of developmental psychology and one which I shall discuss separately.

RELATIONSHIP BETWEEN THE PSYCHOLOGY OF LEARNING AND THE PSYCHOLOGY OF DEVELOPMENT

The second general systematic issue—the relationship between the psychology of learning and the psychology of development—could perhaps have been included in the discussion of the subdivisions of psychology. But because I believe that this relationship deserves special attention, I have treated it as a separate topic. Some developmental psychologists believe that learning and development are the same and that eventually learning will account for all of psychology. Others claim that learning can deal adequately with a limited segment of development, and important problems of the field must be treated from a cognitive or psychoanalytic point of view.

The postulates of a natural science approach are at odds with both these contentions. The natural science approach insists that learning is a part of development and that the other components of developmental analysis which are not learning can also be treated adequately within this framework (Skinner, 1953, 1961). An examination of the similarities and differences between learning and development may help elaborate this thesis.

You will recall that I described development as that branch of psychology concerned with delineating lawful relationships, in progressive changes in the interactions between a physiologically changing organism and environmental events. Such changes, it was said, involve the genetic endowment of the individual, present circumstances, and historical interactions. I would like to add that "present circumstances" are analyzed as stimulus and response functions, setting events, and mediating factors (Kantor, 1959).

Now to the meaning of learning—this is difficult, not because it is a complex construct, but because it has many and varied meanings. First, let me say what learning, from a natural science point of view, is not. Learning is certainly not a causal condition. The statement that a class of behavior occurs in high frequency because of previous learning is worthless, since such a statement does not refer to specific events. We are no further ahead in an analysis of psychological events when we say, "Little Jimmy tells lies because he has learned to do so," than we are when we say, "Little Jimmy tells lies because he is five years old." Learning is not a hypothetical construct, conceptual or physiological; it does not, by definition, refer to observable conditions. Learning is not a change in performance, since such a definition stresses unduly the response aspect of a psychological event and is too inclusive. Learning events most certainly involve changes in behavior, but so do nonlearning events. What is left? Learning can, with good reason, refer to experiments and theory on the relationships in the strengthening and weakening of stimulus and response functions. As such, it would be a generic term which would include behavior changes under antecedent stimulus control or respondent conditioning, as well as behavior under consequent stimulus control or operant conditioning. In my opinion the science of human behavior has advanced to the point at which the term "learning" has become obsolete, except as it may be used to designate an area of psychology which specializes in an analysis, under contrived conditions, of the variables that strengthen and weaken stimulus and response functions.

On the basis of this distinction, the developmental psychologist and the learning psychologist can be discriminated from one another by the problems and the variables each studies. The developmentalist is concerned with *all* the conditions that contribute to developmental changes in behavior; the learning psychologist is concerned with the strengthening and weakening of stimulus and response functions. Zigler (1963) has suggested that the learning investigator is the one who will probably make the most significant future contributions to development. If this turns out to be the case, and I believe it will, I would contend that Zigler made an accurate prediction not because the learning investigator was well informed about learning theory, but because he was well trained in research methodology.

CRITERIA OF DEVELOPMENTAL STAGES

So much for the two external problems of developmental psychology. I turn now to the first major internal problem—the criteria of developmental stages.

As was pointed out, one of the contributions that developmental psychology can offer psychology as a whole is a set of concepts of developmental stages that integrate readily with the general field. One could ask: "How are we progressing in this task?" Much has been written about stages, and many psychologists, as well as poets, novelists, and philosophers, have had their say as to the number of developmental stages and their general characteristics. Among psychologists, the number has ranged widely, but the criteria for each have been limited to three: (a) time since birth, or age; (b) hypothetical constructs, in combination with actual or assumed environmental events; and (c) empirical constructs, based on biological, physical, and social interactions. I shall review each briefly.

Time since Birth, or Age, as a Criterion for Stages

Time since birth, or age, as the criterion for developmental stages has, of course, been most closely associated with Gesell and his co-workers. It is interesting to note, in passing, that although Gesell created many hypothetical variables (for example, Gesell, 1954), he did not use them as criteria for establishing stages. As has been said many times over, this is not a variable but a dimension for recording events. Time as a gross indicator of physiological maturation, or of cumulative interactions with the environment, or both, does nothing to advance a functional analysis of behavior and development. In an empirically oriented cause-and-effect system physiological changes and interactions with the environment must be specified so that they can be evaluated by experimental procedures. If research demonstrates that the specific conditions do play a part in determining change in behavior, further research is required to delineate their parametric characteristics. A system of psychology using time as a causal variable can only describe interresponse relationships, can only describe behavior as a function of a conglomeration of unanalyzed conditions—biological, physical, and social. Most of us recognize this, yet now and then we find lapses in the literature. Behavior is attributed to an age or stage, and a birthday is given magical qualities. There are, for example, educational films called: "The Terrible Two's and the Trusting Three's" and "Frustrating Four's and Fascinating Five's."

Normative accounts of development can contribute to a functional analysis of development by pointing out recurring patterns of behavior and thereby suggesting experimental study of the independent variables. Many psychologists perform experiments that explore the conditions under which a phenomenon occurs.

The final point I wish to make about time is this: To say that time is inadequate as a criterion for stages does not mean that time cannot be a serviceable variable in a functional system for other purposes. However, its use requires a demonstration of a high correlation between time and some cycle—feeding, sleeping, and so forth. With this technique one should keep in mind that the ultimate relationship involved is between biological events and the changes in the behavior observed.

Hypothetical Constructs, in Combination with Actual or Assumed Environmental Events, as Criteria for Stages

The practice of using hypothetical constructs as criteria for stages, which is becoming increasingly prevalent, is typical of the cognitive and psychoanalytic approaches of Piaget and Erikson. Although each theorist makes different assumptions and concentrates on different aspects of development, the logic of each is the same: behavior at a given stage is "explained" in terms of the current situation and the properties of the hypothetical constructs. The latter are innate but modifiable, within certain ranges, by experience. A serious objection to this practice is not that the theories proposed apply to only a segment of development, but that hypothetical constructs are mixed with a loosely defined set of concepts and principles because there should be something "mental" that grows and develops the way the biological components grow and develop, and the way the behavior patterns are elaborated and extended. And after mental constructs are inserted into a system of determining conditions, research is said to be needed which separates the mental terms from the others. Thus Piaget (1966) has said:

> The comparative studies in the field of genetic psychology are indispensible for psychology in general and also for sociology, because only such studies allow us to separate the effects of biological or mental factors from those of social and cultural influences on the formation and the socialization of individuals [p. 3].

To accept hypothetical constructs as criteria for stages is to put stress on evaluating hypothetical terms and relationships. Hence we find a host of studies concerned with whether certain behaviors do, in fact, occur at a particular stage as claimed and, if so, whether their onset can be modified. Preoccupation with hypothetical "growth" variables could have adverse effects on the field of developmental psychology, first, because it discourages researchers from taking a good hard look at all the possible circumstances (not merely those claimed to be important in determining stages) which may affect behavior; second, because it deemphasizes the current need to develop, through experimental procedures, new methods of quantifying behavioral and environmental variables.

To emphasize the need to establish stages on hypothetical constructs is to misunderstand where the science of development is in its evolutional history. I believe that the field of developmental psychology is now in transition from a behavioral-descriptive phase to a functional-analytic period. We are just beginning to see efforts devoted to the analysis of behavioral-environmental relationships and the formulations of functional laws. If this is so, there are too few serviceable empirical laws, as of now, to attempt to integrate them into more general concepts and laws by means of hypothetical terms. The speed with which we reach this advanced stage of theory building will depend, in part, on the rapidity with which we resolve some of the problems discussed here.

I cannot leave this topic without this final note. Whether we are considering the feasibility of hypothetical constructs as criteria for developmental stages or evaluating some other theoretical aspects of developmental psychology, I believe with Zigler (1963) that the field does not any longer need the grand theoretical designs proposed by Piaget, Freud, Erikson, Gesell, and Werner.

Biological, Physical, and Social Interactions as Criteria for Stages

We turn now to the third and last approach to conceptualizing developmental stages, that of basing criteria on empirically defined biological, physical, and social interactions. One example of this approach, long advocated by J. R. Kantor, in fact, as long ago as 1924, segments the stream of development into three gross periods, each with several subdivisions. The first stage—the universal stage of infancy (Bijou and Baer, 1965)—is conceived of as the period which extends from birth to the onset of functional verbal behavior. It is the period in which psychological and biological behaviors are closely related, and, in a sense, the latter imposes restrictions and limitations on the former. It is also the period of the initial elaboration of respondent and operant behaviors (Lipsett and Kaye, 1964; Siqueland and Lipsett, 1966) and the period of initial evolution of ecological or exploratory behavior, a form of operant behavior (Rheingold, Stanley, and Cooley, 1962). The second stage—the basic stage—extends to early childhood and is characterized as the first period free from gross biological limitations. Here interactions with particular people and things become central and prevalent, and the child has an opportunity to build up response equipment that characterizes him as a particular individual. Although the interactions acquired and maintained during this period are basic to the child's behavioral make-up, they are amenable to changes when significant environmental alterations occur (Harris, Wolf, and Baer, 1964; Kantor, 1959). The third stage—the societal stage—is the interval from early childhood to the end of the developmental cycle. It is thought of as the interval saturated with intimate interpersonal and group conditions. Behaviors developed during this period "mark the individual as a member of a large number of cultural communities whose behavior he has evolved in shared performances with other members (Kantor, 1959, p. 168)." Included in this period are the substages of late childhood, adolescence, maturity, and advanced age.

The criteria suggested in this approach are, of course, open to successive revisions as indicated by data. Hence future research along the lines of empirically constructed stages would be expected to refine the descriptions and transitions and to point to further subdivisions. Such research would also be expected to accelerate the formulation of empirical laws with increasingly longer chains within and between developmental periods. From a systematic point of view, such statements are urgently needed.

THE CONCEPT OF THE LONGITUDINAL METHOD OF INVESTIGATION

It is obvious to all that developmental psychology has long been intimately associated with the longitudinal method. Some have argued that studies which do not use the longitudinal methods are not developmental. With the same logic, they have claimed that experimental studies can deal with the cross-sectional problems. I contend that these conceptions of the longitudinal method and developmental psychology are far too limited. Because the longitudinal method has been closely associated with the field of development from its inception, it has come to mean for many a procedure for describing sequential changes in the behavior of individuals and groups over long periods. (See, for example, Kagan's 1964 review of American longitudinal research on psychological development.) The longitudinal method can continue to refer to procedures yielding behavioral-descriptive accounts, or, hopefully, it can become a term to mean procedures for producing data on long-range interactions between a biologically changing organism and sequential environmental events. If it remains the former, the longitudinal method can have only limited use in the future work of developmental psychology; if it is modified to mean the latter, it can retain its role as the central methodology of the area. As indicated previously, I make this statement on the assumption that this field, along with others in psychology, is in the evolutionary stage of moving out of a behavioral description period and into a functional analytic phase.

To delineate functional relationships, data from many methods are required, but the emphasis must be on the experimental method. Drawing inferences about past events from reactions at the time of investigation (for example, from responses to interviews and questionnaires) or from correlations serves only limited purposes. There is no substitute for experimental research which constructs segments of history. It should be acceptable to all that the longitudinal method, like the cross-sectional method, can involve descriptive, correlational, and experimental procedures. Which procedure is used in a study depends on the objective of the study, the training and boldness of the investigator, and the situation in which the study is conducted.

Experimental longitudinal studies, like experimental cross-sectional investigations, have in the past been performed in the laboratory, field, or some combination of both. Well-known examples include Dennis' (1941) study with fraternal twins who were restricted in social stimulation and limited in motor activities during a fourteen-month period, Gesell and Thompson's (1929) study on the effects of training on stair climbing and cube building in twins, Watson and Rayner's (1920) study on the experimental development of reactions to aversive stimulation in a young child, McGraw's (1935) long-range study on the effects of special stimulation with fraternal twin boys, and Hilgard's (1932) group study with preschool children on the effects of academic training on achievement.

Within the past five years there has been a resurgence of interest in the application of the experimental method accompanied by revisions and extensions in techniques. Examples are the field studies on changes in social behavior in preschool children by Harris et al. (1964) and the laboratory studies in concept formation by Bijou (1965) and Sidman (1966).

A final note on the longitudinal method. In 1965 Wohlwill chaired a Society for Research in Child Development symposium entitled "Approaches to Experimental-Developmental Research in Child Psychology." Kendler, Gollin, and Bijou in the same symposium addressed themselves to the question: "How does the experimental child psychologist handle developmental changes that occur with respect to the phenomenon he is interested in?" Wohlwill concluded the discussions by saying he thought future progress in the field can profit from a two-pronged pragmatic attack.

> One phase would follow closely along the lines of the strategies of Tracy Kendler and Eugene Gollin, the aim being to specify as incisively as possible the nature of the developmental processes which are operating to produce the age differences observed for some particular aspect of behavior. The second phase would carry the ball from there, designing experiments and manipulating variables whose effects are theoretically derivable from the specification of the developmental processes achieved in phase one. In order to enhance the developmental

significance of such research, these effects might be measured, not simply in terms of acquisition curves, error rates or the like, but rather in terms of variables directly relating to the developmental changes, such as rate of change, asymptotic level, stability, situational generality, etc.

With this conclusion I am only in partial agreement. I believe that a one-pronged attack will suffice, a belief that is substantiated by the current work in several child laboratories.

RELATIONSHIP BETWEEN DEVELOPMENTAL PSYCHOLOGY AND THE APPLIED AREAS OF CHILD REARING, EDUCATION, AND REMEDIATION

The third and final internal problem in developmental psychology pertains to the relationship between developmental psychology and the applied areas of child rearing, education, and remediation. There should be nothing special about this relationship. It should be the same as that which exists between any basic and applied science—a relationship of research and development. To the question: "How well have we been doing in research and development?" the answer is uncomplicated: "Not well!"

Child developmentalists have always been responsive to practical problems. I merely echo the words of Baldwin and others when I say that in the early days of the institutes of child development great efforts were devoted to research on applied problems before there were principles to apply. In general, problems were analyzed either on a "common sense" basis or in terms of a variation of psychoanalytic theory. Recent history has shown that neither procedure has been fruitful. The common sense approach has not led to new procedures for dealing with recurrent problems, since it attempted merely to objectify the conditions that seemed to be involved and to apply statistical tests of significance. I say "attempted merely to objectify" because many of the quantifying procedures employed, such as questionnaires, structured interviews, projective techniques, and interpreted observations, produced data that were far removed from descriptions of actual interactions. In many instances, admittedly inferior quantification methods were used to study a practical problem "as it existed." Research which attempted to arrive at practical recommendations through the testing of psychoanalytic theory fared no better, not only because of the use of similar methods of measurement and research design, but because, in addition, the concepts and principles tested were elusive.

So, as we all know, reviews by Vincent (1951), Stendler (1950), Senn (1957), and Wolfenstein (1953) indicated that recommendations presumably evolved from such studies carried the tenor of the era in which they were conducted. There was strong suspicion that the recommendations were based more on opinion than fact. Lest you think this reference applies only to research in the remote past, I recommend that you read the conclusions of 8 of the 10 chapters in Volume 1 of Hoffman and Hoffman's *Review of Child Development Research* published in 1964. For example, Caldwell's review of the effects of infant care ends with 10 conclusions. Her first, third, fifth, seventh, and ninth topic sentences, taken as samples, read as follows: (1) "The breast-bottle dilemma must remain exactly that [p. 73]." (3) "Studies concerned with oral gratification and oral activities provide little support for the hypothesis that sustained gratification leads to drive satiation [p. 80]." (5)"The relationship between parent attitudes and parent behavior is still insufficiently explored and imperfectly understood [p. 80]." (7) "The number of variables in which investigators have shown interest is small indeed [p. 81]." (9) "It is imperative that there be a general improvement of methodology in studies concerned with the effects of infant care [p. 81]."

One thing seems sure: there is not much promise in applied research which aims to determine whether treatment method A is better than treatment method B or to test the validity of some aspect of a vaguely defined personality theory. To repudiate both strategies is not to sound a retreat from applied research until we have at hand a large stockpile of basic principles. There is an alternate constructive course: apply known basic principles and do this in a way that the data would indicate, in a continuous manner, the effectiveness of our efforts at application.

I believe that we should approach the problems of technological research in developmental psychology with modesty and optimism. After all, the specific skills of the good mother, teacher, counselor, and therapist have preceded the developmental psychologist's application of known empirical laws. Recognizing this, we can go about the task of technological development in the hope that our efforts will catch up with the practices of the practitioners and will eventually lead to revisions in the handling of old problems and to new ways of dealing with new ones. Our optimism need not rest only in the promise of improving practical problems. Genuine technological research, especially programs with a functional-analytical orientation, will also enrich basic research.

SUMMARY

Developmental psychology is expanding rapidly and is widely supported. Continuous growth of the field is believed to be dependent on the way the five issues discussed here will be resolved. Two of these problems concern external systematic issues.

The relationship between general psychology and developmental psychology From a natural science orientation developmental psychology is viewed as a branch of general psychology which abstracts from general analysis that segment concerned with the influence of near and remote historical events on current interactions. It is concerned with behavior changes within stages and with the influence of interactions between stages. Since practically all psychological events have historical components, developmental psychology has much to contribute to general psychology and, in my opinion, will always remain an integral yet specialized part of the science of behavior.

The relationship between psychological development and psychology of learning The view that the psychology of learning and development are the same, except that the latter is concerned with changes over longer periods, is categorically rejected. Psychological development is believed to be concerned with progressive changes in interaction, the psychology of learning with changes in stimulus-response and response-stimulus relationships in contrived situations. The functional relationship between stimuli and responses plays a central role in any analysis of interaction between a biologically evolving and devolving organism and changing environmental events.

The three additional issues pertain to the following internal systematic problems:

The problem of criteria for developmental stages This issue rests largely on what is believed to be a promising way of formulating criteria. Time and hypothetical constructs in combination with empirical constructs are rejected as criteria in favor of empirical constructs.

The concept of the longitudinal method The study of changes in organism–environment interaction over long-range periods, which is the heart of developmental psychology, should not be limited to methods that yield only descriptive accounts of behavior. Methods providing data which bear directly on cause-and-effect relationships must also be included. Research has already demonstrated its feasibility.

The relationship between developmental psychology and the applied fields of child-rearing, education, and remediation It is reaffirmed that the relationship between the basic and applied developmental psychology should be one of research and development. For the most part, applied research in the past has not been of this sort. Rather, it has been theory testing, treatment-methods testing, and correlational analyses. Since efforts along these lines have not been fruitful, it is suggested that applied research turn toward applying currently known empirical principles generated from laboratory studies. It is argued that the functional analytic approach is particularly suited to engage in technological research because of its insistence on functional terms and principles. A genuine research and development program holds promise of demonstrating improvements in child developmental practices; it can also supply valuable ideas for basic research.

REFERENCES

Bijou, S. W. "Systematic instruction in the attainment of right-left form concepts in young and retarded children." In: Holland, J. G., and B. F. Skinner (eds.), *An analysis of the behavioral processes involved in self-instruction with teaching machines.* Research Report, Grant No. 7–31–0370–051.3, United States Office of Education, Department of Health, Education and Welfare, 1965.

Bijou, S. W., and D. M. Baer. *Child development: A systematic and empirical theory.* Vol. 1. New York: Appleton-Century-Crofts, 1961.

Bijou, S. W., and D. M. Baer. *Child development: The universal stage of infancy.* Vol. 2. New York: Appleton-Century-Crofts, 1965.

Dennis, W. "Infant development under conditions of restricted practice and of minimum social stimulation." *Genetic Psychology Monographs* 23: 143–189, 1941.

Gesell, A. "The ontogenesis of infant behavior." In: Carmichael, L. (ed.), *Manual of child psychology.* (2d ed.) New York: Wiley, 1954.

Gesell, A., and H. Thompson. "Learning and growth in identical infant twins: An experimental study by the method of co-twin control." *Genetic Psychology Monographs* 6: 1–124, 1929.

Harris, F. R., M. M. Wolf, and D. M. Baer. "Effects of adult social reinforcement on child behavior." *Young Children* 20: 8–17, 1964.

Hilgard, E. R. "Learning and maturation in preschool children." *Journal of Genetic Psychology* 41: 40–53, 1932.

Hoffman, M. L., and L. W. Hoffman. *Review of child development research.* Vol. 1. New York: Russell Sage Foundation, 1964.

Kagan, J. "American longitudinal research in psychological development." *Child Development* 35: 1–32, 1964.

Kantor, J. R. *Principles of psychology.* Vol. 1. Bloomington, Ind.: Principia Press, 1924.

Kantor, J. R. *Interbehavioral psychology.* (2d ed.) Bloomington, Ind.: Principia Press, 1959.

Kendler, T., E. Gollin, and S. W. Bijou. "How does the experimental psychologist handle developmental changes that occur with respect to the phenomenon he is interested in?" In: Wohlwill, F. J. (chm.), *Approaches to the experimental-developmental research in child psychology.* Symposium presented at the meeting of the Society for Research in Child Development, Minneapolis, March 1965.

Lipsett, L. P., and H. Kaye. "Conditioned sucking in the human newborn." *Psychonomic Science* 1: 29–30, 1964.

McGraw, M. B., *The growth: A study of Johnny and Jimmy.* New York: Appleton-Century, 1935.

Piaget, J. "Nécessité et signification des recherches comparatives en psychologie génétique." *International Journal of Psychology* 1: 3–13, 1966.

Ross, L. E. "Classical conditioning and discrimination learning research with the mentally retarded." *International Review of Research in Mental Retardation* 1: 21–54, 1966.

Rheingold, H. L., W. C. Stanley, and J. A. Cooley. "Method for studying exploratory behavior in infants." *Science* 136: 1054–1055, 1962.

Senn, M. J. E. "Fads and facts as the basis of child-care practices." *Children* 4: 43–47, 1957.

Sidman, M., and L. T. Stoddard. "Programing perception and learning for retarded children." *International Review of Research in Mental Retardation* 2: 152–208, 1966.

Siqueland, E. R., and L. P. Lipsett, "Conditioned head-turning in human newborns." *Journal of Experimental Child Psychology* 3: 356–376, 1966.

Skinner, B. F. *Science and human behavior.* New York: Macmillan, 1953.

Skinner, B. F. *Cumulative record.* (rev. ed.) New York: Appleton-Century-Crofts, 1961.

Skinner, B. F. "Behaviorism at fifty." *Science* 140: 951–958, 1963.

Stendler, C. B. "Sixty years of child rearing practices." *Journal of Pediatrics* 36: 122–134, 1950.

Vincent, C. E. "Trends in infant care ideas." *Child Development* 22: 199–209, 1951.

Watson, J. B., and R. A. Rayner. "Conditioned emotional reactions." *Journal of Experimental Psychology* 3: 1–4, 1920.

Werner, H. *Comparative psychology of mental development.* (rev. ed.) Chicago: Follett, 1948.

Wohlwill, J. F. (chm.). *Approaches to experimental-developmental research in child psychology.* Symposium presented at the meeting of the Society for Research in Child Development, Minneapolis, March 1965.

Wolfenstein, M. "Trends in infant care." *American Journal of Orthopsychiatry* 23: 120–130, 1953.

Zazzo, R. "Introductory report to the symposium on longitudinal studies." In: Zazzo, R. (chm.), *Diversity, reality and fiction of the longitudinal method.* Symposium presented at the Eighteenth International Congress of Psychology, Moscow, September 1966. (Etudes longitudinales du developpement psychologique de l'enfant. *Enfant* 2: 131–136, 1967.)

Zigler, E. "Metatheoretical issues in developmental psychology." In: Marx, M. H. (ed.), *Theories in contemporary psychology.* New York: Macmillan, 1963.

Part Two

Physiological Bases
of Behavior and Development

In the early years of child psychology a great deal of effort was expended on accumulating data on the physical growth and motor development of children. Normative, cross-sectional data accumulations like those of the Gesell studies at Yale and longitudinal approaches like the California Growth Studies were useful and important enterprises. But overconcern with norms, averages, and age-related development created a negative reaction to such material. Child psychologists moved to other concerns and the physical realm was little studied after the 1930s.

Neglect of body structure and functional influences on behavior has not been complete, of course, but the editors regard the relative neglect as unfortunate and deleterious to understanding development and behavior. It is, after all, the physical body that does the behaving, and the nature and capacity of the organism is one of the determiners of the *what*, *when*, and *why* of much behavior.

Why, specifically, should we be concerned about the person as a physical organism? At least three reasons seem apparent. First, the person is a stimulus for others, and thus influences their behavior toward him. Second, his own self-concept and life-style arises in part from the kind of organism he is. Finally, the person is himself a responding organism.

The individual's stimulus characteristics are unique and important. His physical appearance elicits differential responses from other persons, of both like and opposite sex. This is particularly marked during youth, but continues to some extent throughout life. His absolute and relative size have an effect on his peer relations, his early social experiences, and thus on both his developing behavior and self-image. Some part of his characteristic behavior is dependent on physical traits, such as his energy supply which limits or characterizes the amount and to some extent the nature of the interaction he has with his environment in general and other persons in particular.

In this part of the book Lester Sontag presents data from some of the Fels Institute studies that he has directed for many years. His report demonstrates possible ties between physical functioning of the body and what we call "personality." The possibilities for exploring this relationship in future research seem bright.

Not so readily observable is the self-image or self-concept the person develops. But this self-concept helps to structure individual behavior— the person acts "as if" he were a certain kind of individual. This internalized self-concept has many sources, but absolute and relative size, quickness or slowness, strength and general vigor, physical appearance and the like are necessary determiners of self-image and thus help to produce a particular kind of person. A relevant paper here examines the relationship between a person's vitality and his happiness in life; this essay is laden with wisdom rather than data for a change. The author is F. L. Lucas, who is neither psychologist nor physiologist, but university reader in English at King's College, Cambridge, where he is a Fellow.

In any field of scientific endeavor, it is somewhat easier to collect data than to make appropriate inferences from data. The problem of influence of data-collecting method on the meaning of evidence is especially acute in developmental psychology. A study reported here by Damon illustrates how apparently straightforward "facts" garnered from measurement of height across a mature and older age span appear to lead to erroneous inference when compared to data gathered longitudinally.

A major part of the research literature of modern psychology has been concerned with various aspects of human responses. Only a portion of the literature examines the unique and individual nature of responses and considers the sources of differential responding. Our immediate reaction to the term "response" is probably in terms of *motor* responses, since it is in this fashion that the phenomenon is so often studied in the laboratory. But other responses, more subtle ones, are also made. These include modification of the system's functioning under conditions of stress, the emotion-generated changes occasioned by interaction with other individuals, typically aggressive or passive patterns of responding to social situations. The person's measurable response characteristics, that is, the rate, duration, range, and direction of response, are typical of him as an individual and are closely related to his learning and thus to all of his behavior. The relationship of his response characteristics to those of peers, both in childhood and in later life, has much to do with their evaluation of him. Thus the basic character of an individual's responses become a key to understanding him. Related to the individual's nature is of course his genetic inheritance.

For a variety of reasons, including improved research tools, the science of genetics has been flourishing in recent years. Early psychology attributed too much otherwise unexplained behavior to "instinct"; the decline in credibility of this source of explanation for behavior and the rise of behaviorist "learning-only" formulations tended to retard the involvement of psychology in the new genetics. But this situation is rapidly changing. Geneticists and psychologists of many persuasions, including developmental scientists, are establishing stimulating communication and research bridging the two disciplines. Two papers presented here provide evidence and ideas that are rewarding in themselves and promising of future revelations.

The author of one of the papers, Lissy Jarvik, was an associate of the late Franz Kallmann, who carried out longitudinal studies of hundreds of mature and aged twin pairs. Jarvik is continuing Kallmann's work at the Department of Medical Genetics, New York State Psychiatric Institute, Columbia University. She discusses the interaction of heredity and environment as it influences intellectual changes in the aging process, as well as the interaction influences on etiology of mental defect and on the variability in normal function.

Theodosius Dobzhansky, a genetics experimenter at the University of California, Davis, presents evidence from fruit fly research on a problem fascinating to both psychologists and geneticists—individual uniqueness. He points out that the science of genetics concerns itself in particular with research on individuality and its causes.

This brief sampling of studies related to physical structure and functioning should at least sensitize the reader to the importance of this aspect of the person.

Seven

Somatopsychics of Personality and Body Function

Lester W. Sontag

"Psychosomatic" or mind-body relationship is usually used to describe a physiological disturbance associated with modified autonomic nervous system function, emotional in origin. Much has been written of personality characteristics which are presumed to predispose to certain types of psychosomatic disfunction under stress. Thus, we speak of the peptic ulcer personality or the hypertensive personality. The emphasis is, therefore, on the effect of specific aspects of personality and emotional adjustment and how they affect the development of malfunction of the organ systems of the individual.

There is, however, a very different aspect of mind-body relationship, namely, the question of how basic physiological processes affect the personality structure, perception, and performance of the individual. This body-mind relationship I shall refer to as the somatopsychics of personality and behavior. Freud, of course, recognized the role that differences in "constitution" played in the development of neurotic patterns of behavior. While his emphasis on heredity as a determinant of constitution and, therefore, as a factor in development of social adjustment is plain throughout his writings, this point is too often forgotten by the therapist or the student of personality. This paper will present the results of some researchers which will emphasize the fact that personality and emotional adjustment are not only the products of environmental stimuli and experiences throughout life. It will imply that the way an individual internalizes and interprets experiences—whether he withdraws from perceived threats—or retaliates—may be due in part to the individual nature of his body machinery.

The research results described here are the product, in most instances,

From *Vita Humana*, 1963, 6, 1–10. Reprinted with permission of author and publisher, S. Karger, Basel/New York.

This research was supported in part by Research Grant M-1260 from the National Institutes of Health, United States Public Health Service.

of a longitudinal or long-term study of the same individuals. It is important to describe this research program and the particular method of operation by which such long-term studies have been accomplished.

The Fels Research Institute was established in 1929 by Mr. Samuel S. Fels of Philadelphia. As a layman, without scientific training, he was constantly amazed at the variety of differences in human beings who, in apparently comparable social environments, behaved and functioned quite differently. He established the Institute to study the origin and elaboration of individual differences in human beings, their growth, personality and functioning. He felt that such a study must begin before birth and must include many fields of science as diverse as biochemistry, psychology, physical anthropology, and medicine. At the start of the program of the Fels Research Institute, about fifteen highly cooperative expectant mothers were inducted each year. Among the observations and measures taken on these mothers and their fetuses for a two-hour period each week during the last three months of pregnancy, were records of three types of fetal physical activity—quick, kicking-type movements; slow rhythmic turning of the body; and fetal hiccoughs in those fetuses who exhibited hiccoughs. At the same time fetal heart rate was timed in 10-beat segments over a period of five or ten minutes during each two-hour recording session. The elapsed time for each 10-beat burst was recorded in seconds, and tenths of seconds, and then converted into heart rate. There was a wide range of difference in the stability or constancy of fetal heart rate. Other measurements and observations made during the fetal period included an appraisal of the mother's nutritional intake, nitrogen balance, emotional state, activities, cigarette smoking, mechanics of labor, and the immediate status and developmental progress of the infant. Out of such studies came immediately such findings as: cigarette smoking by the mother increased the fetal heart rate (Sontag and Wallace, 1935); the fetus was susceptible to certain sound stimuli, disturbed maternal emotional states, such types of vibration as was produced by a 1935 washing machine, and the like, and responded by strong kicking movement (Sontag and Wallace, 1935).

As the original research population of fifteen mothers and their newborn infants was augmented each year by an additional fifteen, a variety of measures of growth, development, and behavior of the infants was instituted. Children were brought in four times during the first year and at six-month intervals thereafter for measurements of height, weight, and other body dimensions; for physical examinations; X-rays of the extremities, soft tissue, chest and heart, jaws, and so forth. At two years, each child spent several weeks each year in our experimental or observational nursery school where various aspects of his behavior and performance were observed, described and rated. Records of his nutrition were kept as well as of his illness, school progress, and so on. Developmental tests, such as the Gesell and the Merrill-Palmer, were begun as early as six months, and after two and one-half years, the research subjects were taking the Stanford-Binet test at six-month intervals. From six to twelve years they were brought in to an observational play camp for a period of two weeks in the summer, where their behavior and performance could again be observed and rated. After eight years, the observations which had been systematically made in their homes were discontinued, and interviews with parents were substituted. In the period from six to eighteen years, these children were subjected to measurements of autonomic response to stress, to a battery of tests such as the Rorschach, T.A.T., Wechsler-Bellevue, and to specially designed tests to evaluate such variables as cognitive style. Amino acid excretion patterns are currently being studied on many of these children.

To accomplish this multidisciplinary study, a staff of approximately sixty-five people ranging from senior research investigators to research assistants is divided into four departments: growth and genetics, psychophysiology-neurophysiology, biochemistry, psychology and psychiatry. The variety of scientific products of such an organization is, of course, considerable. It includes, for example, studies of the stereotypy of autonomic responses to stress; of the genetics of patterns of appearance of ossification centers, of changes in I.Q. over periods of years and their personality correlates, of patterns of excretion of amino acids, of parental identification as a determinant of achievement motivation, and of the predictability from early behavioral measurements of adult personality. The research results presented in this paper are the product of a very tiny part of the data gathered over the years.

While I have described the longitudinal program of the Fels Research Institute—the continuous and repeated study of the same individuals over the period of their life times—it should be mentioned that the Institute's program is not solely limited to a study of these children. We have, since the inception of the program, emphasized the fact that while many important factors in the development of the personality, in growth patterns and in the functions of individuals can only be studied adequately by the use of longitudinal methods, we believe there are also unlimited numbers of important problems in human development which can be studied by other methods. We have emphasized that the problem is the all-important thing, and that the Institute must employ whatever research material or method is necessary to solve that problem. Therefore, for the study of some problems, we use classes in public schools. For others, we maintain a small primate colony, a rat colony, and a cat colony. We utilize the parents of our children in various studies of aging processes.

In short, the Institute's program may be considered to be devoted to problems in the field of human development, to be eclectic, multidisciplinary, and both longitudinal and cross-sectional in its methods.

THE RELATIONSHIP OF FETAL ACTIVITY
TO NURSERY SCHOOL BEHAVIOR

Quick fetal activity (arms and legs) was calculated for twenty-four male fetuses on the basis of the average number of quick movements per minute, measured over two-hour periods, once a week during the last two months of pregnancy. Time of day and physical conditions were kept constant. This same group of subjects was studied at nursery school at two and one-half years of age by observers who were not even aware that fetal activity records had been made. Social apprehension, defined as hesitancy to join groups, anxiety in the face of threatened peer aggression, reluctance to enter the nursery school car, and so forth, correlate positively at the .05 level of confidence with amount of quick activity during the last two fetal months. The same type of relationship is found in the study of fifteen females, but at a level which is not statistically significant, probably because the number of cases is too small. The more active the individual was as a fetus, the more likely he was to show social apprehension at two and one-half years.

There is a negative relationship between level of quick fetal activity and peer aggression at two and one-half years in males, at a level of confidence of .05. The very active fetuses, while not inactive at nursery school age, expressed their activity in unaggressive ways; they tended to avoid conflict. The relationship in females is the same but not at a significant level.

Speculation on the meaning of such findings must, at this time, *remain* speculation. A number of tenuous hypotheses can be offered, but without evidence to substantiate them. For example, we know that various emotional episodes in the mother's life, through some humoral mechanism, stimulate excessive physical activity in the fetus. Is it possible that a fetus by such humoral mechanism can experience a series of "physiologically induced apprehensions," and that these traumatic physiological experiences are succeeded during postnatal life by excessive susceptibility to perceived social threats? Or—since we have demonstrated that maternal anxiety and tension produce increases in fetal activity—is the same maternal tension having its effect on the two and one-half years old, and is he showing its effect as social apprehension? Without suggesting that either of these hypotheses is valid, I only want to illustrate the diversity of possible explanations for our statistical findings.

Another part of our longitudinal study has yielded a result which adds interest to these findings. In our study of the relationship of the behavior of nursery school children to their *adult* personality patterns, social apprehension measured at two and one-half years was predictive of social apprehension at age twenty-two and twenty-five years. While the demonstrated relationship between amount of fetal movement and social apprehension at nursery school age is based upon a different group of research subjects, the implications cannot be ignored. There are measurable characteristics in the eighth-month fetus which appear to be predictive of certain aspects of his preschool social behavior. This same preschool social behavior is, in turn, predictive of a similar type of social behavior in the adult.

A typical history of such social apprehension, taken from our files, is as follows:

"At two and one half years S gave the appearance of being frightened at first. When taken downstairs for the physical examination, following the mental test, S would not allow the doctor to take his hand to guide him. He was shy and remained very timid throughout the procedure. He cried when the examination began and cried at new items in the procedure. Upon S's first visit to the Nursery School, it took the entire period for him to make the adjustment. He was very apprehensive and insecure during his first days. He cried a lot and stood about looking sad in the interim between his howls. Initially, he did not enter into any of the play, but was entirely concerned with his unhappiness. He objected to coming and cried in the car. Once at the school, he stood about weeping or sobbing and followed the teacher around for comfort. He was not in a tantrum of anger, but appeared to be showing fear and apprehension. For the first two weeks, S made no attempts to get into any of the groups. He was shy with the other children and timid with the new equipment."

This boy had a lonely high school career, with no dating and almost no participation in extra-curricular school functions. He attended college for two years where he was lonely and depressed. He has done very little dating. He is now, at 26, still single and working as a bookkeeper in a small business, still lonely and avoiding people.

STABILITY OF HEART RATE FROM THE FETAL PERIOD TO FOURTEEN–EIGHTEEN YEARS

Anyone who has put a stethoscope to a chest is aware of the fact that in some individuals irregularity of heart rate is excessive—that bursts of acceleration not accountable for on the basis of physical activity are quite common in some individuals and very rare in others. This characteristic of spontaneous increases and decreases in rate, under resting conditions, 1 shall refer to as cardiac lability-stability. A part of this rate fluctuation or lability is associated with respiratory phase. The remainder is not. Lacey and Lacey (1958) of our laboratory have studied such phenomena in a variety of children and young adults and have published a number of their findings. Lacey, Kagan, Lacey and Moss (1962) have studied the possible relationship between cardiac lability-stability and various aspects of behavior. Briefly, they found that in interviews, the level of dependency material was much higher in cardiac labiles than in stabiles. Cardiac labile males were more reluctant to depend on love objects, they had more conflicts over dependency, and they had more intense strivings for achievement. They showed more anxiety over erotic activity and more compulsive behavior. They vacillated with decisions and were more introspective. I have no glib explanation for these statistically established personality correlates of cardiac rate instability: or why it is that the boys with a fluctuating heart rate showed more conflict over dependency, more compulsive personalities, and the other characteristics mentioned. I am simply calling attention to the fact that the work of Lacey, Kagan, Lacey and Moss demonstrated that this is so.

Such findings have, of course, possible implications for the psychotherapist. They also bring into focus the question, what part of the incorporation of life experiences is a matter of the experiences themselves, and what part is due to gene-determined or prenatally acquired, physiological processes? We know that in the botanical world, some types of plants can stand a great deal of water, but others cannot. Some strains of corn and wheat are much more drought-resistant than others. It seems probable that the difference in autonomic function is a constitutional one, gene-determined, or the result of fetal environment. It may be one indicator of the differences in physiological operation which are accountable for a part of the differences in behavior patterns and personalities of individuals.

Now I should like to go one step further and present some research findings which are not conclusive but are, nevertheless, to me quite exciting. I have suggested that autonomic lability-stability was a constitutional characteristic. It might well be asked, why is it not the product of the life experiences of the individual? I indicated earlier that one of the measures taken on the fetuses was fetal heart rate in 10-beat segments. Thus, each fetus had recorded during the last two months of fetal life, perhaps 400 short samples of fetal heart rate. By calculating the standard deviation of the heart rate of each fetus from these 10-beat samples, one arrives at a measure of its variability. This measure is, to a considerable degree, comparable to the variability of heart rate measured on the adolescents in the Lacey and Lacey study. Insofar as we have data on the same individuals, it is possible, therefore, to compare the variability of heart rate of an eighth-month fetus with the variability of his heart rate at rest at the age of fifteen years. At the time the adolescent heart rates were recorded, the study presented here had not been anticipated. Therefore, the number of subjects for whom we have both fetal and adolescent stability rates is limited, and consequently, the conclusions which can be drawn are limited. Nevertheless, the scatter diagram of the lability of heart rate of the fetus plotted against the lability of heart rate of the individual at adolescence, although it is based only upon twelve cases, is most interesting.

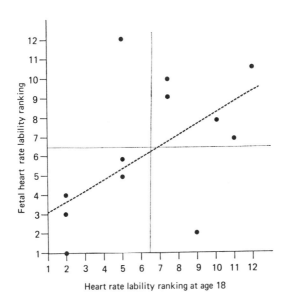

Heart rate lability ranking at age 18

The 5–5 and 1–1 distribution is striking. Because the number of cases is small, the rho of .52, significant at the .05 level of confidence, is suggestive rather than conclusive in indicating that cardiac labiles during fetal life were the cardiac labiles at adolescence. Nevertheless, is it unreasonable to surmise that this characteristic of autonomic function, shown to be related to personality in postadolescence, may be a constitutionally determined factor? And that this characteristic is statistically predictive, even during the fetal period, of some aspects of behavior in later life? ("Constitution" is defined as the sum of those characteristics of an individual that are determined either by genes or by prenatal environment.)

Another individual difference in autonomic nervous system function is degree of cardiac rate response to a stress situation. Research at our Institute has found that the cardiac response of individuals to stress—cold pressor test, mental arithmetic, intense light, loud noise, and others—did not vary appreciably from stress to stress. The degree of response from individual to individual did vary, however. Kagan and the Laceys have found that there are also personality correlates—in this instance, differences in cognitive style, or ways of thinking and perceiving—between the strong reactors and the weak. The following are a few examples.

In a variety of picture test situations, the strong cardiac reactors, those with high heart rate response, were prone to perceive situations imaginatively or as having *affect*. For example, when shown, through tachistoscopic procedure, a picture of a man sitting at a desk writing, they would comment not only on the fact that the man was writing, or that he was a man, or that he was sitting. They would comment rather, or in addition, that the man looked sad, or happy, or angry. Shown a picture of a child drawing at a blackboard, the comment would be that the child was happy, or sad, or was about to draw an elephant, or a bear, or a flower. It would not simply be that it was a child. The nature of the perception was quite different, both in affect and imagination, from that of the low cardiac reactors who, under similar situations, saw the child, or the blackboard. There appears to be a greater projection of the subject's emotions into the picture in the case of the high reactors, and little or no projection, on the part of the low reactors. Furthermore, personality assessment of another group of subjects has suggested a high degree of emotional and behavioral control in all kinds of situations, on the part of the low reactors, whereas the high reactors were more likely to act out their emotions. In other words, people with unstable cardiac rates were also unstable in their social behavior.

Obtaining a fetal cardiac response to a stress situation under *controlled* conditions is not possible. Nevertheless, cardiac response to stress can be obtained. During the last few months of pregnancy, a fetus will respond with an immediate series of limb movements and increased heart rate to the rattle of a doorbell buzzer against a block of wood held against the mother's abdomen (Sontag and Wallace, 1935). This is the typical Moro or startle reflex of the newborn. Some years ago we recorded heart rate changes to such sound stimuli in a number of fetuses. Here, again, we found strong reactors (with a large increase in heart rate) and weak reactors. We are currently beginning to analyze these data to determine whether degree of cardiac rate response to stress at the eighth fetal month is predictive of what it will be at five years and at fifteen or twenty years. If such a relationship is found, it will suggest even more intriguing possibilities in relating fetal autonomic behavior with young adult autonomic behavior, with young adult cognitive and perceptual behavior—with adult personality.

SUMMARY

The Fels Research Institute is conducting a longitudinal multidisciplinary study of human beings from before birth to death. Level of quick fetal activity during the eighth and ninth months is statistically predictive of social apprehension in a nursery school situation at two and one-half years; it is also negatively related to degree of aggression at that age. Social apprehension at two to five years is predictive of social apprehension at twenty-two to twenty-five years. Based on a small number of cases the degree of fetal heart rate instability is predictive of heart rate instability at fourteen to eighteen years. Resting heart rate instability in eighteen-year-olds is statistically related to problems of dependency, achievement striving, compulsive behavior, and anxiety over erotic activity. In eighteen-year-olds the degree of cardiac response to stress is related to degree of affect projected in test pictures. Studies of fetal heart rate response to stress are being made.

REFERENCES

Lacey, J. I., and B. C. Lacey. "The relationship of resting, autonomic activity to motor impulsivity." In: *The brain and human behavior*, Vol. XXXVI, Proc. Assoc. Res. Nerv. Mental Disease. Chapter V, pp. 144–209. Baltimore: Williams and Wilkins, 1958.

Lacey, J. I., J. Kagan, B. C. Lacey, and H. A. Moss. "Situational determinants and behavioral correlates of autonomic response patterns." In: Knapp, P. J. (ed.), *Expression of the emotions in man.* Int. University Press, 1962.

Sontag, L. W., and R. F. Wallace. "The effect of cigaret smoking during pregnancy upon the fetal heart rate." *Amer. J. Obstet. and Gynec.* 29: 77–82, 1935.

Sontag, L. W., and R. F. Wallace. "The movement response of the human fetus to sound stimuli." *Child Developm.* 6: 253–258, 1935.

Eight

Vitality

F. L. Lucas

For thousands of years, sages have discoursed on the vanity of human wishes; sometimes forgetting that life without wishes would become vainer still. It is easy to argue that wealth can be a demoralizing nuisance; fame, a bauble (who reads his own epitaph?); and power a poison (who in his senses could covet the career of Stalin?). Even understanding is sometimes an affliction that makes a man dismal as a wet hen; learning, a burden that can leave him heavy as a hogshead; and beauty not only brief, but often perilous (for the husband of Helen of Troy herself may be no more enviable than the poor Dragon that guarded the Golden Apples). And yet, watching the world, one may wonder that so many men seem blind and heedless about a source of happiness much less dubious or dangerous than these—simple vitality.

It is not merely that even philosophers lose much of their virtue, as Horace observed, when they have streaming colds; or, as Shakespeare echoed, when they have toothache. Vitality is, in my experience, not only a precondition of most happiness, but itself one of the surest sources of it. For many, indeed, the Universe is too grim to be contemplated, let alone enjoyed, without a certain degree of intoxication. Yet most forms of intoxication are hazardous. But there remains this one happy form of it that leaves no headaches—intoxication with the vital rush of the blood in one's own veins. The happiness of Wordsworth, to put it a little differently, was that he so wisely contrived to be drunk with brook water from the fells of Westmoreland. I doubt if, when all was over, Omar Khayyam had got nearly as much happiness from all the grapes of Shiraz.

What else, indeed, but vitality is the main cause for the conspicuous and ubiquitous happiness of young things? If kittens and puppies, lambs and chicks and children are constantly delighted, and delightful, it is,

Reprinted with permission of The Macmillan Company from *The Greatest Problem and Other Essays* by F. L. Lucas. © by F. L. Lucas, 1960. Also reprinted by permission of Cassell and Company Ltd., Publishers.

above all, because they have a surplus of vitality—more than they strictly need, more than they know what to do with.

No doubt for human beings, even in youth, this complete lightheartedness is less easy. *Their* fears, griefs, and jealousies can commence early in childhood; and can cut deep, because they already begin to discover the hardness and complexity of life before they have acquired the power and knowledge to cope with it; and so may feel themselves impotently battling with something vast, mysterious, and cruel. Hence small children can suffer immense miseries—the more immense because they feel so small. Yet for the adolescent, even so, their vitality usually comes to the rescue, with quick forgetfulness and resilient hope. Not the subtlest of metaphysics could reconcile them one-tenth as well with existence as the mere throb and flush of twenty summers.

But afterwards? There is no elixir of life, except vitality itself. Yet, as life goes on, vitality grows ever harder to keep. Some squander it in frantic overwork, like Balzac or Dickens. Some waste it on diversions far more futile—on books not worth reading, or pedantries not worth pursuit; on stupid official functions; on dinners where the speeches are enough to drive one to drink; on social parties where no one can hear what anyone says, and nothing is worth hearing anyway, since people are merely chattering what they do not mean in the least, to people for whom they do not care twopence.

The mature can no longer expect to play like spendthrifts with that gorgeous surplus of energy which blesses youth. And, from mere lack of vitality, many people may miss happiness simply by despairing of it when some temporary mishap has destroyed it. For happiness may need a number of reiterated attempts. But the devitalized lack the dogged persistence to make them. The gift of accepting the inevitable is no doubt also an important part of happiness; but the devitalized accept as inevitable what is not really so, and resign themselves to sit on glumly in the hole, or rut, into which they have fallen.

No doubt such pertinacious energy is mainly a matter of temperament; and men cannot remake their own temperaments. But it depends also, quite a lot, on physical condition. Therefore it seems to me one of the wonders of the world that men should so seldom treat their bodies with even one tenth of the consideration they show for their cars, or their bank balances. Yet it is a platitude that, while cars can be replaced, and bank balances replenished, one will never (unless one believes in transmigration) find a new body.

If I sought a symbol for happiness it would perhaps be a mountain-spring gently, but unfailingly, overflowing its basin with living water. There seems to me nothing in life more vital than to keep always this slight surplus of energy. One should always overflow. If human energy could be bought, like electrical, I do not know on what a wise man would sooner spend his money; even though it meant something near to beggaring himself.

If after five wounds, gassing, and seventeen months of hospital in the First War, yet I have never had an illness since; if I was able during the Second War to work six days a week and fifty weeks a year, from 4 P.M. to 1 or 2 A.M., in addition to Home Guard service, with only one day's absence in six years (from influenza), this good luck was, I think, mainly because between the wars I usually spent at least one month in twelve walking, and kept a car mainly in order to walk, in addition, one day every week or fortnight, for six or seven hours, in country less depressing than the immediate outskirts of Cambridge; and because, even in the Second War, I managed to dig in the mornings before going down to the War Department in the afternoons.

Most of those who lead intellectual lives are not particularly intelligent in the way they lead them, from not realizing that the brain easily becomes a kind of cancer to the body; and that physical condition can do more for men's happiness and usefulness than all the philosophers since Pythagoras. Valetudinarians are a curse to themselves and others; but most people seem to me not sufficiently vitality-conscious—bad stewards of their bodily energy.

I remember, years ago, hearing the late Lord Keynes remark, "I think I have killed the exercise bug." (That is, he had learned to live without it. I seem to recall that Joseph Chamberlain had a similar idea.) At the time I wondered. And I wonder still if Keynes need have worked himself heroically to death in the public service during the Second War, had he adopted in earlier years a way of life less purely intellectual.

So tyrannical, indeed, is the physical, so vitally important its energy, that when I merely shift from the vertical to the horizontal, and think in bed, my whole character and outlook on life often seem changed. I have come to believe that ever to think of any harassing question in bed is idiocy. One may sleep on problems; but one should never lie awake over them. Somehow the helplessness of a recumbent position, like a tortoise on its back, the numbing impotence of drowsiness, and the depressing effect of darkness, combine to make every pebble in one's path seem a rock of offense, every pismire a tiger, every pinprick a bleeding wound. Then when one rises in the morning, back to the vertical, one wonders what all the trouble was about. "I make a point of never lying awake," said Wellington. Enviable man, to be so much his own master! I have learned to moderate this imbecility, but not to overcome it. Another reason why one should tire the body enough in the daytime to drug that fretting mole, the mind.

"When the belly is full," runs the Arab proverb, "it says to the heart, 'Sing, fellow!'" That is not always so; the belly may get overfull. Such a proverb clearly comes from a race familiar with bellies painfully empty. Yet it remains true, I think, that when the body is in radiant health, it becomes extremely difficult for it not to infect the mind with its own sense of well-being.

But more vivid still in its expression of this simple truth which, however obvious in theory, proves in practice so far from obvious to many, is the tale in *The Arabian Nights* of Haroun-al-Raschid and his fool.

And when the Caliph had finished drinking, Bakloul said to him: "And *if*, O Commander of the Faithful, this glass of water you have just drunk refused to issue from your body, because of some retention of urine in your honorable bladder, what price would you pay for the means to make it come?" And Al-Raschid replied: "By Allah, in that case I would indeed give the whole length and breadth of my empire." Then said Bakloul, grown suddenly full of sadness: "Ah, my Lord, an empire that weighs no more in the balance than a glass of water, or a jet of urine, ought not to cost all the cares it causes you, and the bloody wars it brings on us." And, hearing him, Haroun burst into tears.

It is twenty years since I first read this blunt wisdom; but I have never forgotten it.

But there is health of mind as well as of body. The havoc a sick body can play with the mind is obvious; but anyone who troubles to read psychology, soon sees that a sick mind can sometimes play equal havoc with heart, lungs, or stomach, and even distort sight, hearing, taste, or smell. Invalids have been happy; blind men have been happy; but I doubt if much happiness can ever reach the badly neurotic.

Unfortunately, however, though naturally, modern psychologists have concerned themselves so much more with mental sickness, that it is seldom easy to extract from them any clear idea of mental health—indeed from their books one might think at times that no such thing existed. But the type of mind that I have come most to admire is one which wastes as little as possible of the energy it needs for battling with the world, in battling with itself; which keeps its hatreds for things, not persons, knowing that men do not make themselves; which is as active as if it were building for eternity, yet resigned, as sensible men must be, to the realization that in a few years we, and all our works, shall be as if we had never been. For one who lives convinced that nothing matters which he does, becomes a slug; one who imagines that anything he performs is likely to matter prodigiously, is likely to become a pompous fool.

On the other hand one must admit that the mind, though it can do a good deal for the health of the body it inhabits, probably cannot do much for its own health unless it is healthy already. "To him that hath shall be given." A warped mentality can seldom, if ever, straighten itself. And whether it is warped or not, must depend mainly, if not wholly, on inborn temperament, upbringing, and circumstance.

For health of mind and character, upbringing seems extremely potent, both to help and to harm. The vital thing, I believe, is to grow up in a happy home (for conflicts between parents mean conflicts in the child) and in an atmosphere so warm with sympathy that few punishments are needed, beyond the sudden chill of affection temporarily withheld. "Warm" does not mean "stuffy"; on the contrary nothing is more vital than early training in independence.

Two Victorian parents come at this moment to my mind, as opposite instances of wisdom and folly. "My father," said Lord Salisbury's school-boy son, "always treats me as if I were an ambassador; and I like it so." "I have known boys," observed Dr. Arnold, the famous headmaster of Rugby from 1828 to 1842, "—boys of eight or nine years old, who did not so much as know what would happen to them after their death." Did he? Words fail me to express my admiration for the first of these attitudes, or my abomination of the second.

But health of mind as of body needs, even if possessed, constantly to be maintained. In either case, the principle is the same—to keep stirring. Activity is difficult without health: health, without activity. One of the wisest remarks about living that I have ever read, is the Italian poet Alfieri's summary of his own experience—that he had always found life empty except when it was filled by both "*un degno amore*" and "*qualche nobile lavoro*"—"worthy love" and "dignified work."

But simply to live on affection alone without work, or on work alone without affection, is trying to hop through life on one leg. Our natures are double—part egoist, part altruist; and both sides must be served—or one of them may die, and poison us.

In the quest for happiness, then, a sufficiency of activity seems vital; being itself a cause, as well as a result, of continued vitality. Few guardian angels are so effective as *le diable au corps*.

Physical differences must play a large part; but it remains amazing how this ardor of energy can sometimes resist even the decadence of age. It is enough to recall Titian still vigorously painting after ninety; or Sarah Bernhardt, vowing to live to a hundred to annoy her enemies, and acting till her death at seventy-nine, even after losing a leg; or Winston Churchill taking up the weight of a six years' struggle to save the world, at an age when most men abandon themselves to an armchair by the fire.

Most explicit of all is the Japanese, Hokusai. "From the age of six," he says, "I had a mania for drawing the forms of things. By the time I was fifty I had published an infinity of designs; but nothing that I produced before seventy is worth considering. At seventy-three I have learned a little about the real structure of nature, of animals, plants, birds, fishes, and insects. In consequence, when I am eighty, I shall have made more progress; at ninety I shall penetrate the mystery of things; at a hundred I shall certainly have reached a marvelous stage; and when I am a hundred and ten, everything I do—even a dot or a line—will be alive.

"Written at the age of seventy-five by me, once Hokusai, today Gwakio Rojin, the old man mad about drawing."

He died, alas, at a mere eighty-nine, declaring that if he could have had only another five years he would have become a great artist. But that, surely, is the way to grow old; which is simply—not to let oneself grow old. Activity to the last!

Nature, men used fancifully to say, abhors a vacuum: but of human nature this is actually true. Even futile activity, whatever the Buddhist view, may be better, at least for western minds, than none at all. One of Napoleon's prisoners, Ouvrard, kept himself sane in solitary confinement by daily scattering on the floor of his cell a number of pins, and groping about till he had found them all again. It is the idle dog that is most pestered by his fleas. He would be far happier worrying a stick. We all need our sticks to worry, even if we realize at moments that they are merely dry sticks; though of course this wisdom fails if, instead of the dog worrying the stick, the stick begins to worry the dog. If the pure extravert is apt to be unintelligent, the pure introvert tends to grow broody. Life does not bear too much thinking on: the best antidote is action.

One of the problems of our modern world is that so many unfortunate beings are forced to spend their lives on work they dislike, or would dislike if they had more sense. The ancient Greek view that much mechanical labor is unworthy of free men because it warps the body (and, they might have added, the mind) is bleak, but true.

The happiest work, I think, must be creative—not in the narrow sense that we should all be painting pictures or writing odes, but in the sense that it calls on the individual for intelligent skill exerted in his own way, even if it is only growing vegetables; that it should not be the mechanical drudgery of slaves toiling at a conveyor-belt, or pouring nails down a pipe.

If I were a bricklayer, I should have no initiative of my own in laying my bricks—the architect would have settled all that; but I should at least want the fun of finding how to lay bricks as fast as they can be well laid, with that economy of effort which in a work of art we call "grace"; but my trade union would not let me, for quite understandable, yet really lamentable, reasons. And I should have to stroll wearily and listlessly about my job, like all the other bricklayers that anyone may see anywhere any day. A detestable conclusion.

The bane of our time, in short, is that instead of doing work we like, and as much of it as we healthily can, and living largely on the joy of that, we do work we dislike, and as little as possible, and expect to live only in our moments of leisure, which we try our utmost to extend. And when we get that leisure many of us spend it, not on activity, but in the lame impotence of passive spectators.

The curse of labor does not lie in labor, but in the unsatisfactory kinds of it to which all except the lucky have always been, and still are, condemned—work that distorts body or mind, that is tedious, or useless, or immoderate. To be satisfactory, indeed, it is not enough that work be of a kind that can become interesting. The human mind has a useful, yet horrible, capacity for growing interested, sometimes, even in the most sterile and futile subjects. The shelves of libraries groan with books and dissertations that ought to have bored their writers too insufferably ever to get written. Ideally, work should not only be interesting to the worker; it should also not make him uninteresting to his intelligent fellow creatures. And it should be useful not only to the doer but to others. It is this that makes, for example, the supreme good fortune of a good doctor. Indeed it is hard for activity to seem lastingly satisfying if it serves no ends but one's own.

Another condition of happy activity is that it should be successful, yet not too easily successful. There should be difficulties enough to challenge brain or sinew, but difficulties that can be overcome; just as the happiest climate for human development is something less soft than Tahiti, less hard than Greenland. Happiness depends more on progressing and succeeding, than on being something, or possessing something. It seems likely enough that a housemaid promoted to housekeeper may feel as much satisfaction as a Foreign Secretary promoted to Prime Minister. Such things are largely relative. If the lift is going up (no matter from which floor to which)—elation; if it is going down—gloom.

It may, however, be a misfortune to succeed too rapidly, to too easily. For most men, it is usually happier to succeed constantly rather than spectacularly; to advance with a steady moderation that leaves future possibilities unexhausted; so that life itself ends before the tide of success and achievement has seriously ebbed.

The essential is always to have something in hand. In Montaigne's phrase, let death take a man still planting his cabbages. No doubt, this stress on activity may strike some as febrile. There was some Indian who observed in his wisdom: "To stand is better than to move; to sit than to stand; to lie than to sit; and to be dead than all that." But this seems to me a dismal and detestable view of life, attributable partly to a debilitating climate. Any race that really adopted such principles would, in this aggressively competitive world, probably soon attain its wish—or its imagined wish—and become extinct indeed.

Without vitality, life becomes as void of gaiety and zest as a wood where no birds sing. It seems to me a folly, except where it is a duty, or a necessity, to lose that active, buoyant *joie de vivre* whose basis is so largely physical, for any of the things that human ambition labors for; still more a folly to lose it for any bauble in the toyshop of the world. Poor Bakloul the buffoon was wiser after all than the golden Caliph of Baghdad.

Nine

Discrepancies between Findings of Longitudinal and Cross-sectional Studies in Adult Life: Physique and Physiology

A. Damon

The thesis that trends in growth, maturation, and aging are best determined by longitudinal rather than cross-sectional studies needs no theoretical defense. Cross-sectional studies permit inferences as to what *might* happen under certain conditions *if* certain assumptions are valid. Only by following the same persons over time can we learn what in fact *does* happen, as well as when, how, and at what rate it happens. This point has been well demonstrated in many longitudinal studies of child growth, including Terman's work on intellectual functions of gifted children and the growth studies at the Harvard School of Public Health, the Brush Foundation in Cleveland, the Fels Foundation in Yellow Springs, Ohio, the Child Study Center in Denver, and the Institute for Human Development in Berkeley, California. Some of these well-known investigations are beginning to extend into adult life.

For obvious practical reasons, longitudinal research on aging has been much slower in getting under way than longitudinal research on child growth. There are now at least fifteen longitudinal studies of adults throughout the country, mostly centered around cardiovascular disease, as well as several in the psycho-social area. The most extensive, in respect to breadth of areas encompassed and also because it includes the adult offspring of the subjects, is the Framingham (Massachusetts) Heart Study of the U.S. Public Health Service. Others are the Tecumseh (Michigan) Community Study of the University of Michigan and the Thousand Aviator Study, now in its twenty-fourth year, conducted by the U.S. Naval School of Aviation Medicine, Pensacola, Florida.

Since longitudinal studies in adult life are so new, most of our knowl-

From *Human Development*, 1965, 8, pp. 16–22. Published with permission of the author and publisher, S. Karger, Basel/New York.

Presented at the American Psychological Association Meeting, Los Angeles, Calif., September 1964.

108

edge—or perhaps "notions" would be a better term—of age changes rests on cross-sectional findings. Strictly speaking, cross-sectional studies can describe only age *differences between groups,* and we assume that these correspond to *age changes in individuals.* Only a little reflection, however, will tell us that differences between older and younger adults at one point in time can have two other explanations beside aging in the individual. Age differences can indeed reflect age changes; but they can also signify (1) membership in cohorts with different life experiences or (2) selective survival.

Let us examine these possibilities with respect to height. All cross-sectional studies show that height decreases from the third decade of life to the seventh or eighth (Stoudt, Damon and McFarland, 1960; Stoudt, Damon, McFarland and Roberts, 1965). This trend is also seen in grip strength and pulmonary functions, but we shall consider only height for the moment. Long-term secular increase in body size is amply documented and could alone account for the observed difference in height, even without invoking shrinkage with age or selective survival of short men. To be specific, we are currently engaged in a Normative Aging Study of 1,500 healthy veterans, at the Boston Outpatient Clinic of the Veterans Administration. As a preliminary step, we studied 133 Spanish-American War veterans with an average age of 81.6 years (Damon and Stoudt, 1963). After exclusion of 14 men with severe kyphosis, the 119 remaining men averaged 66.3 inches tall. As compared with estimates for American men in general, they were 0.2 inches shorter than men in their seventies, 0.5 inches shorter than men in their sixties, and so on until they were 2.4 inches shorter than men in their twenties.

Secular increases in height have been thoroughly documented, and their magnitude could more than account for this 2.4-inch difference. Bowles (1932) found that in Harvard men measured between 1892 and 1920, the sons measured 1.3 inches taller than their fathers on entry to Harvard. Between 1920 and 1946, American soldiers in the Second World War were 0.7 inches taller than those in the first World War (Davenport and Love, 1921; Newman and White, 1951). Since 1946, there has been a further increase of height among young men, reported as 0.5 inches (Karpinos, 1961). These three stature increases of American men between 1890 and 1960 add up to a total of 2.5 inches, more than enough to account for the difference between Spanish-American War veterans in their eighties and young men of today in their twenties *without having to invoke aging in the individual at all!* Incidentally, Newman (1963) has shown that tall young men of today continue to grow between the late teens and early twenties, refuting Morant's (1947) contention that secular increase of height in young men is an illusion resulting from their earlier attainment of adult stature.

The gist of the foregoing argument is that secular or long-term trends, reflecting the different life experiences of the several age cohorts represented in a cross-sectional survey, can account for the observed differences. The possibility that older people are small because small people live longer is seriously advanced by pediatricians and other physicians. For example, the eminent cardiologist, P. D. White, urges a halt, largely on these grounds, to what might be termed forced feeding of the young. Only longitudinal or cohort studies followed throughout entire lifespans can settle this question, but the issue of selective survival should be kept in mind when we try to account for age differences in height.

All of the foregoing discussion is indirect. It *is* known that the height of individuals decreases late in life; but when does this begin, at what rate, to what extent, and with how much variation from one person to another? This we can learn only from longitudinal studies. The two longitudinal studies that we know about indicate that *age changes* in height by no means follow the pattern indicated by cross-sectional *age differences*. Cross-sectional studies would lead one to expect a slow but steady decrease beginning in the fifth or sixth decade of life. But Lipscomb and Parnell (1954) found no height decrement by age 72 among 44 retired British servicemen remeasured after fifty years, and Damon and Sheldon (unpublished) found no decrement among 187 healthy Columbia College men re-studied after 37 years. Of the 187 Columbia men, only 27 showed any degree of height loss between the mean ages of 18–19 and 56 years. The net difference for the whole group was a *gain* of 1.62 ± s.e. 0.13 cm. To be sure, men grow between the ages of 18–19 and 22, when adult stature is attained; the amount of such growth, in 1958, being 1.3 cm (Karpinos, 1961). Nevertheless, cross-sectional studies would lead one to expect an appreciable decrement by age 56.

Equally striking discrepancies between cross-sectional and longitudinal findings have been found in the two physiological measures of grip strength and vital capacity. Cross-sectional studies show a gradual rise in grip strength and vital capacity to a peak between the ages of 25 and 30, and thereafter a steady decline with age (Fisher and Birren, 1947; Norris et al., 1956; Shock, 1962). The secular trend could be a partial explanation here too, by virtue of the correlation between height and each of the two physiological measures. Coefficients of correlation (r) between height and grip strength have ranged from 0.19 among young physical education students (Sills, 1950) to 0.36–0.46 among four older groups of men (Damon, 1961, 1962, 1964; Zwerdling et al., 1963), while r's between height and vital capacity are higher, ranging from 0.56 to 0.78 among four groups of men (Kory et al., 1961; Brody et al., 1963; Zwerdling et al., 1963; Damon, 1964). Removal of the effect of age yields partial correlations between height and each of the two physiological measures negligibly lower than the zero-order coefficients.

In the study of Columbia College men just mentioned, mean grip strength and vital capacity of the same subjects had not decreased after 37–38 years. Post-training grip measurements were obtained for 108 men observed twice during their first year in college—once on entry in the fall and again several months later, after a regimen of physical training. Their right-hand grip strength averaged 52.6 ± s.e. 0.9 kg at age 18.6 years, and 53.2 ± s.e. 0.7 kg at 57.1 years of age. Adding 62 men observed only on entry to college, the mean difference in right-hand grip between the ages 18–19 and 56–57 for 170 men was 2.8 ± s.e. 0.8 kg, in favor of the older men.

For 158 of these men, the mean difference in vital capacity between the same ages was 169 ± s.e. 41 cc., with the larger measurement again at the older age. In this comparison, the larger of the two "early" measurements was used. For 35 men observed only in the fall, before physical training, only the single observation was available. Using the second or post-training vital capacity for the remaining 123 men, their vital capacity at mean age 57 was 190 ± s.e. 42 cc. greater than at age 18–19.

Dill (1963) is currently undertaking detailed physiological follow-up studies after 20–25 years, but to date has not published comparative data on vital capacity (Dill et al., 1964). We await his findings with great interest.

The discussion to this point has concerned clear discrepancies between findings from cross-sectional and longitudinal studies. There are two other areas where large questions exist, if not discrepancies. The first area comprises changes which certainly occur with age, but with unknown time of onset, rate of change, or sometimes even the precise amount. In addition to height, an example is the increased anteroposterior diameter of the thorax, part of the complex called "senile chest."

The second area includes characteristics where change with age is variously reported in the cross-sectional studies. Skinfold thickness, an excellent index of body fat, was found to increase with age in Canadian men (Pett and Ogilvie, 1956) and in London men (Appenzeller, 1963), but not in three series of American men (Lee and Lasker, 1956; Brozek et al., 1963; Damon and Stoudt, 1963).

The conclusion we draw from the foregoing discussion is that longitudinal studies in adult life, though expensive, laborious, and long-lasting, are more than merely desirable to study the aging process—they are indispensable. It is no longer a matter of refining or confirming what we think we know about aging from cross-sectional studies. What we think we know may be in serious error.

SUMMARY

Whereas cross-sectional surveys suggest that height, grip strength, and vital capacity should all decrease between the ages of 18–19 and 56 or later, no such decrease was observed in longitudinal studies. Possible explanations for the apparent decrease shown cross-sectionally are, in addition to the usual assumption of individual changes during aging, (1) the secular trend toward increasing height and (2) selective survival of small persons. In addition to such discrepancies, there are other problems which can be settled only by longitudinal study: changes known to occur with individual aging but with unknown time of onset, rate, and magnitude (for example, stature and the anteroposterior chest diameter); and contradictory findings of several cross-sectional studies (for example, skinfold thickness). Longitudinal studies of aging are not merely desirable —they are mandatory.

REFERENCES

Appenzeller, O. "Skin-fold thickness in the aged. Measurements in a sample of the London population." *Brit. J. Prev. Soc. Med.* 17: 41–44, 1963.

Bowles, G. T. *New Types of Old Americans at Harvard and at Eastern Women's Colleges.* Cambridge, Mass.: University Press, 1932.

Brody, A. W., H. J. Wander, P. S. O'Halloran, J. J. Connolly, Jr., and F. W. Schwertley. "Correlations, normal standards, and interdependence in tests of ventilatory strength and mechanics." *Amer. Rev. Resp. Dis.* 89: 214–235, 1964.

Brozek, J., J. K. Kihlberg, H. L. Taylor, and A. Keys. "Skinfold distributions in middle-aged American men. A contribution to norms of leanness-fatness." *Ann. N.Y. Acad. Sci.* 110: 492–502, 1963.

Damon, A. "Constitution and smoking." *Science* 134: 339–341, 1961.
"Notes on anthropometric technique: Stature against a wall and standing free." *Amer. J. Phys. Anthropol.* 22: 73–78, 1964.

Damon, A., H. K. Bleibtreu, O. Elliot, and E. Giles. "Predicting somatotype from body measurements." *Amer. J. Phys. Anthropol.* 20: 461–474, 1962.

Damon, A., and H. W. Stoudt. "The functional anthropometry of old men." *Human Factors* 5: 485–491, 1963.

Davenport, C. B., and A. G. Love. *Army Anthropology,* Vol. 15, Medical Department of the U.S. Army in the World War. Washington, D.C.: Government Printing Office, 1921.

Dill, D. B. "The influence of age on performance as shown by exercise tests." *Pediatrics* 32: 737–741, 1963.

Dill, D. B., W. H. Forbes, J. L. Newton, and J. W. Termann. "Respiratory adaptations to high altitude as related to age." Chap. 5, pp. 62–73, of *Relations of Development and Aging,* Birren, J. E. (ed.). Springfield, Ill.: Charles C Thomas, 1964.

Fisher, M. B., and J. E. Birren. "Age and strength." *J. Appl. Psychol. 31:* 490–497, 1947.

Karpinos, B. D. "Current height and weight of youths of military age." *Human Biol. 33:* 335–354, 1961.

Kory, R. C., R. Callahan, H. G. Boren, and J. C. Syner. "The Veterans Administration—Army Cooperative Study of Pulmonary Function. I. Clinical spirometry of normal men." *Amer. J. Med. 30:* 243–258, 1961.

Lee, M. M. C., and G. W. Lasker. "The thickness of subcutaneous fat in elderly men." *Amer. J. Phys. Anthropol. 16:* 125–134, 1958.

Lipscomb, F. M., and R. W. Parnell. "The physique of Chelsea pensioners." *J. Royal Army Med. Corps 100:* 247–255, 1954.

Morant, G. M. "Anthropometric problems in the Royal Air Force." *Brit. Med. Bull. 5:* 25–31, 1947.

Newman, R. W. "The body sizes of tomorrow's young men." Ch. 8 in *Human Factors in Technology*, Bennet, E., J. Degan, and J. Spiegel (eds.). New York: McGraw-Hill, 1963.

Newman, R. W., and R. M. White. *Reference anthropometry of Army men.* Rept. No. 180, Env. Protection Branch, U.S. Army Quartermaster Research and Development Center, Natick, Mass., 1951.

Norris, A. H., N. W. Shock, M. Landowne, and J. A. Falzone, Jr. "Pulmonary function studies: Age differences in lung volumes and bellows function." *J. Geront. 11:* 379–387, 1956.

Shock, N. W. "The physiology of aging." *Sci. Amer. 206:* 100–110, 1962.

Sills, F. D. "A factor analysis of somatotypes and of their relationship to achievement in motor skills." *Res. Quart. 21:* 424–437, 1950.

Stoudt, H. W., A. Damon, and R. A. McFarland. "Heights and weights of white Americans." *Human Biol. 32:* 332–341, 1960.

Stoudt, H. W., A. Damon, R. A. McFarland, and J. Roberts. "Height, weight, and selected body dimensions of U.S. adults, 1960–1962," Nat. Health Exam. Survey, U.S.P.H.S. Washington, D.C.: Government Printing Office, 1965.

Zwerdling, M., J. R. Goldsmith, and F. Massey, Jr. "Maximal expiratory flow tests in a population sample, with an application of component analysis." *Amer. Rev. Resp. Dis. 87:* 69–87, 1963.

Ten

Biological Differences
in Intellectual Functioning

Lissy F. Jarvik

Biological differences in intellectual functioning as well as variations in the capacity to handle life situations are considerable. We need only compare the utter helplessness of the human infant with the sophisticated achievements of his adult counterpart to indicate the vast biological changes that take place in a lifetime.

Yet, some capacity for handling life situations is an essential attribute of all organisms. Survival itself implies the interaction of living matter with its surroundings. During the long process of evolution—from the simplest unicellular organism to complex human beings—certain adaptive mechanisms, with selective advantages over others, have been perpetuated in our species. In this way we can account for the basic similarities among all men, in terms of growth, maturation, physiological homeostasis, psychological processes, and other constitutional and behavioral attributes which characterize human beings as such. At the same time, hereditary mechanisms permit such great diversity in genic constellation that each person is truly an individual, unique, and distinct from all others.

Individuals with different hereditary endowments (different genotypes) are then exposed to a wide variety of circumstances. Personal adjustment to life situations is a function of both the hereditary make-up and the environmental conditions which are encountered. The genotype does not determine the response to a given set of stimuli, but merely limits the range within which the organism can respond. Thus, for example, individual variations in the ability to learn from others and to profit from experience are determined by genic elements, while the utilization of this potential ability is governed largely by socio-cultural factors and is not directly under genic control.

From *Vita Humana*, 1962, 5, pp. 195–203. Reprinted with permission of author and publisher, S. Karger, Basel/New York

Presented at the Symposium on Intellectual Development in Post-Maturity, as part of the program of the annual meeting of the American Psychological Association, St. Louis, Missouri, August 31, 1962.

The intricate interaction of hereditary and environmental agents, which ultimately results in a complex functioning human being, is difficult to analyze. One-egg twins give us two separate individuals with the same genotype. Thus we know that their genes are identical and any differences observed in the subsequent development of such twins must be attributed to differences in genotypic expression resulting from variations in intrinsic and extrinsic milieu.

Obviously, no conclusions can be drawn from the comparison of just two persons and the main value of twin studies derives from the existence of two types of twins. One-egg (monozygotic) twins are derived from a single fertilized ovum which splits into two halves and gives rise to two individuals with the same complement of genes. By contrast, two-egg (dizygotic) twins are the product of two separate ova, fertilized by two spermatozoa. They are genically no more alike than are two ordinary siblings born at different times, although their environments may be somewhat more similar than those for siblings of different ages. The comparative differences between one-egg and two-egg twin partners give a rough indication of the relative efficacy of the hereditary and environmental factors which interact in the production of observable characters.

In representative samples of twins, the average difference between two members of a twin pair—the so-called mean intra-pair difference—is smaller for one-egg than for two-egg pairs, if genic elements exert a measurable influence upon the trait. The operation of hereditary factors has been confirmed in this way for variations in normal characters such as height, weight, and intelligence. It does *not* follow that ultimate stature, either literally or figuratively, is predetermined at conception, but the limits of possible variations are indeed set at that moment.

The problem of the relative contributions of innate endowment and subsequent training was approached by means of twin studies as early as 1883. Galton concluded at that time that ordinary environmental differences are not sufficient to make "similar" twins unlike. By the same token, "dissimilar" twins show no tendency to become more alike under the influence of similar surroundings. These conclusions have since been confirmed by numerous psychometric studies of school-age twins.

At the adult and especially the post-maturity level, however, we encounter a more difficult situation. The opinion has often been expressed that genic elements, though governing intellectual behavior during childhood, cease to be of importance in later life. Until recently there was little evidence either for or against this belief. The data now available are derived from a psychometric study of aging twins which was organized in 1946 as part of a long-term investigation of the hereditary aspects of aging and longevity in a series of about 2000 twins over the age of sixty (Kallmann and Sander, 1948, 1949). The criteria by which subjects for psychological testing were selected from this large sample were described in detail elsewhere (Feingold, 1950).

The pertinent features may be summarized as follows: The subjects were residents of communities in New York State (not institutionalized), literate, white, English-speaking, in a satisfactory state of health, and at least sixty years of age at the time of first testing (1946–1949). As discussed elsewhere (Jarvik et al., 1962), 268 twins, forming 134 same-sex pairs, received the test battery consisting of five subtests (Digits Forward, Digits Backward,[1] Digit Symbol Substitution, Block Designs, and Similarities) from the Wechsler-Bellevue Intelligence Test, Scale 1 (Wechsler, 1944), the Vocabulary test from List 1 of the Stanford-Binet (Terman, 1916) and a simple paper-and-pencil tapping test (Feingold, 1950). The same battery was administered again to 207 of these subjects after a period of approximately two years, and for a third time to 78 survivors six years later. A fourth test round was arranged after an additional two-and-a-half years and could still be administered to 17 subjects in this sample.

The results of the initial test round revealed that the scores of one-egg twin partners were more similar than those of two-egg twins even after the age of sixty, indicating the persistence during adult life of gene-specific differences in mental functioning. No significant change in the intra-pair correlations was demonstrated during the follow-up period.

Aside from the demonstration of measurable genic effects during senescence, our psychometric twin study provided some interesting information on intellectual changes in relation to the aging process. The applicability of data derived from twin studies to the single-born population has often been a topic of discussion. Last revived by reports in the European literature, various investigators ascribe a lower level of intelligence to school-age twins than to singletons (Mehrotra and Maxwell, 1949; Sandon, 1957; Tabah and Sutter, 1954; and Zazzo, 1952). In the age group over sixty, however, the average test scores were found to be similar for twins and for single-born aged persons, so that it is appropriate to utilize twin data in the evaluation of longitudinal intellectual changes taking place during senescence.

Details regarding such intellectual changes observed during a twelve-year follow-up period were published periodically and revealed an overall tendency toward decline in test scores with advancing age, although the rate of decline was smaller than had been inferred from cross-sectional studies. Moreover, different tests showed varying rates of decline (Jarvik et al., 1957, 1962). The general trend may be illustrated by results obtained from 17 twins who were tested on four consecutive occasions (Table 10.1). A definite decline was evident on the two speeded motor tests only—Tapping and Digit Symbol Substitution. The statistical significance of the decline was established for the larger group of subjects

TABLE 10.1   Mean Scores of 17 Aged Subjects on Four Successive Testings (1947–1957)

	Number of subjects	First		Test session Second		Third		Fourth	
Vocabulary	14	30.4	(5.9)*	31.1	(5.9)	31.9	(4.4)	30.0	(5.3)
Tapping	16	73.9	(15.2)	74.5	(16.8)	56.4	(15.1)	61.4	(16.6)
Digits Forward	15	6.4	(0.9)	6.3	(1.2)	6.4	(1.3)	6.5	(1.4)
Digits Backward	15	4.8	(1.4)	4.9	(1.0)	4.6	(0.9)	4.3	(1.0)
Digit Symbol	15	30.3	(7.6)	28.1	(10.7)	27.2	(8.3)	25.9	(8.7)
Similarities	17	11.1	(4.1)	10.5	(3.9)	11.4	(4.6)	11.1	(3.6)
Block Designs	13	15.4	(4.8)	16.0	(6.6)	12.5	(5.1)	13.2	(3.2)
Age	17	67.5	(6.8)	68.3	(6.8)	74.2	(6.1)	76.3	(6.3)

* Standard deviations in parentheses.

tested on three occasions (Jarvik et al., in press). On the other five tests there was a limited decline only, even though over nine years elapsed between the first and the last testing.

Nevertheless, a positive relationship emerged between test score and survival for most of the tests (Jarvik et al., 1957, 1962). In other words, the original test scores were lower for subjects who died before 1955 than for those still alive at that time. Similar indications of association between survival and intellectual performance have since been reported by Sanderson and Inglis (1961) and by Kleemeier.[2] The former noted a higher mortality rate in patients with than in those without memory disorder, while the latter postulated an "imminence of death" factor to account for the sharp decline in intellectual performance prior to death.

A recent analysis of our own data revealed a positive relationship between five-year survival after the last test session and stability of scores on Vocabulary, Digit Symbol Substitution, and Similarities tests. A "critical loss" on two or all three of these tests occurred with greater frequency among those who died during the five-year interval than among those who survived it (Figure 10.1). "Critical loss" was defined for the vocabulary test as a score on the last testing that was below the lower score on either of the first two testings. For the Similarities and Digit Symbol tests the additional stipulation was made that the annual rate of decline had to exceed ten and two percent, respectively (Jarvik and Falek, unpublished). The deceased group, selected on the basis of "critical loss," did not differ in age from the surviving group (73.8 versus 76.9 years, respectively).

The three tests which may reflect changes indicative of subsequent survival include two tests with opposite age trends. Vocabulary tends to

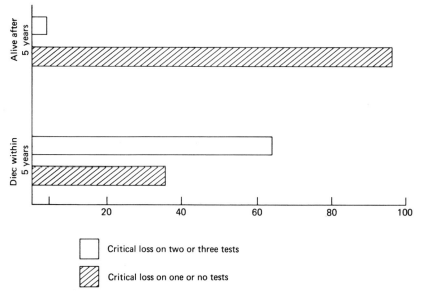

Figure 10.1   Survival and critical loss on digit symbol, similarities and vocabulary tests (34 subjects).

show a positive correlation with age, and even on a longitudinal basis only 32 percent of the aged subjects decreased their scores. Digit Symbol, on the other hand, tends to show a negative correlation with age. In our 12-year follow-up group, over 63 percent of the subjects declined in scores (Figure 10.2). Apparently, separate mechanisms are responsible for "critical losses" and decline on speeded tasks. The latter may represent general concomitants of aging, while the former may reflect more specific cerebrovascular changes associated with high five-year mortality.

The assumption that both mechanisms are biologically rooted rests upon indirect evidence as does the role assigned to genic elements in the determination of individual differences in intellectual functioning. There is as yet no direct proof of differences in hereditary material as the substrate of differences within the normal range of intellectual abilities, either in children or in aged persons. Yet, an analogous situation existed with respect to mongolism until a few years ago. In 1953 Kallmann inferred from twin studies of mongolism that, in spite of the well-known relation to maternal age, genic factors had to be of considerable importance in the etiology of this type of mental defect. Similar conclusions were reached after expansion of the studies (Allen and Kallmann, 1955, 1957; Allen and Baroff, 1955). In 1959 Lejeune and his associates, Gautier and Turpin, described the presence of an extra chromosome in a French child with mongoloid idiocy (Lejeune et al., 1959). This chromosome has since been

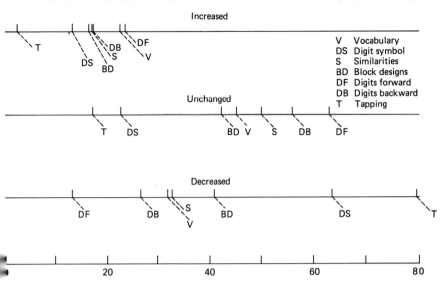

Figure 10.2   Percent of subjects with increased, unchanged, or decreased retest scores.

identified as No. 21, and the anomaly, visible under the microscope, leads to the well-known physical and mental stigmata of the disorder. Neither the nature of the excess genetic material, nor the mechanism of action, are known but this should not let us despair. Hereditary defects are no less curable than acquired ones. In fact, of the two forms of mental deficiency which respond to medical management, provided it is undertaken early enough, one—phenylpyruvic oligophrenia—is clearly hereditary and the other—cretinism—is frequently so. There is every reason to hope, therefore, that identification of the crucial elements located on chromosome No. 21 will enable us to devise a rational program for the prophylaxis of the deleterious effects of mongolism.

For psychologists, the new advances in cytogenetics are pertinent inasmuch as nearly all of the chromosomal abnormalities are associated with an increased frequency of disturbed or subnormal mental functioning. Even sex-chromosome irregularities result in a higher rate of mental retardation, possibly related to the number of excess X-chromosomes (Grumbach, Marks and Morishima, 1962; Hamerton, 1961; Schuster and Motulsky, 1962).

In conclusion, it seems justified to stress the importance of an awareness of the role of heredity not only in the etiology of mental disease and mental defects but also in delineating normal range of variability in anatomical, physiological, and psychological functions. Heredity supplies the raw materials which are molded by external influences into the final shapes to which we refer as physical and mental traits. Realization that the nature of these raw materials differs from one individual to the next, should help in the optimal environmental manipulation of each case. With the study of human chromosomes still in its early stages, it is to be expected that further research in cytological and biochemical genetics will substantially clarify the biological substrate of intellectual functioning.

SUMMARY

To gauge the extent of the influence of genic elements in intellectual functioning during later life, data were obtained from a psychometric study of aging twins organized in 1946 as part of a long-term investigation of the hereditary aspects of aging and longevity. Greater similarity was observed between the scores of one-egg than of two-egg twins even after the age of sixty, indicating the persistence during adult life of gene-specific differences in mental functioning. Average test scores were as high for the twins as those reported for single-born aged persons in the general population. During a 12-year follow-up period statistically significant decline occurred only on speeded motor tasks and this decline may represent general concomitants of aging. By contrast, the formula for *critical loss*, as described in the paper, may reflect specific cerebrovascular changes associated with high five-year mortality.

NOTES

[1] Digits Forward and Digits Backward together make up Wechsler's Digit Span.

[2] Kleemeier, R. W. *Intellectual changes in the senium or death and the I.Q.* Presidential Address, Division on Maturity and Old Age, American Psychological Association, New York, September 1961.

REFERENCES

Allen, G., and G. S. Baroff. "Mongoloid twins and their siblings." *Acta Genet.* 5: 294–326, 1955.

Allen, G., and F. J. Kallmann. "Frequency and types of mental retardation in twins." *Amer. J. Hum. Genet.* 7: 15–20, 1955, "Mongolism in twin sibships," *Acta Genet.* 7: 385–393, 1957.

Feingold, L. *A psychometric study of senescent twins.* Doctoral dissertation, Columbia University, 1950.

Galton, F. *Inquiries into human faculty and its development.* New York: Macmillan, 1883.

Grumbach, M. M., P. A. Marks, and A. Morishima. "Erythrocyte glucose-6-phosphate dehydrogenase activity and X-chromosome polysomy." *Lancet 1*: 1330–1332, 1962.

Hamerton, J. L. "Sex chromatin and human chromosomes." *Int. Rev. Cytol. 12*: 1–68, 1961.

Jarvik, L., and A. Falek. "Intellectual stability and survival in the aged." *J. Geront.* (in press).

Jarvik, L. F., F. J. Kallmann, and A. Falek. "Intellectual changes in aging twins." *J. Geront. 17*: 289–294, 1962.

Jarvik, L. F., F. J. Kallmann, A. Falek, and M. M. Klaber. "Changing intellectual functions in senescent twins." *Acta Genet. 7*: 421–430, 1957.

Jarvik, L. F., F. J. Kallmann, I. Lorge, and A. Falek. *Longitudinal study of intellectual changes in senescent twins.* Fifth Congr. Intern. Ass. Geront.; San Francisco, 1960. In: Tibitts, C., and W. Donahue (eds.), *Social and psychological aspects of aging.* New York: Columbia University Press, 1962.

Kallmann, F. J. *Heredity in health and mental disorder.* New York: Norton, 1953.

Kallmann, F. J., and G. Sander. "Twin studies on aging and longevity." *J. Hered. 39*: 349–357, 1948. "Twin studies on senescence." *Amer. J. Psychiat. 106*: 29–36, 1949.

Lejeune, J., M. Gautier, and R. Turpin. "Les chromosomes humains en cultures de tissus." *C. R. Acad. Sci. 248*: 602–603, 1959.

Mehrotra, S. N., and J. Maxwell. "The intelligence of twins. A comparative study of eleven-year-old twins." *Pop. Stud. 3*: 295–302, 1949.

Sanderson, R. E., and J. Inglis. "Learning and mortality in elderly psychiatric patients." *J. Geront 16*: 375–376, 1961.

Sandon, F. "The relative numbers and abilities of some ten-year-old twins." *J. Roy. Stat. Soc. 120*: 440–450, 1957.

Schuster, J., and A. G. Motulsky. "Exceptional sex-chromatin pattern in male pseudo-hermaphroditism with XX/XY/XO mosaicism." *Lancet 1*: 1074–1075, 1962.

Tabah, L., et J. Sutter. "Le niveau intellectuel des enfants." *Ann. Hum. Genet. 19*: 120–150, 1954.

Terman, L. M. *The measurement of intelligence.* Boston: Houghton Mifflin, 1916.

Wechsler, D. *The measurement of adult intelligence.* Baltimore: Williams and Wilkins, 1944.

Zazzo, R. "Situation gémellaire et développement mental." *J. Psychol. Norm. Path. 45*: 208–227, 1952.

Eleven

Of Flies and Men

Theodosius Dobzhansky

One of the assertions which have gained acceptance by dint of frequent repetition is that science is competent to deal only with what recurs, returns, repeats itself. To study something scientifically, this something must be made representative of a class, group, or assemblage. A single *Drosophila* fly is of no interest whatsoever. A fly may merit some attention only if it is taken as a representative of its species. An individual person may, to be sure, merit attention. However, it is allegedly not in the province of science, but of insight, empathy, art, and literature to study and understand a person in his uniqueness.

I wish to challenge this view. Individuality, uniqueness, is not outside the competence of science. It may, in fact it must, be understood scientifically. In particular, the science of genetics investigates individuality and its causes. The singularity of the human self becomes comprehensible in the light of genetics. You may, of course, object that what science comprehends is not really a singularity but a plurality of singularities. However, an artist, no less than a biologist, becomes aware of the plurality because he has observed some singularities.

In the main, genetics is a study of differences among living beings. Genetics would be superfluous if all living beings were exactly alike. If all members of a species were exactly alike genetics could do very little. Since Mendel, the most powerful method of genetics is to observe differences among individuals in the progenies of parents which differed in some ways. Heredity and variation are the two sides of the same coin. Geneticists are always on the lookout for genetic diversity. Variety is said to be the spice of life. It is a staple necessity to geneticists. (This applies, of course, to Mendelian genetics proper. The great discoveries of the role of

Reprinted from *The American Psychologist*, 1967, Vol. 22, pp. 41–48. Copyright of 1967 American Psychological Association, and reproduced by permission of author and copyright holder.

Invited address presented at American Psychological Association, New York, September 1966. Some of the experimental work referred to in the text supported under Contract No. AT-(30-1)-3096, United States Atomic Energy Commission.

chromosomes in the development, and the relationships between DNA, RNA, and protein synthesis could conceivably have been made even if Mendel's laws remained unknown.)

That every person differs from every other person is so obvious that this is taken usually for granted. What continues controversial is to what extent the human differences are due to genetic and in what measure to environmental variations. Though in a new guise, the old nature-nurture problem is still with us. Now, the individuality of flies is rather less evident than human individuality. I do not claim to recognize every *Drosophila* by her face. The drosophiline individuality is nevertheless easier to analyze, and this analysis helps to throw some needed light on human individuality.

The theory of genetic individuality is simple enough. It stems directly from Mendel's second law, the law of independent assortment. An individual heterozygous for n genes has the potentiality of producing 2^n genetically different kinds of sex cells. Two parents, each heterozygous for the same n genes, can give rise to 3^n genotypes among the progeny, and parents heterozygous each for n different genes may produce 4^n genotypes. To be sure, not all of these genotypes are equally probable, because the linkage of genes in the same chromosome limits their independent assortment. Linkage disequilibrium delays but does not prevent eventual realization of the genetic variety. More important is the problem how large is n, that is, for how many genes an average individual is heterozygous, or how many genes are represented each by two or more variants in the populations of a species, such as man or a *Drosophila*.

The disagreement among geneticists on this point is rife. Those who espouse the classical theory of population structure believe that most genes are uniform, not only in all individuals of a species but even in different species not too remote in the biological system. The unfixed genes are a minority, perhaps of the order of same tens. Moreover, among the unfixed genes one variant, one allele, is normal and adaptively superior, while others are inferior and are maintained in populations by recurrent mutation. Though adherents of the classical theory are reluctant to admit this, the theory is a product of typological thinking. Lurking behind the facade of the variability, they like to envisage the Platonic archetype of the Normal Man, homozygous for all good with no bad genes.

The balance theory of population structure would assume numbers of variable genes of the order of hundreds, perhaps even thousands. An appreciable part of this variety is maintained in populations by several kinds of balancing natural selection. The kind most often discussed is the heterotic selection, operating because of hybrid vigor. There are, indeed, genetic variants which are adaptively favorable when heterozygous and unfavorable when homozygous. The gene which in homozygous condition causes sickle-cell anemia in man is a classical example; in heterozygous condition it confers a relative immunity to *falciparum* malaria. Perhaps even more important in evolution is diversifying natural selection. This can be explained most simply by pointing out that every living species faces not just one environment but a variety of environments. Human environments are certainly diverse, and moreover the diversity is growing. It is improbable that genes can be found to show optimal performance in all environments. More likely, different genes will be relatively more adaptive in different environments. Genetic variety is a method to cope with variety of environments.

Theoretical arguments cannot settle the questions for how many genes is an average individual heterozygous, and what proportion of the genes are represented by different alleles in different individuals of a species. Geneticists are busy working on these matters. I can cite here only the brilliant work of Lewontin and Hubby (1966), of the University of Chicago, as an example. Since the total number of genes is unknown, but is surely too large to have the whole set examined one by one, Lewontin and Hubby have decided to study what they believe is a random sample of genes. They chose a battery of enzymes that can be detected by electrophoresis in single individuals of the fly, *Drosophila pseudoobscura*. Some of these enzymes did and others did not show detectable genetic variations. The authors, after making a thorough examination of the possible biases and pitfalls, came to the conclusion that an individual fly was heterozygous for on the average between 10 percent and 15 percent of the genes in their sample. The numbers of kinds of genes in a sex cell can hardly be less than 10,000; an average fly may, then, be heterozygous for a number of genes of the order of 1,000.

Do these results have any bearing on man? Although man is not an overgrown *Drosophila*, he must have as many or more genes than *Drosophila* does. If the degree of heterozygosity in man is anything like it is in *Drosophila*, brothers or sisters are quite unlikely to inherit from their parents the same genes. The likelihood that any two unrelated persons are genetically identical is practically nil. Only identical twins may be genetically identical, since they arise by asexual division of a sexually produced fertilized egg. Even there the possibility of mutation and of cytoplasmic difference must be reckoned with. Human nature is, then, not unitary but multiform; the number of human natures is almost as great as the number of humans. Every person is unique, unprecedented, and unrepeatable.

The demonstration of the genetic uniqueness of individuals only opens, rather than solves, the problem as far as behavioral and social sciences are concerned. There seems little point in belaboring the truism that behavior as such is not inherited. Only genes can be inherited, in the sense of being handed down from parents to offspring. Even so, I have mostly division products, true copies of the genes I have inherited from my parents, rather than these genes themselves. The skin color is not inherited either, because the skin pigment is not carried in the sex cells. However, I am yet to meet anybody who would contend that one's genes have nothing to do with one's skin color. Human, as well as animal, behavior is the outcome of a process of development in which the genes and the environment are components of a system of feedback relationships. The same statement can be made equally validly with respect to one's skin color, the shape of one's head, blood chemistry, and somatic, metabolic, and mental diseases.

There are some authors who go so far as to question the existence of problems of genetics of behavior, distinct from genetics of anything else. They are right only inasmuch as there is not likely to exist a special brand of DNA concerned with behavior, different from that in other kinds of genes; moreover there are no genes "for" behavior, as there are no genes "for" the shape of one's nose. The problem is more subtle. It is the problem, or rather problems, of the genetic architecture of behavioral differences. We want to know how many genes are usually involved in such differences, the magnitude of their effects, the nature of their interactions, the parts played by mutation pressure, hybrid vigor, environmental heterogeneities, and by all forms of natural selection in the formation, in maintenance, and in normal and pathological variations of behavior. In this sense, the genetics and the evolution of behavior may well be different from, let us say, the genetics and the evolution of blood chemistry, or of metabolism, or of chromosomal polymorphism, or of concealing colorations, or mimetic resemblances. And in this sense, which is the only meaningful sense, the genetics of behavior, especially the experimental genetics of behavior, is not yet even a fledgling field, although it has recently begun to chirp rather lively.

In this article I can discuss only one example of a study of genetics of behavior, that made by my colleague B. Spassky and myself on phototaxis and geotaxis in Drosophila pseudoobscura. Hirsch and his students (Erlenmeyer-Kimling, Hirsch, and Weiss, 1962; Hirsch, 1962; Hirsch and Erlenmeyer-Kimling, 1962) have constructed a classification maze, and selected populations of Drosophila melanogaster which were clearly positively and others negatively geotactic in their behavior. They showed furthermore that the genetic basis of this behavior was polygenic, the three large pairs of chromosomes all influencing the result. Hadler (1964a, 1964b) made a similar maze for selection for phototaxis, and succeeded in obtaining positively and negatively phototactic strains of Drosophila melanogaster. Dobzhansky and Spassky (1962), using Hirsch's maze, selected positively and negatively geotactic strains of Drosophila pseudoobscura. Their starting population was polymorphic for some inverted

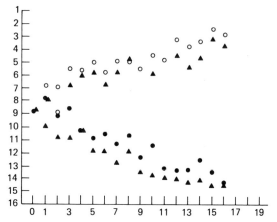

Figure 11.1 Selection for negative (light symbols) and positive (dark symbols) phototaxis. (Circles—females, triangles—males. Abscissa—generations of selection, ordinate—the phototactic score.)

sections in the third chromosomes, and one of the variant chromosomes proved to favor negative geotaxis, while chromosomal heterozygosis favored positive geotaxis.

The results of newer experiments on selection for positive and negative phototaxis and geotaxis in *Drosophila pseudoobscura* are presented in Figures 11.1 and 11.2. The ordinates show the phototactic or the geotactic scores, that is, the averages of the 16 terminal tubes of the mazes into which the flies distribute themselves. On the geotaxis maze the tube No. 1 is the uppermost and No. 16 the lowermost, on the phototaxis

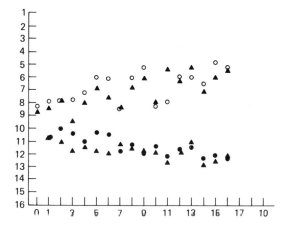

Figure 11.2 Selection for negative (light symbols) and positive (dark symbols) geotaxis. (Circles—females, triangles—males. Abscissa—generations of selection, ordinate—the geotactic score.)

maze No. 1 is reached by 15 choices of dark passages and No. 16 by 15 choices of light passages. The selection is made by running through the maze 300 females or 300 males; the 25 most positive, or most negative, individuals of each sex are selected to be the parents of the next generation. The initial populations in our experiments were photo- and geotactically neutral on the average. Or, to be more precise, these initial populations had positive, neutral, and negative individuals in such proportions that the average scores were between 8 and 9 (an average of 8.5 is exact neutrality). After 15 generations of selection, the positively phototactic line had average scores 13.4 and 14.5 for females and males respectively, the negatively phototactic line 2.4 and 3.1, the positively geotactic line 12.1 and 12.5, and the negatively geotactic line 4.7 and 6.1. The frequency distributions overlap only slightly in the middle, that is, only few flies of the selected strains end up in the terminal tubes Nos. 8 and 9.

Is it, then, the heredity which makes a *Drosophila* walk towards lights or darkness, climb up or descend? Even with flies, not only with men, the situation is more complex than that. From the effects of the selection in the first generation, the heritability of the photo- and geotactic responses can be calculated to lie between .15 and .20. This is somewhat oversimplifying the issue, but one can say that, as a first approximation, the genetic component of the behavior of the fly in our mazes is only 15 percent to 20 percent, while random chance and environment is responsible for 80 percent to 85 percent. Nor is this all. Taking the data for the 15 generations of selection as a whole, we can compute the so-called realized heritability, that is to say the efficiency of the response to the selection. This turns out to be very small, only about 9 percent for the phototaxis, and only about 3 percent for the geotaxis. In other words, a prediction of what the selection could accomplish in 15 generations, based on the initial heritability figure, would be a gross overestimate. There are several factors responsible for this situation, among which I shall single out just one, which seems most interesting.

In our first experiments (Dobzhansky and Spassky, 1962) we made selection in three populations of *Drosophila pseudoobscura* during 18 generations for positive and for negative geotaxis. After the positive and the negative populations have diverged about as much as the populations on Figure 11.2, the populations were split each into two. In one member of each pair the selection was reversed, that is, a population formerly selected for the positive was now selected for a negative geotaxis, and vice versa. In another subpopulation the selection was relaxed, that is, the subpopulation was propagated without selecting either the positive or the negative individuals. The selective gains obtained through 18 generations of the original selection were almost erased in 6 generations of the reverse selection. The simple relaxation of the selection resulted in a loss of about half of the selection gains.

A partial, or even complete, loss upon abandonment of selection of what had been gained by previous selection is a phenomenon well known to breeders of agricultural plants and animals. Lerner (1954) has called this the genetic homeostasis. Very simply, the average height, weight, speed of maturation, and many other characteristics of a population which are determined by cooperation of numerous polygenes, are held by natural selection at levels near optimal for the population in the environments in which that population usually lives. When a breeder selects toward higher or toward lower levels of certain characteristics, he does so for his benefit, not necessarily for the benefit of the animal or the plant in its original environments. In other words, the artificial selection is often pitted against natural selection. As the artificial selection progresses it becomes more and more frustrated by natural selection. When the artificial selection is stopped, natural selection is given an opportunity to undo what the artificial one had gained; and reverse selection is highly effective because the artificial and the natural selections then work in the same direction, in alliance rather than in opposition.

Biologically, adaptively, this is an excellent strategy for evolution to follow. It combines high adaptedness to the existing environmental conditions with high adaptability to environmental changes. This strategy is, however, not at all what the classical theory of genetic population structure envisages. If the environment were uniform, constant, and favoring phototactically and geotactically neutral *Drosophilae*, then the simplest solution of the adaptive problem would seem to be to make all members of the species homozygous for the genes favoring photo- and geotactic neutrality. "Normal" or "typical" flies would then be neutral, and positive and negative ones would be abnormal or atypical. But this is not what is observed. The populations, though neutral on the average, contain also positive and negative genetic variants.

The availability in the populations of this genetic variance confers upon them evolutionary plasticity. A change in the environment that favors a positive or a negative photo- or geotaxis makes the population respond rapidly by adaptive genetic changes. Such responses might occur also in a genetically uniform and homozygous population, but they would be much slower. They would have to wait for the occurrence of mutations. These mutations would have to produce genetic variants which were unfavorable in the old but adaptive in the new environments. The rapidity of the genetic adjustment is, however, not the whole story. A genetically polymorphic population not only responds adaptively to environmental challenges, but in so doing it does not, so to speak, burn the bridges for retreat. It is hedged against the contingency that the environmental change to which it is adapting may only be a temporary one. If it is indeed temporary, and the original environment returns, the population can readapt itself speedily, by returning to its former genetic composition.

And yet genetic homeostasis does not stand in the way of permanent, irreversible, progressive evolutionary changes. If a new environment or a new way of life endures, a new genetic system becomes stabilized. This genetic system will be buffered against the vagaries of the new environments, but no longer able to retrace its steps to the conditions of the bygone age. If these conditions returned, the species would probably become adapted to them in some new way. One of the most interesting lessons that evolutionary biology teaches us is that there may be many more than a single method to eke out a living from an environment. Major evolutionary changes are irreversible and unrepeatable.

This point is so central that it must be reiterated: Man is not just an overgrown *Drosophila*. We reject the belief that man is nothing but an animal. Yet he is, among other things, also an animal. Like *Drosophila*, he is a sexually reproducing, outbreeding species, and his populations are abundantly provided with genetic variability. The genetic diversity affects all kinds of traits—morphological, physiological, and behavioral. The discrete, clear-cut, and usually pathological genetic variations of behavior, such as the so-called Mongoloid idiocy or phenylketonuria, need not be considered in the present discussion. The genetic variations among healthy persons are no less interesting, though much harder to study. The same situation exists also in *Drosophila*: sharp, easily distinguishable, and poorly viable mutants of classical genetics, versus slight, quantitative, polygenic variations. The difficulty, in human as well as in *Drosophila* genetics, arises because in the phenotypic variance of the second kind of traits the genetic and the environmental influences are intermingled.

Neither in the most highly selected, nor in the unselected, photo- or geotactic lines of *Drosophila* is the behavior of an individual rigidly determined. We have seen that the heritability of these behavioral traits is rather low. Whether at a given point of the maze an individual climbs upwards or downwards, takes a light or a dark passage, is in part a matter of environment, or simply of chance. The evidence is nevertheless conclusive that the genotype does bias the choices. Some flies are inclined to walk more often upward and others downward. Are the behavioral traits in human populations also conditioned by genetic variations? I shall be among the first to insist that the evidence is incomplete, and that more data must be collected. Yet the existing evidence, for a variety of traits ranging from I.Q. measurements to smoking habits, indicates that at least some genetic conditioning is involved, of course relatively more for some traits and less for others.

It is no secret that the study of the genetic conditioning of human behavior is hampered by the emotional reactions which this issue elicits in many people. Some wish to give an aura of scientific respectability to their race and class biases. Differences in material well-being and in social position are represented as just and necessary outcomes of the genetic differences. Others cling obstinately to the old tabula rasa theory. Man is a product of his environment and social conditions, and his genes are simply irrelevant. I submit that, irrespectively of your preconceptions, modern biology makes it necessary to state the problems of genetic conditioning of behavior in terms rather different from the traditional ones. This is because one of the most significant changes in the biological theory in the recent decades has been a shift from typological to populational models and concepts. This conceptual reformation has been discussed with admirable clarity and discernment, particularly by Simpson (1961) and by Mayr (1963), making it possible to state what is essential for us here very briefly.

To a typologist, what is real and important is the species or the race to which an organism belongs. Differences among individuals of the same species and race are, of course, too obvious to be denied. A typologist regards them, however, as merely a kind of troublesome noise in the biological system. He tries, as it were, to recognize the melody obfuscated by the noise; he seeks to identify, classify, and name the species and the races. He hopes that once he can determine to which species and race an individual belongs, that individual is thereby adequately described.

A populationist, on the contrary, regards the individuals and their diversity as the prime observable reality. The biological validity of species and races is not thereby refuted (although some extremists, try to do just that, in my opinion ill advisedly). Species and races are, however, derivative from individuals, not the other way around. Species and races are Mendelian populations, reproductive communities of sexually reproducing organisms, forms of adaptive ordering of systems of individuals, evolved because they have made the evolutionary feedback processes between the organisms and their environments most efficient and successful.

Man in the street is a spontaneous typologist. To him, all things which have the same name are therefore alike. All men have the human nature, and an alleged wisdom has it that the human nature does not change. All Negroes are alike because of their negritude, and all Jews are alike because of their jewishness. Populationists affirm that there is no single human nature but as many human natures as there are individuals. Human nature does change. Race differences are compounded of the same ingredients as differences among individuals which compose a race. In fact, races differ in relative frequencies of genes more often than they differ qualitatively, one race being homozygous for a certain gene and the other lacking it entirely. The extremists who deny that races exist are disappointed typologists who have discovered for themselves the gene gradients between race populations. They fail to understand that such gradients elucidate the nature of race as a biological phenomenon; the facts warrant the conclusion that Platonic types of races do not exist, not that races do not exist.

The typological and populational operational approaches are characteristically different. A race of typology is described in terms of means or averages of height, weight, cephalic index, intelligence, and so on. Populationists regard variances at least as important as means. Genetic variance characterizes not only the status but also the evolutionary possibilities of a population. The *Drosophila* populations with which Hirsch and his colleagues as well as ourselves began our experiments were photo- and geotactically neutral on the average. Yet the experiments have shown that the average neutrality did not mean that all individuals were neutral. Selection has attested the presence in the populations of genetic elements for positive and negative photo- and geotaxis. This does not quite mean that the original populations contained individuals as sharply positive and negative as are individuals of the selected strains. Natural and artificial selection do not act as simple sieves which isolate genotypes which were there before selection. Selection creates novelty, because it compounds genotypes the origin of which without selection would be altogether improbable.

All human beings have certain universally recognized rights because they are members of the species *Homo sapiens*. Members of other species do not have the same rights. Cows are sacred to Hindus, but even in India cows are not treated exactly like humans. An imaginative French writer, Vercors, has given a thought-provoking discussion of legal and other problems that might arise if a hybrid of man and some anthropoid species were produced. Anyway, membership in a group, be that a species or a race, does not define all the characteristics of individuals. The notion that it does is implicit in race pride, exclusiveness, and bias.

Racists busy themselves attempting to scrape up any kind of evidence that Race X has a lower mean I.Q., or smaller mean brain volume, or greater emotionality than Race Y. How large is the genetic component in such differences is questionable. The partitioning of the genetic and environmental variances obtained through studies on monozygotic and dizygotic twins cannot be used as a measure of the genetic and environmental components of the group differences. The basic assumption of the twin method is that the environments of the cotwins are uniform. This is obviously not true when different social classes, castes, and races are compared. Even if we had much more complete data on twins than are actually available, this would still leave the question of the magnitude of the genetic component in the group differences wide open. The argument that about one-half of the interracial variance in I.Q. must be genetic because this appears to be so among cotwins is a misinterpretation when it is not an intentional obfuscation.

To say that we do not know to what extent group differences in psychological traits are genetic is not the same as saying that the genetic component does not exist. It is a challenge to find out. If individuals within populations vary in some character, be that blood grouping, or stature, or intelligence, it is quite unlikely that the population means will be exactly the same. What matters is how great is the intrapopulational variance compared to the interpopulational variance. This is different for different characters. Skin pigmentation is individually variable in probably all races, but the interracial variance is evidently larger. Although precise data are not available, it is at least probable that the relation is reversed for psychological traits. In simplest terms, the brightest individuals in every class, caste, and race are undoubtedly brighter than the average in any other class, caste, or race. And vice versa—the dullest individuals in any of these groups are duller than the average of any group. There are sound biological reasons why this should be so. Very briefly, in the evolution of mankind the natural selection has worked, nearly always and everywhere, to increase and to maintain the behavioral plasticity and diversity, which are essential in all human cultures, primitive as well as advanced.

True enough, an individual taken from a population with a higher mean of some trait, say a higher intelligence, has a higher statistical probability to possess this trait more developed than an individual from a population with a lower mean. When we select *Drosophilae* for stronger or weaker photo- or geotaxis, we generally breed the high and the low selection lines separately. Spassky and myself have, however, some experiments in progress, in which pairs of populations exchange migrants in every generation. The migrants are selected for high or for low photo- or geotaxis or for some other genetically conditioned trait. This may be considered to represent to some extent an experimental simulation of social mobility in human populations. The preliminary results of these experiments are, at least to us, fascinating. Genetically selective social mobility seems to be a powerful evolutionary agent.

A day may conceivably arise when mankind will embark on some all-out eugenical breeding program. This day is not yet in sight, because mankind has not reached a level of wisdom when it could decide with anything approaching unanimity what combination of genetic qualities should the ideal man have. It is rather easier to agree what qualities he should not have. As for positive ideals, we can only recommend that a diversity of tastes, preferences, abilities, and temperaments should be preserved and perhaps even increased. Anyway, when we consider the social implications of the human genetic diversity we are not usually preoccupied with eugenical breeding programs. The genetic diversity is, for example, most relevant to educational problems. The students are, however, selected for study, not for stud.

Insofar as the genetic component is concerned, the intelligence, or temperament, or special abilities of the parents have little predictive value for these qualities in an individual child. This does not mean that such genetic components do not exist, as some authors have overhastily concluded. It means two things. First, the heritability is fairly low, as it is low in the photo- and geotactic behavior of our flies. In other words, the environmental variance is high, and in man the parent-offspring similarities in behavioral traits may well be due more to the cultural than to the biological inheritance. Second, one cannot too often be reminded of the fact that we do not inherit the genotypes of our parents but only one half of their genes. The genes do not produce their effects in development each independently of the others. The genes interact; the genetic "nature" of an individual is an emergent product of the particular pattern or constellation of the genes he carries. This is often the reason why a child is sometimes so strikingly dissimilar to his parents in some traits, even if the environment is kept constant.

How can I summarize the contents of this article, which is itself a summary of thinking concerning a variety of issues? Perhaps the best way is to say that genetics bears out John Dewey's emphasis on "the infinite diversity of active tendencies and combinations of tendencies of which an individual [human] is capable."

REFERENCES

Dobzhansky, T., and B. Spassky. "Selection for geotaxis in monomorphic and polymorphic populations of Drosophila pseudoobscura." Proceedings of the National Academy of Science 48: 1704–1712, 1962.

Erlenmeyer-Kimling, L., J. Hirsch, and J. M. Weiss. "Studies in experimental behavior genetics. III. Selection and hybridization analyses of individual differences in the sign of geotaxis." Journal of Comparative and Physiological Psychology 55: 722–731, 1962.

Hadler, N. "Genetic influence on phototaxis in Drosophila melanogaster." Biological Bulletin 126: 264–273, 1964. (a)

Hadler, N. "Heritability and phototaxis in Drosophila melanogaster." Genetics 50: 1269–1277, 1964. (b)

Hirsch, J. "Individual differences in behavior and their genetic basis." In: Bliss, E. L., Roots of behavior. New York: Harper, 1962, pp. 3–23.

Hirsch, J., and L. Erlenmeyer-Kimling. "Studies in experimental behavior genetics. IV. Chromosome analyses for geotaxis." Journal of Comparative and Physiological Psychology 55: 732–739, 1962.

Lerner, I. M. *Genetic homeostasis*. Edinburgh and London: Oliver & Boyd, 1954.

Lewontin, R. C., and J. L. Hubby. "A molecular approach to the study of genic heterozygosity in natural populations." *Genetics 54:* 595–609, 1966.

Mayr, E. *Animal species and evolution*. Cambridge: Harvard University Press, 1963.

Simpson, G. G. *Principles of animal taxonomy*. New York: Columbia University Press, 1961.

Part Three

Perceptual and Cognitive Development

There is no topical area in contemporary developmental psychology—or general psychology for that matter—livelier than cognition. The term is used broadly to include much that has been traditionally called learning, memory, language, and thinking, as well as several newer additions, including assimilation and accommodation. The title "perceptual and cognitive development" is used here because disentangling "perception" from "cognition" is a difficult task, frequently more of a semantic problem than a matter of content. In general terms, however, perception refers to the processes of receiving (or obtaining) and organizing sensory information from the environment, and cognition refers to the processes by which an individual comes to know and to understand the world in which he lives.

Why a particular concept or topic becomes fashionable is an intriguing question in the history and sociology of science. Without more study than is appropriate here, we can only speculate. The diminishing returns of quantitative studies (that is, psychometric tests) in intelligence may have played a part in the turn to more qualitative considerations of cognition. Certainly no improvement in prediction of performance has occurred in the past four decades. Traditional neo-behaviorist animal learning seems to have reduced appeal for the most recent generation of psychologists. Gradual but increasing awareness of the implications of Jean Piaget's work has had great influence on American psychology. It is of interest to note that the first major compendium of research in child development, Murchison's 1931 volume entitled *A Handbook of Child Psychology*, contained a chapter written by Piaget, "Children's Philosophies." Subsequent American handbooks paid little heed to his work; however, the most recently issued handbook (*Carmichael's Manual of Child Psychology*, 1970), published 40 years after Murchison's initial volume, contains a chapter authored by Piaget. At any rate, how an individual thinks has become a tremendously popular topic for speculation and research.

While much cognitive theorizing and research concerns itself with the mind of abstract *Homo sapiens* (usually the ubiquitous college sophomore), the topic has a natural developmental flavor. Observing and measuring the intellectual capacities of children and adolescents, noting the emergence of new qualities, or the changes in later life, tell something about the structure, content, and functioning of thought. For practical as well as theoretical purposes, changes in thinking that occur through maturity and old age need delineation. The papers presented here are all developmental, that is, concerned with change in functioning, rather than with theory or data of cognition per se.

Research stemming from traditional learning theory has not, for the most part, paid much attention to developmental considerations. An exception to this general rule has been the work of Robert Gagné, who has long studied the relation of learning to development. In his paper here, Gagné presents his theoretical position, which maintains that any developmental cognitive sequence can be interpreted as a hierarchic organization of isolable, learned behaviors. Professor Gagné is an educational psychologist now teaching and researching at Florida State University, after several years' work at the University of California, Berkeley.

As noted earlier, there is a great deal of research activity in learning; studies of memory are a part of this activity. Few, however, are developmental in nature. An exception is the study of Inglis, Ankus, and Sykes at Queen's University, Kingston, Ontario. They use the popular dichotic listening task (involving the presentation of stimuli separately to each ear), and employ this device to explore the relationship between age and memory processes.

If developmental studies of learning are scarce, longitudinal (that is, repeated testing of the *same* individuals over time) learning studies are virtually nonexistent. An exception is the series of self-examinations carried out by the late Madorah Smith. Dr. Smith had a long and distinguished career in psychology, but one of her most unique contributions was her measurement of retention of material overlearned in childhood and retested in middle and old age—in 1935, 1951, and in the study presented here, 1963. It is regrettable that more individual psychologists do not add to such evidence.

The essay by William Fowler first attacks the deficiencies of the traditional learning-theory approach to the study of development. It then discusses the attractive features of a stage analysis of developmental change, but it also points out some of the problems inherent in adopting the stage perspective. Perhaps Fowler's most important contribution here is his call for greater attention to "cognitive styles," that is, individual characteristics and uniquenesses in thinking, in our investigations of cognitive development. Dr. Fowler is now at the Ontario Institute for Studies in Education.

Another content area exciting a great deal of theoretical and research activity is that of language. Much of this work is developmental in nature since the evolution of verbal competence in the child tells much about language per se. Except for vocabulary data from intelligence testing, little concern has been exhibited for language functioning at the end of the age scale. Klaus Riegel, a psychologist at the University of Michigan, is unique because of his involvement in this neglected area. He attempts, with theory and data, to explain some verbal phenomena that change with age. Riegel's article is a challenging one, but it deserves careful study by the reader, for the model presented therein is the only one in existence that attempts to account for language development across the entire course of the lifespan.

As we have noted previously, very few theoretical positions have incorporated a truly lifespan orientation. However, one attempt at such a theory is that proposed by K. Warner Schaie. Schaie has produced an intriguing view of cognitive development by casting it into a Lewinian field-theory framework. A central concept in this theory is *rigidity*, which refers to the degree of flexibility in one's thinking and problem-solving activities. Professor Schaie has for many years served as chairman of the Department of Psychology at West Virginia University, which supports an ongoing program of research specifically oriented toward lifespan development issues.

Twelve

Developmental Readiness

Robert M. Gagné

[An] essential element that sets the stage for an act of learning is the readiness of all the cognitive states and processes that enter into such an act. Limitations of the performance of a human being often can be directly related to processes of growth. At a given age, for example, the child may be unable to perform some particular act because he is not tall enough or not strong enough. Similarly, it appears that a certain amount of neurological growth must occur before he can walk or talk. If one attempts to have him learn to carry out these activities before he is maturationally ready, one finds it impossible. From observations such as these on young children has come the idea of developmental readiness and the associated notion of "stages of development."

Is it possible that the child must reach a particular stage of developmental growth before he can learn? Many studies have indicated that simple types of learning can occur with the very young infant, and a few investigations have shown signal learning to occur before birth. For practical purposes, it may apparently be assumed that if the conditions for learning of simple connections are properly established, learning of this variety will occur, even in the youngest children.

There are obvious differences in the intellectual "power" displayed by children of different ages. It appears to be much easier for a child of seven, say, to learn the meaning of a new word than it is for a child of four. Similarly, a youngster of eleven may readily learn to use the abstract rules involved in inferring the kinship structure of a primitive society, whereas the seven-year-old may not. When attempts are made to have children learn intellectual skills that are beyond their stage of development, the result is highly predictable. Although some learning may indeed occur, the objectives are typically not achieved within any reasonable periods of time.

Does the existence of these limitations of developmental readiness imply that the more complex forms of learning (concept learning, rule learning, problem solving) are conditional upon stages of neurological growth? There are many who hold this view. Piaget (1952), for example, proposes the stages of cognitive operation called sensori-motor, preoperational, concrete operational, and formal operational. The differences in intellectual performance he observes in various tasks at various ages are, according to his theory, reflections of differences in the intellectual skills of logical thinking. At a particular age, the growing child may be able to perform certain kinds of intellectual tasks because he has attained a developmental stage in which he can think logically when dealing with concrete events. At the same time, he may be unable to display the operations of logical thought that are necessary to solve problems involving symbolic representations of events.

The alternative view to be described here is, very simply, that differences in developmental readiness are primarily attributable to differences in the number and kind of *previously learned intellectual skills*. At any given age, a child may be unable to perform a particular intellectual task because he has not acquired the specifically relevant intellectual skills as prerequisites to that task. According to this view, limitations of intellectual growth do not prevent a young learner from solving an abstract problem, or from learning new higher-order rules that are symbolically represented. Such learning may be readily accomplished if the learner has acquired, or will undertake to acquire, the intellectual skills that are prerequisite to the task. Prerequisite skills may be derived according to the method described in the previous chapter. It is recognized, of course, that learning such skills may take some time. Can the abstract rules of calculus, such as those of minima and maxima, be learned by fourth-graders? The answer is yes, if they first attain the developmental readiness implied by the skills prerequisite to such learning, which, of course, include many of the concepts and operations of algebra. (This hypothetical example does not imply, of course, that such learning *should* be undertaken; only that it *could* be.)

CUMULATIVE LEARNING

Developmental readiness for learning any particular new intellectual skill is conceived as the presence of certain specifically relevant subordinate intellectual skills. If the individual is to engage in problem solving to acquire a new higher-order rule, he must first have acquired some other, simpler rules. These in turn depend for their acquisition on the recall of other learned entities, rules, or perhaps concepts. As we have seen in previous chapters, concepts in their turn depend for their learning on the recall of other prerequisites, the discriminations which are specifically related to them, and so on. Viewed from the other end, the learning history of the individual is *cumulative* in character. The Ss → R connections and chains that are learned form a basis on which concepts are built. Concepts contribute positive transfer to the learning of rules; and the latter support the learning of more complex rules and problem solving. Among the rules acquired are those special kinds that are called cognitive strategies, which enable the individual to attack many kinds of problems, regardless of their particular content.

Learning has the specific effects of establishing the particular capabilities necessary for the performance of any intellectual task. It also has *cumulative* effects. When an individual has learned a particular rule, for example, he has established a capability that can transfer to the learning not only of a single higher-order rule but also to several others as well. As a specific instance, learning rules regarding the factors of numbers up to 100 may be shown to contribute to the learning of higher-order rules, say, in adding fractions. The factor rules are also prerequisite to other mathematical tasks, such as completing ratios or simplifying equations. When learning of the latter tasks is undertaken, these subordinate rules of number factors do not have to be learned all over again. They are already available in the learner's memory. Learning is cumulative, then, because particular intellectual skills are likely to be transferable to a number of higher-order skills and to a variety of problems to be solved. As the individual develops, he continually increases his store of intellectual skills. This means that the possibilities of combination of these learned skills, in transferring to the learning of higher-order capabilities, increases exponentially. These cumulative effects of learning are the basis for observed increases of intellectual "power" in the growing human being.

An Example of Cumulative Learning

The effects of cumulative learning may be illustrated in a task of liquid conservation, similar to that used by Piaget (Piaget and Inhelder, 1964) and by other investigators who have been interested in studying aspects of his theory (see Bruner, 1966). The task to be performed is illustrated in Figure 12.1. When the liquid in a container shaped like A is poured into container B (top row), many children below age seven are inclined to say that the taller container has more liquid. In another variant of the task, children at this age level tend to say that the volume in the shallower container (second row), exhibiting a larger surface area, is "more." Thus a few children at age seven are "conservers" in this task, while most are "nonconservers." To Piaget, these differences in performance reflect a critical point in intellectual development, marking the difference between a preoperational phase of thought and a concrete operational phase. The kind of logical operation particularly relevant to such a task is considered to be the *multiplying of relations* (recognizing that an increase in height may be compensated by a decrease in width). Such operations are in turn dependent on those that appear at an earlier stage of development, particularly *reversibility* (carrying out an inverse operation).

From the standpoint of a cumulative learning theory, performance of this conservation task may be accomplished when the individual has acquired the specific intellectual skills that are relevant to it. The set of

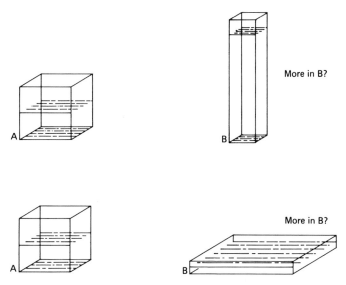

Figure 12.1   Two tasks of "conservation of liquid" in rectangular containers. (From R. M. Gagné, "Contributions of learning to human development," *Psychol. Rev.*, 1968, 75, Figure 2, p. 183. Copyright 1968 by the American Psychological Association and reproduced by permission.)

intellectual skills required may be derived in the manner described in the previous chapter, to yield a learning hierarchy shown in Figure 12.2.

It may be noted, first of all, that this particular hierarchy has been derived under the assumption that the children to be tested are uninstructed in mathematical concepts of volume, specifically in the relation volume equals height times width times length. Obviously, one *could* construct a learning hierarchy which proposed that children learn this exact mathematical relation in order to perform the final task. This possibility was not followed in the present instance in order that fewer assumptions about prior learning could be made.

According to the figure, reading from the top down, the child needs to have learned the rule that volume of a liquid in rectangular containers

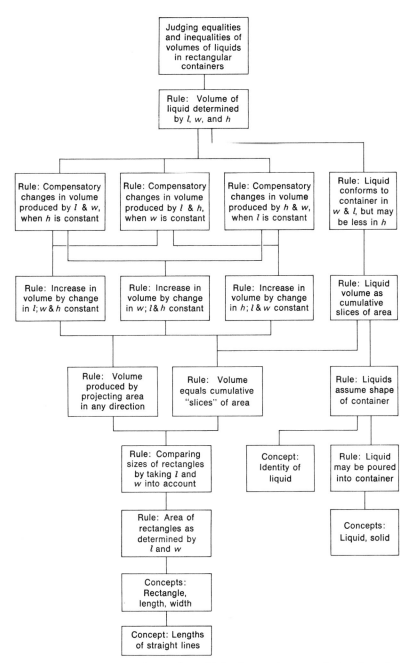

Figure 12.2 A learning hierarchy which shows the cumulative learning of intellectual skills leading to the task of judging equalities and inequalities of liquid volume in rectangular containers. (From R. M. Gagné, "Contributions of learning to human development," *Psychol. Rev.*, 1968, 75, Figure 3, p. 184. Copyright 1968 by the American Psychological Association and reproduced by permission.)

is determined by length, width, and height. Volume will be changed if any of these is altered. Proceeding one step further in the analysis, we find three rules about compensatory changes in two dimensions when the third dimension remains constant. (The fact that Piaget considers compensatory multiplying of relations essential to this task is noteworthy.) If the width of a liquid remains the same in two different containers, the two can have the same volume if a change in height in one is compensated by a change in length in the other.

In order for a child to learn these complex rules, the figure says, he must have learned three other rules, relating to change in only one dimension at a time. For example, if height is increased while width and length remain constant, volume will increase. The learning of these rules is in turn supported by the prior learning of still other rules. One is that volume of a container is produced by accumulating "slices" of the same shape and area. A second is that volume can be projected from area in any direction, up or down, to the front or back, to the right or left. Following these, one can identify considerably simpler rules, such as those of comparing areas of rectangles by compensatory action of length and width, and the idea that length and width determine area. If one chooses to trace the hierarchy still further downward, he encounters the concepts of rectangle, length, width, and an even simpler one, the concept of length of a line.

In its right-hand branch, the hierarchy describes the intellectual skills having to do with liquids in containers. Included here are rules having to do with changes in shape of liquid in containers and liquid volume as cumulative slices of area. This branch is necessary because at the level of higher-order rules, the child must distinguish the volume of the liquid from that of the container. It is notable that this branch also includes the concept of liquid "identity"—that is, matching a liquid poured from one container to another as "the same liquid." This sort of concept of identity is generally considered to be arrived at very early in the intellectual development of the child, as shown, for example, by Bruner's (1966, pp. 183–192) evidence.

These, then, are the subordinate intellectual skills, each of which can be learned, that are hypothesized to lead to the capability of judging equalities and inequalities in rectangular containers. It is worth noting that studies of children's performance in such tasks have sometimes undertaken to train the children directly on the final task, and have generally reported limited success. Some more recent studies, however, have followed the suggestion of the cumulative learning model by setting out to have children learn subordinate skills (see Kingsley and Hall, 1967; Le François, 1968; Bearison, 1968). Such investigations have found that children can learn relevant subordinate skills and that such learning makes possible transfer of learning to conservation tasks. While more evidence is surely needed, it is particularly noteworthy that "ability to conserve" has been clearly related in these studies to the mastery of prerequisite intellectual skills.

The cumulative learning theory may be seen to propose a conception of "what is learned" (or "what develops") in marked contrast to Piaget's theory. The latter states that performing conservation tasks depends on the development of logical processes, such as "reversibility," "seriation," and "compensation." The proposal of the cumulative learning theory, in contrast, is that development results from the learning of relatively specific intellectual skills having to do with liquids, containers, volumes, areas, lengths, and heights. Specifically, it is supposed that the child does not acquire an ability like "compensatory multiplying of relations" all at once at some particular stage of his development. Instead, he acquires through learning the specific intellectual skill of "identifying equal volumes given compensatory changes in length and width, height remaining constant." In addition, and at approximately the same time, he may acquire a number of other intellectual skills such as those shown in Figure 12.2. If the necessary specific capabilities are learned, by being taught in some systematic fashion, the child will be able to perform the conservation task.

Do the intellectual capabilities of the child remain as specific as this? Of course they do not. The reason is because of the existence of the well-known property of learning, transfer of learning. Suppose, for example, a child has learned to judge equalities and inequalities of liquids in rectangular containers. Suppose, further, that one undertook then to have him learn, in addition, to judge volumes of liquids in cylindrical containers. Presumably, such learning could be based upon another different learning hierarchy, which would have certain skills in common with that of Figure 12.2, but would also have some new ones dealing with areas of circles and volumes of cylinders. Having learned to judge volumes in *both* kinds of containers, a learner would then have some intellectual skills of great value for *generalizing* to still other problems. One might, for example, try him on the entirely new problem of judging volumes of liquids in irregularly shaped containers. Since he would already know how to equate volumes in rectangular and in cylindrical containers, it would perhaps not be too difficult for these skills to transfer to the new problem, composed of shapes that were partly and roughly rectangular and partly and roughly cylindrical. Transfer of previously learned skills to the entirely new problem seems highly predictable.

Once a learner has acquired a repertoire of learned specific skills, it is not difficult to realize that his performance of conservation-type problems would become progressively easier, because an increasing number of these subordinate skills would exhibit transfer of learning to any new conservation task one might devise. At some stage in his development, one might then choose to speak of the child as a conserver, or to say that he has acquired a general principle of conservation. To categorize his stage of development in this manner may have some value. However, it does not appear to be at all a necessary inference, in contrast to a finding that the child has learned a variety of relatively specific conservation skills applicable to volume, substance, weight, number, and other physical properties.

It is surely to be expected, then, that intellectual skills of whatever type, although they may be learned as relatively specific entities, will generalize through the mechanism of learning transfer to the learning of many other skills and to the solving of many previously unencountered problems. Readiness for new learning is thus a matter of a stage of development, to be sure. But the stage of development for any learner depends first upon what relevant prerequisite skills he has already learned, and, second, upon what capabilities he has yet to acquire in order to meet the particular objective set for him. Stated simply, the stage of developmental readiness of any learner is determined by what he already knows and by how much he has yet to learn in order to achieve some particular learning goal.

REFERENCES

Bearison, D. J. The role of measurement operations in the acquisition of conservation. Paper presented at Annual Meeting, Eastern Psychological Association, April 1968.

Bruner, J. S. "On the conservation of liquids." In: Bruner, J. S., R. R. Olver, and P. M. Greenfield. Studies in Cognitive Growth. New York: Wiley, 1966.

Kingsley, R. C., and V. C. Hall. "Training conservation through the use of learning sets." Child Development 38: 1111–1126, 1967.

Le François, G. "A treatment hierarchy for the acceleration of conservation of substance." Canadian Journal of Psychology 22: 277–284, 1968.

Piaget, J. The Origins of Intelligence in Children. New York: International Universities Press, 1952.

Piaget, J., and B. Inhelder. The Early Growth of Logic in the Child. New York: Harper & Row, 1964.

Thirteen

Age-related Differences in Learning and Short-term Memory from Childhood to the Senium

J. Inglis, Mary N. Ankus, and D. H. Sykes

There are many studies which show that human performance on learning tasks changes with age; there are, however, few that have studied a wide range of age groups. A decrease in efficiency with advancing years has been found on standardized learning tests by Shakow, Dolkart, and Goldman (1941) and Wechsler (1945), on paired-associate learning tests by Ruch (1934), Gilbert (1941), Korchin and Basowitz (1957), Gladis and Braun (1958) and Canestrari (1963), and on rote-learning tests by Kubo (1938), Bromley (1958) and Eisdorfer, Axelrod, and Wilkie (1963), to name but a selection of studies. The question of what altered *processes* may underlie such changes in performance (Inglis, 1965a) is, however, still undecided. It has been pointed out by Jerome (1959) that the poorer performance of old, as compared with young persons might, for instance, result from insufficient motivation to learn on the part of the elderly. It has also been pointed out by Schaie (1965) that experiments aimed at the study of the effects of age on learning usually fail to disentangle age *changes* from age *differences*. Thus, in the typical cross-sectional study comparing young and old groups of subjects, the effects of different life duration are confounded with the effects of different life experience.

The investigations to be reported in this paper have tried to specify and examine one part of the learning process which seems to be particularly sensitive to the effects of age, and to explore age differences in terms of results which cannot easily be seen as due to motivational factors. The

From *Human Development*, 1968, 11, pp. 42–52. Reprinted with permission of the authors and publisher, S. Karger, Basel/New York.

The studies described in this paper were carried out at Queen's University, Kingston, Ontario with the assistance of Ontario Mental Health Grants Nos. 25 and 59. This aid is most gratefully acknowledged as is the help of the public school system in Kingston.

147

range of ages of the subjects studied is also wider than has been usual in such experiments.

In one analysis of the essential aspects of learning and memory function, Welford (1958) has described a number of critical phases. These comprise a sequence that involves perception, short-term storage, evolution of a durable trace, endurance of such a trace, recognition, recall or retrieval and finally, the use of recalled material. It is evident that if these are several stages of a sequential process, with "learning" or "memory" as their product, then changes at any one point in the system would necessarily affect the whole later succession of the learning chain. Welford (1956) has also stated that short-term storage is perhaps the stage most likely to be vulnerable to age changes.

For the closer analysis of this latter part of the learning sequence, the method of dichotic stimulation developed by Broadbent (1958) has shown much promise. He has suggested that the ability successively to reproduce a series of digits when this has been presented as two simultaneous half-sets to each ear, depends upon some short-term storage mechanism. In one experiment Broadbent (1957) showed that when digit-span stimuli were relayed at a speed of two per sec through headphones, one half of the span to one ear and, simultaneously, the other half of the span to the other ear, subjects tested under these conditions could reproduce the digits sequentially. Furthermore, in such reproduction, elements from one half of the span were rarely alternated with elements from the other half. The first half-set recalled, however, commonly contained fewer errors than the second half, producing a kind of serial-order effect. Broadbent has suggested two kinds of mechanisms which may underlie such performance on this modified digit-span test: First, a "p-system" which can only pass information successively; secondly, an "s-system" which can store excess information arriving, for example, when the p-mechanism is already fully occupied in transmitting information from another channel. In each case the half-set of digits recalled first, passes directly through the p-system while the half-set recalled second spends some time in storage.

There is now a growing body of evidence to suggest that human memory may depend, in part at least, upon the efficiency of this short-term storage stage of the learning process. Inglis (1960), for example, has suggested that the defect of acquisition which has been found in elderly psychiatric patients suffering from memory disorder might be based upon a breakdown of this kind of storage mechanism. Such patients should therefore show disturbance of recall in those half sets of digits reproduced second in the dichotic stimulation situation relative to the performance of a matched group of patients without memory disorder. This expectation has been confirmed by Inglis and Sanderson (1961) for the case in which the two channels for simultaneous stimulation were the two ears. Caird and Inglis (1961) have confirmed these results and extended them to the case in which the ear and eye were together presented with different digits.

Inglis (1962) has pointed out that since increasing age beyond the middle years also seems to affect learning capacity, and since such impairment may likewise depend on changes in short-term storage, it might be that responses to dichotic stimulation would vary with normal aging. If age primarily affects storage, the reproduction of the first half-set recalled should not be affected by advancing years. If the second half-set recalled must pass through the storage process, the recall of these digits should be affected by age.

This expectation has been confirmed in a number of studies. A study by Inglis and Caird (1963) and a study by Inglis and Ankus (1965) each examined the changes in sequential responses to simultaneous stimulation in 120 subjects between the ages of eleven and seventy years. In another study, MacKay and Inglis (1963) examined the responses of 160 subjects between the ages of eleven and ninety years. The results of these experiments, carried out on different subjects by different experimenters, proved to be in very close agreement. As age increases there is little or no significant differential impairment in the ability to recall the half-spans reproduced first. Progressively and significantly greater difficulty is, on the other hand, shown in the reproduction of the second half-spans as age advances; the longer the span to be recalled the greater the difference between the first and second half-spans. Further confirmation of these findings has now come from similar, independent studies by Broadbent and Gregory (1965) and by Craik (1965).

One important characteristic of changes seen with senility or with advancing age on the dichotic listening task is that they cannot readily be explained away in terms of differences in motivation. When either senile patients or elderly persons perform poorly on any test it can usually be argued that they could not be bothered, or were not able to try the task in hand. In the case of recall after dichotic listening, however, it was predicted in the studies cited above that in one respect the senile groups would perform at the same level as the controls (that is, in recall of the first half-spans) and in another respect, gauged at almost the same point in time, they would be markedly inferior (that is, in recall of the second half-spans). It is difficult to see how any general hypothesis couched in terms of motivation level could account for the appearance of both similarities *and* differences in simultaneously secured measures. The same is true of the effects of normal aging. A general lack of motivation in the elderly could not plausibly explain a differential decline which appears in only one part of the recall of dichotic digits, another part being unaffected. It is also difficult to see how differences in life experience could account for such results.

The data obtained thus far on the effect of age on short-term memory function, as this is reflected in performance in dichotic listening tasks, seem consistently to show that this process progressively decreases in efficiency after the third decade of life. It also seemed of interest to try to discover what changes may take place over a wider span of years. This paper is an attempt to portray differences which appear in subjects between five and seventy years of age.

It was anticipated that the relation between age and measures of learning processes over this wide age range would approximate the form of an inverted U-shape. In the first decade of life it would be expected that learning abilities would develop; then, at or about the third decade, a maximum capacity would be reached, with an involution of ability appearing thereafter. It was expected that part, at least, of this curvilinear relation might be due to age changes in short-term memory storage capacity.

METHOD

Subjects

Two sets of subjects were examined by two different investigators. Sykes (1967) tested 120 children from the public schools of Kingston, Ontario, between five and ten years of age. There were twenty children in each year group: ten boys, ten girls. Ankus (1965) tested 120 persons between eleven and seventy years of age. None of these was known to be suffering from any gross mental or physical handicap; none had taken part in any previous studies. There were twenty persons in each decade group: ten male and ten female. They were taken at random (but not, strictly speaking, "randomly") from the population of Kingston, Ontario.

Procedure

All the subjects, with the exception of the five-year-olds, were given a serial anticipation rote-learning test of the kind used by Bromley (1958) in his study of the effects of age on learning. The score on the rote-learning task was obtained as follows. Twelve nonsense syllables of 80 percent association value (for example, TAS, DIT, PEL) were taken from the lists provided by Hilgard (1951) and printed in large letters on separate cards. The experimenter exposed the syllables one by one in serial order at a rate of approximately one every two seconds. After the first complete presentation, the subject went through the list trying to anticipate each syllable, correcting himself or being corrected by the experimenter whenever he made a mistake. The score obtained was the number of correct anticipations on the eleventh trial. If a subject managed to learn the complete series accurately before the eleventh presentation, he was given an extra credit for each unused trial. This test was given in order to secure a more orthodox measure of "learning ability," to see if it would be affected by age.

The apparatus used to present the dichotic stimuli was a "Roberts 990" tape recorder with different sets of digits recorded on two channels of stereophonic tape. The test series were played over earphones at the rate of one digit every two-thirds of a second. For the subjects aged five to ten years there were eight of each length of span, with from one to three digits per half-span, making 24 spans in all. For the subjects aged eleven to seventy years there were four of each length of span, with from one to three digits per half-span, making twelve spans in all.

This test material was administered as follows: Immediately after completing the serial learning test, each subject was given two practice exercises to get him used to the experimental situation. He was told, "You are going to hear a number. Tell me what you hear." The spoken digit 3 was then played on channel 1. If the subject reproduced this correctly, the same procedure was repeated for the digit 7 on channel 2. If the subject failed to respond or gave the wrong number the stimulus was repeated and, if necessary, the volume increased until the correct response was made. Secondly, the subject was told, "Now you are going to hear two numbers together, one in the right ear and one in the left ear. You will hear them both at the same time. Tell me what you hear." The two channels then played the spoken digits 3 and 7. If the subject reproduced the correct digits (in either order 3–7 or 7–3) the experimenter went on to the test series; if not, this item was repeated until both numbers were correctly reproduced. For the test series each subject was told, "Now you are going to hear n numbers, $n/2$ in each ear" (where n was 2, 4 or 6). "Tell me what numbers you hear."

Responses were scored as follows: The first digit repeated determined in each case which channel was taken to be the half-span recalled first. The score obtained was the average number of correct responses for each half-span of digits, taking each digit's position in the series into account. If, as sometimes happened, the first digit recalled was not in fact the first term of either half-span then this sequence was regarded as spoiled and excluded from the scoring calculations.

RESULTS AND DISCUSSION

First consideration may be given to the nonsense-syllable rote-learning test. The scores made by the different age groups are represented in Figure 13.1, together with data derived from Bromley's (1958) study.

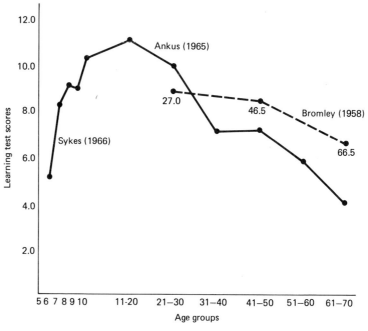

Figure 13.1   Age and serial rote-learning scores.

For Sykes' (1967) and Ankus' (1965) data, separate trend analyses of variance (Edwards, 1960) fully confirm what is apparent by inspection of the figure. There is a very significant increase ($p < 0.005$) in the learning scores between the ages of six and ten, and a significant decrease ($p < 0.005$) in the decade groups thereafter; the difference between the eleven to twenty-year-old and the twenty-one to thirty-year-old group, however, is not in itself significant. It will be noted that the age decrement in Ankus' (1965) groups is steeper than the still significant decline shown by Bromley's (1958) subjects. This may possibly be because Bromley only tested intellectually superior subjects in the older age groups. The apparent discrepancy between the points on the graph represented by Ankus' twenty-one to thirty-year-old group and the group of mean age 27.0 tested by Bromley may be because the *mean* age of Ankus' group was only 24.65.

So far the results are in accord with expectation. It still remained to be determined, however, at what stage or stages of the learning process some of the crucial change with age may take place.

The principal expectation derived from the hypothesis that short-term memory storage is the phase most susceptible to the effects of age was that the material recalled second in sequence, after simultaneous dichotic stimulation, should also increase in accuracy up to the third decade and decrease thereafter. The material recalled first in sequence, not having been held in store, should be unaffected by age.

The mean scores of the different age groups for the first and second half-spans recalled after dichotic stimulation may be represented in the form of the graphs shown in Figure 13.2.

These plots show very clearly the effects of age on the dichotic digit recall for spans of different lengths. As age increases from five to ten and from eleven to seventy, there is no significant differential change (Edwards, 1960) in the ability to recall the first half-spans with one or two digits per half-span. Between five and ten, however, there is a slight but significant increase in accuracy for recall of the first half-span for three digits; this score does not, however, change with age between eleven and seventy.

In accord with expectation there is, on the other hand, a significant change with age in the form of an increase in accuracy in the first decade and a decrease from the third decade to the seventh in all of the material recalled second in sequence which, by hypothesis, has spent some time in short-term memory storage.

Evidence on the effect of maturation in childhood on successive responding to simultaneous stimulation has previously been described by Kimura (1963). She showed in 145 normal children between four and nine years of age that the digits arriving at the right ear were more accurately reported than digits arriving at the left. This finding she interpreted to mean that those digits arriving at the ear opposite the dominant cerebral hemisphere were *perceived* more efficiently. A similar conclusion concerning the performance of young adults was reached by Dirks (1964). As has been pointed out elsewhere, however, (Inglis, 1965b) unless both laterality *and* order of recall are taken into account in such experiments the observed errors of reproduction might be caused either by failure of part of the input to enter the system (that is, a perceptual defect), or by a decay of input after it has entered the system (that is, a storage defect) or both. Further analyses of the data obtained from the 120 children described above have been carried out by Inglis and Sykes (1967). When order of recall was left out of account, significant differences were found between reproduction of material originally read to the right as compared to the left ear in only two comparisons out of 18 (six groups × three conditions). As Figure 13.2 shows, however, when laterality of recall was

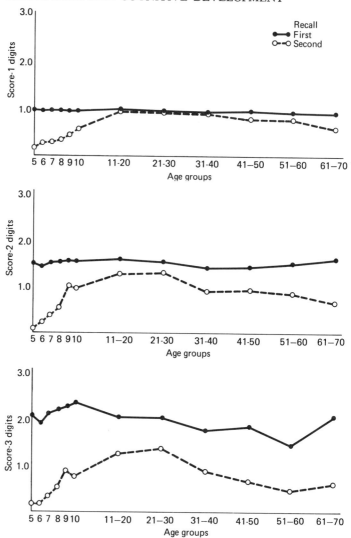

Figure 13.2 Age and recall of first and second half-spans of dichotic digits with from one to three digits per half-span.

ignored there was, in every case, significantly less accurate reproduction of the digits reproduced second in sequence. Since the main source of differential accuracy of the reproduction of dichotic digits in both children and adults is not laterality but order of recall, it would seem more plausible to regard such performance as a function of short-term memory storage capacity rather than as an index of perceptual ability. It seems possible that Kimura's (1963) and Dirk's (1964) finding of right-ear superiority when order was left uncontrolled might have been due to some preference in their subjects for first recall of the material read to the right ear. Although Kimura's (1963) interpretation of her data may be disputed, she also found that the performance of children in dichotic listening improved over the years four to nine.

The results secured in the present investigation, then, show the expected curvilinear relationship between age and performance from childhood to the senium on a serial anticipation rote-learning task. It has been suggested that the process of human learning requires that the material to be learned should initially and briefly be held in some short-term store. The results described above suggest that it may be the changing efficiency, over the years, of this kind of storage which accounts, in part at least, for the age-related changes found in grosser learning performance.

SUMMARY

This investigation was concerned with age-related changes in learning and short-term memory. A total of 240 subjects between the ages of five and seventy were tested on a rote-learning and on a dichotic listening task. It was anticipated that there would be a curvilinear relation, in the form of an inverted U-distribution between age and serial rote learning. This proved to be the case. A similar relation between age and accuracy in one aspect of the sequential recall of simultaneous stimuli suggests that changes in short-term storage processes may underlie variation in other learning performance.

REFERENCES

Ankus, Mary N. *The effect of age on short-term storage and learning.* Unpublished M.A. dissertation, Queen's University, 1965.
Broadbent, D. E. "Immediate memory and simultaneous stimuli," *Quart. J. Exp. Psychol.* 9: 1–11, 1957.
Broadbent, D. E. *Perception and communication.* London: Pergamon, 1958.
Broadbent, D. E., and Margaret Gregory. "Some confirmatory results on age differences in memory for simultaneous stimulation," *Brit. J. Psychol.* 56: 77–80, 1965.
Bromley, D. B. "Some effects of age on short-term learning and remembering," *J. Geront.* 13: 398–406, 1958.
Caird, W. K., and J. Inglis. "The short-term storage of auditory and visual two-channel digits by elderly patients with memory disorder," *J. Ment. Sci.* 107: 1062–1069, 1961.

Canestrari, R. E. "Paced and self-paced learning in young and elderly adults," *J. Geront.* 18: 165–168, 1963.

Craik, F. I. M. "The nature of the age decrement in performance on dichotic listening tasks," *Quart. J. Exp. Psychol.* 17: 227–240, 1965.

Edwards, A. L. *Experimental design in psychological research.* New York: Holt, Rinehart and Winston, 1960.

Eisdorfer, C., S. Axelrod, and F. L. Wilkie. "Stimulus exposure time as a factor in serial learning in an aged sample," *J. Abnorm. Soc. Psychol.* 67: 594–600, 1963.

Gilbert, J. G. "Memory loss in senescence," *J. Abnorm. Soc. Psychol.* 36: 73–86, 1941.

Gladis, M., and H. W. Braun. "Age differences in transfer and retroaction as a function of inter-task response similarity," *J. Exp. Psychol.* 55: 25–30, 1958.

Hilgard, E. R. "Methods and procedures in the study of learning." In: Stevens, S. S. (ed.), *Handbook of experimental psychology.* New York: Wiley, 1951.

Inglis, J. "Dichotic stimulation and memory disorder," *Nature* 186: 181–182, 1960.

Inglis, J. "Effect of age on responses to dichotic stimulation." *Nature* 194: 1101, 1962.

Inglis, J. "Problems in the detection of memory disorder," *O.P.A. Quarterly* 18: 85–89, 1965.

Inglis, J. "Dichotic listening and cerebral dominance," *Acta Oto-laryng.* 60: 231–238, 1965b.

Inglis, J., and Mary N. Ankus. "Effects of age on short-term storage and serial rote learning," *Brit. J. Psychol.* 56: 183–195, 1965.

Inglis, J., and W. K. Caird. "Age differences in successive responses to simultaneous stimulation," *Canad. J. Psychol.* 17: 98–105, 1963.

Inglis, J., and R. E. Sanderson. "Successive responses to simultaneous stimulation in elderly patients with memory disorder," *J. Abnorm. Soc. Psychol.* 62: 709–712, 1961.

Inglis, J., and D. H. Sykes. *Some sources of variation in dichotic listening performance in children* (in preparation, 1967).

Jerome, E. A. "Age and learning-experimental studies." In: Birren, J. E. (ed.), *Handbook of aging and the individual.* Chicago: University of Chicago Press, 1959.

Kimura, Dorcen. "Speech lateralization in young children as determined by an auditory test," *J. Comp. Physiol. Psychol.* 56: 899–902, 1963.

Korchin, S. H., and H. Basowitz. "Age differences in verbal learning," *J. Abnorm. Soc. Psychol.* 54: 64–69, 1957.

Kuro, T. "Mental and physical changes in old age," *J. Genet. Psychol.* 53: 101–118, 1938.

Mackay, H. A., and J. Inglis. "The effect of age on a short-term auditory storage process," *Gerontologia, Basel* 8: 193–200, 1963.

Ruch, F. L. "The differentiative effects of age upon human learning," *J. Genet. Psychol.* 11: 261–286, 1934.

Schaie, K. W. "A general model for the study of developmental problems," *Psychol. Bull.* 64: 92–107, 1965.

Shakow, D., M. B. Dolkart, and R. Goldman. "The memory function in psychoses of the aged," *Dis. Nerv. Syst.* 2: 43–48, 1941.

Sykes, D. H. *Children's performance on dichotic listening tasks.* Unpublished M.A. dissertation. Queen's University, 1967.

Wechsler, D. "A standardized memory scale for clinical use," *J. Psychol.* 19: 87–95, 1945.

Welford, A. T. "Age and learning: Theory and needed research." In: *Symposium on experimental gerontology.* Basel: Birkhäuser, 1956.

Welford, A. T. *Ageing and human skill.* London: Oxford University Press, 1958.

Fourteen

Delayed Recall of Previously Memorized Material after Fifty Years

Madorah E. Smith

A. INTRODUCTION AND PURPOSE

This report is a follow-up of two studies (Smith, 1935 and 1951) made over ten and twenty-six years ago. In the first report the method used, one of distributed practice and frequent reviews in the memorization of the 107 answers to the questions in the Westminster Shorter Catechism was explained in detail. Although the answers to all the questions were perfectly recited at one sitting sixty years ago before the child's thirteenth birthday, there was considerable incidental practice during the next ten years of the earlier portion of the catechism. The latter part of the book received no such practice so its present recall can be considered to be after sixty years but the earlier portion after approximately fifty years. Aside from the attempts at repetition, ten and twenty-six years ago, only the answers quoting the Bible, as in naming the Commandments, and the very first answer have received any even incidental practice.

TABLE 14.1 **Record of Number of Answers Known in**

	1934	1950	1960
Remembered	54	53	41
Prompted once	44	39	32
Partly forgotten	9	15	34

When repetition of the answers was undertaken recently, some forgetting of previously recalled answers took place. As shown in the table, more answers were partly forgotten during the last ten years than in the preceding forty or fifty years. At least one cue was required for 66 answers as against 53 on the first trial; while 34 now needed more than one

From *Journal of Genetic Psychology*, 1963, 102, 3–4. Reprinted with permission of the publisher.

prompting as against nine on the first trial. However, on none of the three trials were all phrases of any answer completely forgotten.

Of the 54 perfectly remembered answers in 1934 only two were not recalled after a single cue was given; but of the 53 other answers, 21 again required but one cue and the rest two or more cues.

In 1934, each answer, except the Commandments, was given a difficulty score of from zero to seven points based on length, lack of help from the number of words in the question that are repeated in its answer, and its place in the numerical order of the catechism. As stated above the direct quotations from the Bible had received some interim practice and these 11 answers were perfectly remembered each time. As before, a larger proportion of the answers with low difficulty scores were recalled. The average difficulty score of those perfectly remembered was 2.7; of those remembered in 1934 but now requiring a single prompting, the score was 2.8; of those requiring such prompting both then and now, the score was 4.2, while those requiring more prompting had an average score of 5.0.

Of the 22 answers in the last quintile, which had had no practice during sixty years, only two of the shortest were remembered and three others with but a single cue. But of the 22 in the first quintile, which had received considerable practice until fifty years ago, 14 were still perfectly remembered and only three of the longest required more than a single prompting.

B. CONCLUSION

Forgetting was more rapid during the last ten years than during the preceding thirty or forty years; that is, after sixty-three years of age than before that age. The answers forgotten were the more difficult, those not overlearned in childhood and that had received no incidental practice during the sixty years after first completing the memorization of the Westminster Shorter Catechism.

REFERENCES

Smith, M. E. "Delayed recall of previously memorized material after twenty years," *J. Genet. Psychol.* 47: 477–481, 1935.

Smith, M. E. "Delayed recall of previously memorized material after forty years," *J. Genet. Psychol.* 79: 377–378, 1951.

Fifteen

Dimensions and Directions in the Development of Affecto-cognitive Systems

William Fowler

The idea that the course of development through the life cycle follows a sequence of stages, each of which is composed of different types of structures and modes of functioning, has been one of the universal currencies of history. It is found in the literature of all cultures; one expression of it widely known to Western culture is found in a passage from Shakespeare on the ages of man. Anthropological evidence repeatedly turns up the existence of one or another *rites de passage*, marking major migrations from one kind of developmental status to another. Among the most frequent of these are ceremonies associated with the onset of puberty and/or the assumption of an adult status (Van Gennep, 1909).

Psychological and educational theories of development, both those rooted in the historical past and those emerging since the institutionalization of scientific study, have freely resorted to explanations based on the assumption that development includes changes in form as well as quantity. Rousseau, Pestalozzi, Froebel, Montessori, and Dewey all elaborated their educational theory on the proposition that the child, especially the young child, is qualitatively different from the adult (Raymont, 1937). In our own era, psychodynamic theories of personality development lean heavily on stage concepts, such as oral, anal, and genital periods, each defined in terms of radically different modes of functioning at different ages. Similarly, in developmental psychology and education, the concept of development as sequentially patterned by stages is a widely held view. Most statements in these fields have remained on a descriptive level, assigning characteristics empirically to ages with little regard for logical order, except for the more obvious sequences of motor development like sitting, crawling, and standing. Nevertheless, the concept of "learning readiness" is often implicitly defined in terms of some schema of traits, upon whose emergence, on an all-or-none basis, successful learning depends (Thompson, 1952).

From *Human Development*, 1966, 9, pp. 18–29. Reprinted with permission of the author and publisher, S. Karger, Basel/New York.

Theories of cognitive development (Werner, 1957; Vygotsky, 1962; Piaget, 1953) typically center upon definitions of fixed sequences of development. Moreover, a considerable body of research is accumulating, much of which tends to support both the fact as well as the order of the stages of cognitive development as Piaget has defined them (for example, Elkind, 1961; Furth, 1964; Smedslund, 1961).

The only major category of theoretical orientation and research which, in its central outlook, has not addressed itself to the notion of stages of development are traditional core areas of psychology, itself, learning theory and experimental psychology. This appears to be largely a function of their interest in formulating general behavior theory which transcends age considerations, as well as a preoccupation with the methodological precision of the physical sciences (Brunswik, 1952). Inevitably, this latter emphasis has funneled research activity into a molecular focus in time and space. It has also led to the isolation of problems in a laboratory setting to gain greater control. The developmental status of the organism has for the most part been treated peripherally, with development seen essentially as straight line learning progression, building additively upon a basic foundation implicitly little different from Locke's original notion of a "tabula rasa."

In spite of this basic outlook, certain concepts obtrude, even within this framework, which take on the flavor of "saltatory jumps" in development. Among these are "stimulus generalization," particularly when in the form of second- and third-order generalization, the concepts of "threshold," Pavlov's notion of a "second signal system," and Hull's idea of "habit family hierarchies." All of these concepts seem to imply more than the simple definition of development as change in size or amount by means of the accretion of equivalent small steps. They all, in fact, appear to suggest the idea of switches in direction and the *emergence* of coordinated response *systems*, producing patterns of behavior different in *kind* from earlier, simpler S-R forms of infant activity. In more recent years, developmentally oriented learning theorists, such as the Kendlers (1962) have been conducting experiments contributing to a growing body of evidence, recently summarized by White (1965), of the existence of a possible developmental shift in the nature of thought processes around the age of five to seven years.

It would appear that the idea that developmental processes include some kind of transformational process on the road from infancy to adulthood has been difficult to exclude wherever life processes have been studied. It is equally true that there are a number of issues involved which have been differently weighted or resolved according to the various ways in which stage concepts have been conceived. The most basic of these is the extent to which differences in types of behavior found or classified for different ages have been attributed to biologically governed maturational mechanisms. At one extreme lie the many age-stage descriptions of Arnold Gesell and his followers (for example, Ames and Learned, 1948; Gesell et al., 1940). These have been implicitly and explicitly attributed to an unfolding of a universal natural human order, despite their uniform emergence from heavily represented, middle-class samples of a single American community and the thin experimental tests to which they were subjected. The latter day observations of Dennis (1950) in infant care institutions in Teheran would underscore the wide variation in timing which can occur with even the supposedly more biologically based processes of early motor development. Delays of months in sitting and a year or two in walking were common. The apparent omission of a crawling stage in many children, moreover, further reduces the proportionate contribution of genetic mechanisms of sequential development which can be postulated.

The extent of maturational participation in the emergence of developmental transformations is difficult to determine. Except for myelination, a process well established by the age of two years, most of the neurophysiological developmental changes in brain structure appear to consist of changes in size (Conel, 1939, 1941, 1947, 1951; Langworthy, 1933). Knowledge of neurophysiological development is extremely primitive, however. How to probe further than the study of dead tissue—anatomy—remains a persisting research problem. The existence of structural changes in development, of itself, sheds no light on their relative origins in biological blueprints or external stimulation—in the absence of experimental demonstration of antecedent-consequent relations.

Learning theory is, of course, generally environmentalist in its basic viewpoint, although often framed in terms of relations to primary drive and reinforcement theories. On the other hand, there is apparently nothing intrinsic to Piaget's elaborate conception of cognitive sequentiality which dictates any close association with maturational timing, despite his original statement of the schema in rough age approximations. It is also clear that more rigorous empirical tests, to which his concepts are now being subjected, reveal a wide range of individual differences in the ages at which stages may occur. A few preliminary studies are suggesting the possibility of general developmental shifts in timing (Braine, 1959; Sigel and Roeper, 1965). The notion that it may be possible to teach certain concepts such as conservation systematically and earlier is appearing in recent studies. Wallach and Sprott (1964) were successful in teaching children at six and seven years and Sigel and Roeper (1965) at age five. In our work on number-concept learning at The University of Chicago Laboratory Nursery School at least three and possibly five, of nine four-year-olds plus one three-year-old and two five-year-olds show definite evidence of learning stable number conservation. All children in the two age groups have also made considerable progress on a variety of dimensions of number concepts as measured by a number-concept scale devised by the writer. It would appear, therefore, that if there exist critical periods with respect to the occurrence or nonoccurrence of developmental transitions, they are quite loosely hinged to biological doors.

The long delay of learning theorists in directly considering the hypothesis of developmental stages in learning, aside from their devotion to general laws of learning and their preference for linear theories of development, probably arises primarily from the fundamentally reductionist level of analysis at which they handle phenomena. This orientation has been well suited to delineating the mechanisms of change (or learning) and the plotting of step-function changes in development, but has been unduly tied to narrow sets of conditions and situational time slices.

The broad gauge level of analysis which has marked the structural approach to problems has been just as admirably suited to outlining the general course of development and discovering and characterizing broad, major transitions. As Zigler (1963) has discussed at length, the resolution of this issue will probably require explanations at a middle range of analysis, interrelating the scope of the developmental structuralists with the more precise operational definitions and mechanisms of the learning theorists.

The question of levels of analysis brings us directly into the issue of explanation. Structuralists have been strong on description, short on explanation. Associationists have been strong on defining antecedent conditions but the time dimension has been so telescoped as to make prediction beyond the immediate situation tenuous. If we are to gain any real confidence in and understanding of stage processes in development, we must confront the issue of defining and controlling the long-range conditions that lead to hypothesized stages of development. Although development does not appear to follow a linear path constructed of equivalent steps or learned units, each simply added to what has gone before, life processes do appear to involve continuity and progression. The knowledge structures and coping styles of any given ontogenetic phase, however new they may be as systems of functioning, are not disjunctive with prior modes. They are probably at most, transformations leading to more complex hierarchical integrations. Newly acquired elements and concepts are pulled together in ever broader, more differentiated, intricate structural networks, facilitating coordination of an individual's functioning. The fact of life processes contained in a single, physically identifiable and separated psychophysiological life system, with a beginning, middle and an end, is perhaps the best testimony for the need to conduct a search for the long-term, cumulative antecedents of development.

This reaffirmation of the obvious may add little to Piaget's broad definitions of developmental processes in terms of organism-environmental adaptive relations, operating through assimilation-accommodation processes. But, as we have suggested, these are explanations only in the most general sense, without specification of mechanisms for change. More cogently, Piaget's theoretical outlook does not seek explanation of the variations in impact which different organism-environmental interactions of an early period might have upon stage development in any later period. It is true that data gathered in different countries and cultures, including some reported by Kohlberg (1964) on *Formosan* children, suggest that stage sequences for cognitive and moral development may be universal, developmental phenomena, probably varying considerably in timing, according to experience and genetic foundations, but not in sequence.

Actually, there is a disarmingly simple principle which governs the logic of much cognitive theory on stages of development. This is the notion of levels of task complexity. Later stages must follow earlier stages, principally for the reason that concepts defining later stages are intrinsically more complex. The strategies of successively more advanced modes not only require but are constructed *with* and *on* the dimensions of the less advanced stages. It is not easy to imagine, for instance, how one could arrive at the abstraction level of Piaget's general stage of logical operations without having first been steeped in first-hand manipulations with the object world, itself. Stages of development accordingly may be viewed as bench marks, defining the level of complexity an individual has attained in the organization of his understanding in dealing with reality. Since this appears to be something on the order of a universal axiom, it is not surprising to discover that the epistomological history of human society finds many parallels in the cognitive development of the individual, as Werner (1957), Piaget (Flavell, 1963), and others have observed.

Let us assume, therefore, that some kind of cognitive staging, similar to the form Piaget has conceptualized, is indeed universal to human development. Are we not still left with only a rough and highly general approximation of the nature of cognitive development? Does the theory not still fail to come to grips with a number of important problems? To summarize, cognitive stages, although perhaps universal, are subject to considerable variation in the rate with which they can be acquired. The kind of investigation necessary to discover the extent of variation possible must focus systematically on the developing and cumulative interactions of the organism with his environment. In short, let us find out what, over the long haul, *produces* these variations in timing so that ultimately, where advantageous, we can predict and control their development.

In the same vein, a general description of the main cognitive levels of complexity through which an individual traverses leaves unspecified the *particular processes* which govern the fact and rate of progress from stage to stage. If the general sequences are, in fact, universal and a function of task difficulty, this does not preclude the possibility of gradients of difficulty which lead up to and prepare the ground for emergent transformations.

General stage theory is also sparse in concepts with respect to the vast range and diversity of types of differences in cognitive functioning between individuals operating on any of the same basic levels. This gap appears especially wide at the upper levels when one attempts to compare the cognitive levels of two scientists, one of whom may be a theoretical physicist of the order of Einstein and the other a high school physics teacher, but both of whom operate at the most abstract level of logical processes. Nor does cognitive stage theory as presently conceived have much to offer in the way of explanations for the unevenness which typifies the development of abilities in the same individual. Catholicity of competence may have attained certain heights in some Renaissance men like Leonardo da Vinci, but even prior to our era Descartes was a poor chess player, and Bach, Shakespeare, and Goya are little remembered for their rudimentary knowledge of mathematics. In the social evolution of cultures, simplistic, magical explanations in one domain have freely coexisted with explanations of a scientific order in another—and typically in the same individuals. It is useful to match the kind of explanations Trobriand Islanders employed in their predictions of weather phenomena with their more realistic orientations toward crop planting or techniques for fishing (Malinowski, 1954). All too similar are current rational views on the operation computers, compared with the magical powers widely attributed to the contemporary "Communist devils," in "explaining" the complex operations of our social system and international affairs (see Park, 1943).

Today, the most obvious fact which inheres in the individual's I.Q. score or achievement test pattern is the often wide component pattern variation appearing. General intellective ability or "g," as measured by I.Q. tests, holds up well on the statistical level, as many general traits do in their distribution over large populations. But predictive efficiency slips badly when attempts are made to predict an individual's competence in particular areas of knowledge or symbol systems, such as icthyology, mathematics, or ordinary language systems. Presumably, for these reasons Spearman (1904) formulated his theories of ability in terms of a two-factor theory, composed of both general and specific factors. Similarly, Thurstone's (1938) dissatisfaction with "g" led him to concentrate his efforts on the development of measures of "s," special or primary ability factors.

Bearing more directly on cognitive stage theory itself are the findings of Uzgiris (1964) and others that the rate of acquisition of general concepts like those of conservation may vary considerably according to the nature of the materials and tasks involved. An even more fundamental question is raised by the work of Galperin (Bauer, 1962) who, contrary to Piaget, not only believes that concepts of length measurement precede those of number, but succeeded in accelerating their development in 55 out of 60 six-year-olds well beyond control children, as a result of experimental training.

There is also another line of conceptual orientation emerging in the realm of cognitive style which offers promise of an additional approach to the question of multiple and alternate systems of directionality in cognitive development. Recent work by Santostefano (1964; Santostefano and Paley, 1964) suggests the possibility that such cognitive styles as leveling-sharpening (R. W. Gardner et al., 1959, 1960), field dependence-independence (Witkin et al., 1962) and nonanalytic-analytic (Kagan, Moss, and Sigel, 1963) might be grouped under the broader concept of cognitive differentiation. Santostefano observes that ". . . each of these dimensions seems to progress toward differentiation with age." Yet a general trend toward developmental differentiation does not rule out basic cognitive-style differences at *intermediate* levels of organization. For example, many of the differences between adult subjects in traits of cognitive style not only do not disappear developmentally, but are sometimes very large. Differentiation in type and style appears to be one of the important rules of cognitive development.

There is still another domain, the affective domain, which cognitive stage theory, although encompassing in a general way, does little to elaborate. Aside from the general proposition that each stage of cognition and mode of functioning has its affective aspect, cognitive theory has little to add. If psychoanalytic theory is preoccupied with affect and flow in a structural context, cognitive structuralists are inclined to focus on a description of organizations and their dimensions with little regard for the flow of energies or motivating forces which bring cognitive reorganizations about. Havighurst's (1953) concept of "developmental task" is one of the few early attempts in developmental theory to subsume in a common frame of reference the socio-emotional domains of personality-culture theory and the fields of intellectual learning and education.

Another question of developmental differentiation in type functioning arises in this context. Cognitive stage theory not only underestimates the role of particular concepts and dimensions intrinsic to specific subject areas and symbol systems in the development of competence, it offers little on the problem of development of alternate coping and ego defense systems of action. It would be useful to explore relations between coping systems and types of cognitive style and control. Is there any relationship or commonality, for example, between paranoid "over-centering" and the analytic cognitive style of Kagan, Moss, and Sigel (1963) or the focusing style of Santostefano and Paley (1964)? Is the first possibly an extreme polarization of these latter styles?

Although cognitive stage theory has proved useful in my own studies of early cognitive stimulation, certain concepts have proved more valuable than others. There is something of an incompleteness to a structural stage orientation in the areas of specificity, affective processes, and coping styles and strategies, type differentiation and causality. In my work on early stimulation I am attempting to formulate concepts which may cast some earlier concepts of cognitive stage theory in a new light. Many of the ideas are, of course, derived from other sources.

Central to my orientation toward strategies and processes of early stimulation is, first of all, the notion of presenting stimulation in a *form* and on a *level* with and at which the child can receive information—that is, stimulation that is appropriate to the sensory-motor and infralogical developmental systems of the infant and young child. All stimulation, in short must be geared and graded according to the learning the child has accumulated up to the given point in his individual history. But, unlike the usual stage theory framework, the focus combines an identification of present levels and styles of functioning with a perspective of advancing the child toward increasingly higher levels of abstraction and logical functioning.

In this scheme of things, early development, while not conceived of as a critical period, is viewed as an optimal period during which basic mental sets, styles, and orientations get established. These serve as primary foundations of affecto-cognitive categories governing all subsequent selection and flow of information which the organism will process.

This advancement is facilitated through *systematic* and sequential programming of concepts through hierarchies of task complexity. The organizing and sequencing of a program begins with selection and prior analysis of a defined reality structure or subject area, for example, a transportation system, community structure, a foreign language, mathematics, and so on. The analysis involves identification and association of structural-functional units at various levels of analysis—of components to wholes and of integral structures to one another in larger ecological systems. Emphasis is placed upon selecting, and initially isolating and simplifying central concepts and elements, which are then presented alternately in and out of context.

Stimulation tactics center on the employment of model discrimination-generalization and problem-solving tasks. Learning activities are developed in plays, games, and dramatic activities and presented in a setting of warm and accepting group and individual teacher-child relationships.

Much attention is devoted to careful pacing as well as grading of stimulus elements, relations, and patterns. A stimulation program proceeds step-by-step, following each child's individual rate, governed and cued by the observed successes of the child. The method is one of guided, means-end exploration of conceptually graded materials, to ensure successful coping and mastery. Progress is envisaged as sequential, cumulative learning along a course, leading to the elaboration of a structural network and hierarchy of concepts and affecto-cognitive styles of problem solving. One such style may be generated (but remains to be elaborated and tested) through the combined analytic-synthctic orientation, that is, through alternating the child's perceptual focus on part-versus-whole and structural-versus-functional relations at each level of unit and ecological analysis. Finally, control of language and other mediational and memory processes is considered essential. These are regulated and developed through a process of graduating in space and time the distance among stimuli to be associated and conceptualized, on the one hand, and between stimuli and the child's verbal responses and problem-solving activities, on the other.

With this conceptual organization, I am continuing to accumulate data on early learning in several projects, including a number at The University of Chicago Laboratory Nursery School. This past year at the Nursery School, in addition to our studies on stimulating mathematics concepts, we have been continuing exploratory work in other areas—biological and natural science concepts, foreign language (Russian), and others. The most carefully conceptualized and programmed effort is on early reading. Using a set of seven linguistically-sequenced primers, written by myself, reading this year has been extended to embrace nearly half or about fifty children in the Nursery School. Nearly all children are now reading with much comprehension at least as far as Book V, with the range covering from Book II on. Although data is still being collected, reading vocabulary has reached about fifty words or more, without including the varying quantities of word-generalizing abilities which many children are learning.

REFERENCES

Ames, Louise B., and Janet Learned. "The development of verbalized space in the young child," *J. Genet. Psychol.* 72: 63–64, 1948.

Braine, M. D. S. "The ontogeny of certain logical operations: Piaget's formulation examined by nonverbal methods," *Psychol. Monogr.* 73: 5 (whole No. 475), 1959.

Brunswik, E. *The conceptual framework of psychology.* Chicago: University of Chicago Press, 1952.

Conel, J. LeRoy. *The post-natal development of the human cerebral cortex.* Vol I–IV. Cambridge: Harvard University Press, 1939.

Dennis, W. "Causes of retardation among institutional children: Iran," *J. Genet. Psychol.* 96: 47–59, 1960.

Elkind, D. "Children's discovery of the conservation of mass, weight and volume," *J. Genet. Psychol.* 98: 219–227, 1961.

Flavell, J. H. *The developmental psychology of Jean Piaget.* New York: Van Nostrand, 1963.

Furth, H. C. "Conservation of weight in deaf and hearing children," *Child Develop.* 35: 143–150, 1964.

Gardner, R. W., et al. "Cognitive controls: A study of individual consistencies in cognitive behavior," *Psychol. Issues 1:* No. 4, 1959.

Gardner, R. W., et al. "Personality organization in cognitive controls and intellectual abilities," *Psychol. Issues 2:* No. 4, 1960.

Gesell, A., H. M. Halverson, Helen Thompson, Frances L. Ilg, B. M. Castner, Louise B. Ames, and Catherine S. Amatruda. *The first five years of life: A guide to the study of the preschool child.* New York: Harper, 1940.

Havighurst, R. J. *Human development and education.* New York: Longmans and Green, 1953.

Kagan, J., H. A. Moss, and I. E. Sigel. "Psychological significance of styles of conceptualization," *Monogr. Soc. Res. Child Develop.* 28: No. 2, Serial No. 86, 1963.

Kendler, H. H., and T. S. Kendler. "Vertical and horizontal processes in problem-solving," *Psychol. Rev.* 69: 1–16, 1962.

Kohlberg, L. *The preschool child's conception of reality.* Paper read at a Meeting of the Women's Board of the University of Chicago, October 29, 1964.

Langworthy, O. R. "Development of behavior patterns and myelination of the nervous system in the human fetus and infant," *Contrib. Embryol. Carneg. Inst.* 24: No. 139, 1–58, 1933.

Malinowsky, B. *Magic, science and religion.* New York: Doubleday, 1954.

Park, R. E. "Magic, mentality and city life." In: Burgess, E. R. (ed.), *The City.* Chicago: University of Chicago Press, 1943.

Piaget, J. *The origins of intelligence in children.* New York: Int. University Press, 1952.

Raymont, T. *A history of the education of young children.* London: Longmans and Green, 1937.

Santostefano, S. G. "A developmental study of the cognitive control 'leveling-sharpening,'" *Merrill-Palmer Quart.* 10: No. 4, 343–359, 1964.

Santostefano, S. G., and Evelyn Paley. "Development of cognitive controls in children," *Child Develop.* 35: 939–949, 1964.

Sigel, I. E., and Annamarie Roeper. *The acquisition of conservation: A theoretical and empirical analysis.* Detroit, 1965.

Smedslund, J. "The acquisition of conservation of substance and weight in children. V. Practice in conflict situations without external reinforcement," *Scand. J. Psychol.* 2: 156–160, 1961.

Spearman, C. "'General intelligence' objectively determined and measured," *Amer. J. Psychol.* 201–293, 1904.

Thompson, G. G. *Child psychology.* New York: Houghton Mifflin, 1952.

Thurstone, L. L. *Primary mental abilities.* Psychometr. Monogr. No. 1, 1938.

Uzgiris, Ina C. "Situational generality of conservation," *Child Develop.* 35: 831–841, 1964.

Van Gennep, A. *Les rites de passage*. Paris: Emile Nourry, 1909.

Vygotsky, L. S. *Thought and language*. Cambridge: M.I.T. Press, 1962.

Wallach, Lise, and R. L. Sprott. "Inducing number conservation in children," *Child Develop.* 35: 1057–1071, 1964.

Werner, H. *Comparative psychology of mental development*. New York: Int. University Press, 1957.

White, S. H. "Evidence for a hierarchical arrangement of learning processes." In: Lipsitt, L., and C. C. Spiker (eds.), *Advances in Child Development and Behavior*. Vol. II. New York: Academic Press, 1965.

Witkin, H. A., et al. *Psychological differentiation*. New York: Wiley, 1962.

Zigler, E. "Metatheoretical issues in developmental psychology." In: Marx, M. H. (ed.), *Theories in contemporary psychology*, pp. 341–369. New York: Macmillan, 1963.

Sixteen

Development of Language:
Suggestions for a Verbal Fallout Model

Klaus F. Riegel

Most spatial conditions can be reliably reproduced and measured, but changing temporal conditions—strictly speaking—cannot. Until rather complex periodic processes had been identified and recording techniques developed, it was impossible to deal with time except on a primitive psychological basis. Moreover, we do not sense the passage of time directly but rely on the changes in stimulation which accompany it. Thus, time perception is intimately linked with memorization. Since memory is never perfect, neither is time perception. If, indeed, we were able to retain every input received during a certain interval, our time perception could be perfect. In this case there would be no psychological gaps or "empty spaces" that could distort the time estimation; the intervening period between two instances of time would be solidly filled with psychological events that could be completely enumerated for time estimates.

Because of the difficulties described, it is not surprising that physicists devoted much attention to providing a theoretical foundation for the time dimension and its measurement. These attempts have led to a reduction of the time variable and to a substitution of time measurements by enumerative procedures of spatial conditions. The analysis of radioactive decay or of irradiative accumulation are characteristic examples of such attempts. Both processes (as well as perfect memory, mentioned above) represent enumerative processes. In order to explicate these attempts a brief consideration of the measurement of time in physics is relevant here. (See also Maria Reichenbach and Ruth A. Mathern, 1959, and Strehler, 1962.)

From *Human Development*, 1966, 9, pp. 97–120. Reprinted with permission of the author and publisher, S. Karger, Basel/New York.

The preparation of the manuscript has been aided by the Grant No. 55–139 from the Foundation's Fund for Research in Psychiatry. The manuscript has been completed under the Grant 1 P01 HD01368 from the United States Public Health Service.

MEASUREMENT OF TIME IN PHYSICS

Let us imagine that a film camera has been mounted above a billiard table across which a ball is rolling. If we obtain a film strip, cut it into single pictures, and randomize them, we could reconstruct the order of the pictures—but we could not state for certain from which side the ball came. All that we could reconstruct would be the sequential properties but not the direction of the sequence.

Next, let us assume that a bar is placed across the billiard table and that one half contains a greater number of randomly moving balls than the other. After the bar is removed, balls will move from one half to the other. Ultimately they will be about equally distributed over the whole table. If we had taken pictures again, we would not only be able to reconstruct the sequences of moves, but could also ascertain the direction of the sequences. The state in which the two sets of billiard balls are still separated is less likely than the state in which they are mixed. The separate states can be reconstructed only through the intervention of some external (ordering) forces. During the normal course of events, the separate states will merge into a mixed one but not vice versa.

Our procedure allows the reconstruction of sequences as well as of their directions, but requires reliance on a periodic system, namely the film camera with its constant speed of exposures. The periodicity of the camera is not a crucial condition for the description of physical systems, however. Although for all practical purposes we prefer a periodic and, in particular, a linear time concept, we might well imagine a film camera which takes pictures at irregular intervals. In spite of this, we should be able to reconstruct sequences and their directions with a reasonable degree of accuracy.

Of greater importance, our procedure reduces indeed the measurement of time to an enumeration of spatially ordered events: First, we determined the sequence of these events by aligning them in a systematic manner. Then, we estimated the direction by separating the less probable from the more probable states. If we want next to derive measurements, we chop off equal numbers of observations from the less likely end of the ordered sequence assuming that under nonlinear conditions, long and short distances between adjacent events will be randomly distributed.

MEASUREMENT OF TIME IN PSYCHOLOGY

Almost exclusively, physical time measures have been applied in psychological research. In perceptual studies, for instance, the time for the identification of stimuli is taken in physical units. In studies of learning, the time for the acquisition of items as well as for the forgetting are measured in physical terms. Noteworthy exceptions exist, however, for instance when the intervening periods are specified psychologically by the number of intervening lists or the amount of activity.

In developmental studies the age of Ss has been specified in physical units of chronological time. Quasi-psychological time-scales have also been applied; the mental age scale of I.Q. tests is the most outstanding example. Interestingly enough, Ss do not necessarily progress equally on the chronological and mental age scales, but may reveal periods of retardation or acceleration in their intellectual growth. Similarly, intervening tasks in learning studies have different effects, depending on whether the material was interspersed early or late during learning.

Our examples have pointed to the usefulness of independent time measures for psychological research. Attempts to derive psychological time scales have concerned either the relatively short time periods of experimental studies, or the measurement of age in developmental investigations. Most psychological studies of the short-term time variable have been restricted to subjective estimates of physical time periods (see Wallace and Rabin, 1960). The primary objective of developing long-term psychological time scales has been to derive more meaningful interpretations of behavior changes than chronological age allows. Most of these scales have been selected for specific biological or psychological reasons and only a few have furthered our theoretical notions about growth and aging. Since the discussion of such theoretical notions will be the main purpose of our presentation, some statements on developmental theories need to be made.

THEORIES OF DEVELOPMENT

In spite of its impressive history, developmental psychology has not provided a clear answer to the question of why organisms grow and age. Some have tried to obviate the problem by declaring that time itself may serve explanatory functions and that, for instance, four-year-olds have a certain height *because* they are four years old. Others have been satisfied by the common attempts to reduce psychological development to changes in nonpsychological conditions, particularly biological factors, and to substitute phenotypical by genotypical descriptions. Such an interpretation is implied, for example, when we explain growth in size by changes in the endocrine system. For many purposes such explanations are fruitful. But they merely remove the problem into another area of study, where the same question arises again and no satisfying answer to why organisms age is provided.

In a more abstract form reductionism is implied when nontemporal factors are substituted for the time or age dimension. This possibility is inherent in all attempts to derive psychological or biological age scales independently from chronological age, for instance, by evaluating the potassium concentration, the calcification of certain organs, or merely the number of items solved on a mental age scale. If the evaluation of psychological or biological age is well substantiated theoretically, powerful explanations of growth and aging can be derived. For clarification of this point, let us consider two biological theories of development (for a comprehensive discussion see Strehler, 1962, and Curtis 1963).

The first theory may be called *waste theory*. Basic metabolic processes of the cells lead to the production of waste products, such as lipofuscin, which cannot be completely removed; these products accumulate over the years until they reach a critical level and produce a slow or sudden decrease in functioning. This theory leads to the important notion that life itself implies growth and aging, *via* the accumulation of metabolic waste. The "waste theory" may be called an intrinsic theory of development: Biological age is determined by the amount of waste accumulated by and within the organism.

The second theory has been called the *mutation theory* and is concerned with the instability of the chromosomes in somatic and gonadic cells. Mutations—which are generally regarded as deleterious for the organism—occur at random. The older a person is, the greater the number of mutations that have occurred and the more likely it is that structural and functional defects will result.

Since the mutation rate depends on the amount of irradiation to which an organism has been subjected, some theorists regard the amount of radioactive fallout as a major determinant of changes during development and aging. This formulation is of importance not only because it suggests ways to manipulate developmental processes in the laboratory, but also because the determinant of age has become an extrinsic, nonbiological entity. Again, the amount of irradiation received (fallout) may be used to redefine the time or age scale. Thus, we may rid ourselves in another manner of chronological time as a primary dimension and at the same moment develop a deterministic interpretation of growth and aging.

SUGGESTIONS FOR A VERBAL FALLOUT
MODEL OF LANGUAGE DEVELOPMENT

Contrary to what some experimentalists like to make us think, language does not consist of short responses in single situations but—as Anderson (1949) has put it—individuals are virtually "bathed in linguistic stimulation." A five-year-old may speak 12,000 words per day and will receive even larger quantities of spoken language. As a first question, we have to ask whether the amount of verbal activity on one hand, and of "verbal fallout" on the other, is about constant throughout life.

In spite of its practical importance, few reliable studies on age differences in the amount of communication are available. Figure 16.1 shows the results of a sample survey on the amount of time spent daily in reading. Even though we might question whether the type of material listed adequately represents the total written input, let us, for simplicity, accept the results as they are presented, let us also agree that the amount of written input is about constant for the adult years. Accordingly, the total number of messages received will increase linearly with age.

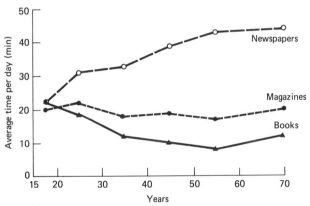

Figure 16.1   Amount of time spent on daily reading by different age groups (Link and Hopf, 1946).

Messages differ in quality or information value, and, on the average, the information value decreases with the length of messages. What is the precise relationship between these two variables?

Fortunately, researchers in literary statistics have worked extensively on this problem and have suggested a number of mathematical models. Thus, Carroll (1938) analyzed the relationship between the number of different words (types) and the total number of words (tokens) in James Joyce's *Ulysses*. Plotting the first occurrence of any word in an accumulative manner against text length, he derived a negatively accelerated trend shown in Figure 16.2. For the present purpose it is conceptually simpler to regard Figure 16.2 as representing the information received by a person *reading Ulysses*. On the first few pages the reader will be faced with many different words (types). After the more common words have occurred, the encounter of new words becomes less likely. Thus, the curve flattens out as the end of the book is approached.

In his study of the relationship between word variability and text

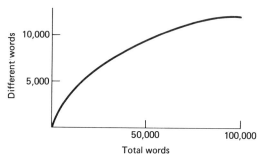

Figure 16.2 Predicted number of different words as a function of total number of words in James Joyce's *Ulysses* (Carroll, 1938).

length Carroll has been followed by a number of other scientists. In an earlier publication (Riegel and Riegel, 1965) we have reviewed the pertinent literature and selected the best fitting mathematical function for such a study. In the present paper we will generalize these interpretations to the study of language development.

GENERALIZATIONS AND EMPIRICAL SUPPORT FOR THE MODEL

If a word count were available on all the books and materials a person has read during his lifespan, we would expect a growth curve for his vocabulary quite similar to the one for the reader of James Joyce's *Ulysses* —only much longer. Moreover, if we maintain our assumption that the amount of daily reading is constant, we could substitute the total number of pages read for the age of the reader or vice versa, leaving as an open question whether we regard the vocabulary as a function of the total amount of material read or of age.

Slicing off equal intervals on the ordinate of the curve, we may also use the enumeration of the different words as an indicator of a person's "reading age." At later ages, when the reader is far advanced in his "book of life" greater amounts of reading are required in order to produce equal amounts of reading growth. But as long as he continues to read, he has a chance—even though a decreasing one—to encounter a word that has not occurred to him before. This word might be a new one, only recently introduced into the language, or it might be a rare word which requires an exceedingly large sample of the language for its occurrence.

Substantial evidence has been accumulated to support our propositions. During the first two years of life, scores on recognition vocabularies (McCarthy, 1954) show a positively accelerated increase. Thereafter, the trend becomes negatively accelerated, but even during the adult and later years of life a high stability or a slight increase in scores has been noted in most of the previous studies (Riegel, 1965). Figure 16.3 includes a mathematical trend for a multiple choice vocabulary given to 500 Ss (Riegel, Riegel, and Meyer, 1966).

The evidence thus far reported is on recognition vocabularies and does not represent the words actively used by Ss. Unfortunately, much less is known about age differences in the active vocabulary of Ss. The writer observed an increase in variability of responses to stimuli on an association test over an age span from seventeen to over seventy-five years of life (Riegel and Riegel, 1964) and similar results have been reported by Riegel and Birren (1966) in their study of syllable associations of young and old Ss. Since all these investigations rely on cross-sectional comparisons, it remains possible that the increase in variability expresses an increase in vocabulary specializations, that is, in inter-, rather than intra- individual variations.

This difficulty is resolved in word counts of diaries and journals written by a woman at a median age of thirty-five years and of letters written by the same person after the age of sixty-eight years [Figure 16.4(A) and (B)]. In the analysis of these records, Madorah Smith (1955, 1957) observed an increase in word variability up to an age of

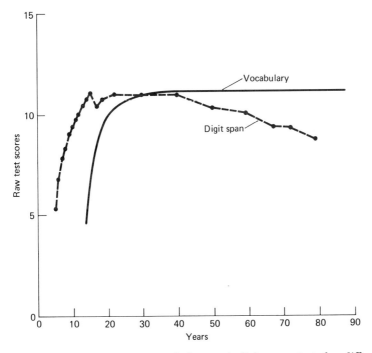

Figure 16.3   Mean scores on a vocabulary and digit span test for different groups (Riegel, unpublished data; Wechsler, 1949; Wechsler, 1955).

about seventy-five years. She attributed the decline thereafter to extraneous factors particularly the declining health of the S.

The data illustrated in Figures 16.3 and 16.4 provide only indirect support for our interpretation. Ideally, complete counts of *all* the words read or written, heard or spoken during a person's life ought to be obtained, but such a study would be an exceedingly expensive if not an impossible endeavor. Fortunately, simplifications of the procedures are possible: First, we may study time samples only. Second, we may stratify our material and restrict our analysis to specific strata, for instance, to nouns or verbs or adjectives. Indeed, we may study only words beginning with a certain syllable or letter of the alphabet. Undoubtedly the type of restriction will affect the results, but by comparing different restrictions, generalizations may be drawn.

If we restrict our study in the indicated manner, we approximate quite closely the method of verbal fluency developed by Bousfield and Sedgewick (1944) in which Ss are asked to name as fast as possible words of a particular type, such as all the words beginning with the letter S or Q but to omit all others. Moreover, with this restriction we compress our procedures to such an extent that we may obtain our records during a single laboratory session.

Our comparison implies that in the verbal fluency task Ss are scanning through their repertoires in the same manner as readers would through the pages of a book in order to find particular words such as different S-words. To the person scanning a book for different S-words, the interspersed words are a nuisance and they become more so, the

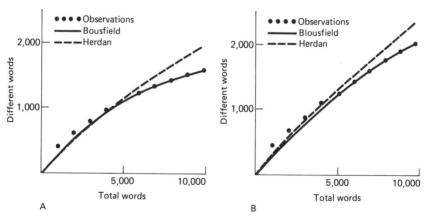

Figure 16.4 (**A**) Accumulated record of different words written by a thirty-five year old S plotted against text length (after Smith, 1957). (**B**) Accumulated record of different words written by a S after her sixty-eighth year of life plotted against text length (after Smith, 1957).

farther he has advanced in the text (because an increasingly greater number of words is interspersed between two new S-words). If the time intervals between the emission of adjacent words in the verbal fluency test are assumed to be proportional to those that elapse until a person scanning a book comes upon new words, and—moreover—to those that elapse until a regular reader encounters new words, we would have derived a method of simulating developmental processes in the laboratory. Let us apply this idea for the purpose of generating some experimental hypotheses.

SIMULATION OF DEVELOPMENTAL PROCESS

Previously, we cited some evidence that older persons have larger active as well as passive vocabularies than young Ss. Consequently, their performance in verbal fluency tests ought to be superior. In order to test this hypothesis and—more generally—the idea of simulating aging processes with the verbal fluency test, we compared the accumulation of S- and Q-words in James Joyce's *Ulysses* with the production of such words by young and old Ss in ten-minute laboratory sessions.

In our study, 31 college students and 23 members of senior citizens' club were asked to write as many words as they could think of beginning with the letter S or Q, respectively. Ss marked the end of each 2-min period on their papers. The number of responses given during consecutive periods was used in the analysis.

According to the results illustrated in Figure 16.5, our prediction that the larger vocabulary of old Ss should lead to a superior performance in the verbal fluency test was not confirmed. The total number of different words emitted during the 10-min periods was found to be much smaller for the old than for the young Ss. Apparently, other factors than vocabulary size affect verbal output and prevent old Ss from making full and efficient use of their large vocabularies. Subsequently, we have to examine possible sources of interference and, generally, to explain the failure of our prediction.

1. Thus far, we have considered only the *physical environment* according to Lewin (1954) and *la langue* according to Saussure (1931). We have dealt only with the different words that occur in the surroundings of a S but not with whether he registers and perceives these words or not. We have considered the words which he can possibly produce under optimal conditions but not with the limitations which a realistic situation imposes upon his performance. Quite different mechanisms representing psycho-biological changes rather than changes in physical or linguistic conditions are involved in perceiving or producing words and sentences.

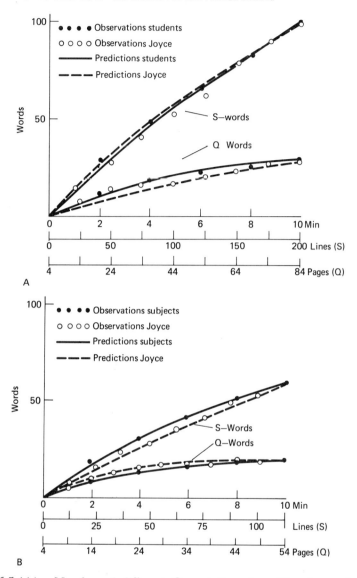

Figure 16.5 (A)   Number of different Q- and S-words written by students and by James Joyce plotted against time or text length. **(B)** Number of different Q- and S-words written by old Ss and by James Joyce plotted against time or text length.

Figure 16.6 shows some characteristic data on age differences in the speed of silent reading and writing. Mature Ss read about ten times as fast as they write, but there is also a differential decline in writing speed beyond the age of about fifty years. The data of Figure 16.6 on silent

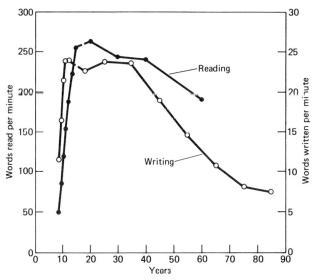

Figure 16.6   Speed of writing and reading at different age levels (Wills, 1938; Birren and Botwinick, 1951; Pressey and Pressey, n.d.; Hall and Robinson, 1942).

reading speed have been collected by Pressey and Pressey (n.d.) for children between seven and fourteen years of age and by Hall and Robinson (1942) for Ss above ten years of age. The data on writing speed have been obtained by Wills (1938) for children between eight and twelve years of age and by Birren and Botwinick (1951) for Ss above sixteen years of age. Further information has been provided by Rader and O'Connor (1957, 1959). The decline in writing speed with age is one important factor that explains why old Ss produce markedly fewer words in the verbal fluency test than anticipated.

2. A writer of a book avoids undue repetition of words, but he is not rigidly restricted in his output. In the verbal fluency test, however, S is prevented from using the same response twice, and thus, he has to check any response he is intending to give against all the others previously emitted. This counter-checking of words causes interference and again differentially affects the performance of old Ss (Riegel and Birren, 1966). Moreover, the checking of words requires a good short-term retention span which, according to Figure 16.3, old Ss are lacking. The data of Figure 16.3 represents the Digit Span Subtest of the Wechsler Intelligence Scales for Children and Adults (Wechsler, 1949, 1955). The differential susceptibility to interference and the decline in retention span are other factors that explain why old Ss produce markedly fewer words in the verbal fluency test than anticipated.

3. The complexity of the sentences produced or comprehended is

another important factor in the analysis of developmental differences in verbal behavior. Old Ss may differ from young in the syntactic or semantic structure which they are using or are able to comprehend at a certain reading rate. The problem of linguistic structures seems far remote from the present interpretation, but, since basic components of such structures are classes and class relations, similarities become conceivable. After all, one major difficulty of the verbal fluency test consists in identifying members of a class and performance is greatly facilitated if categories, such as S-words or words denoting particular groups of objects, pieces of furniture, animals, and so on, are readily available to Ss.

In order to outline the role of linguistic structures for verbal performances, we have to analyze how classes are acquired, how their relations are recognized, and how transformations of these relations are performed. Although a detailed discussion of such problems is beyond the scope of the present paper, we may show that age differences do indeed exist and are at variance with changes deduced from our model.

Figure 16.7 provides pertinent information on age differences in reading comprehension and sentence length. The data on the length of spoken sentences are taken from a review by McCarthy (1954) and represent the average results of four studies. The data on the length of written sentences of children and adolescents are taken from the same source and represent averages of two studies. The data on the adult Ss have been made available through the courtesy of Dr. Madorah Smith (unpub-

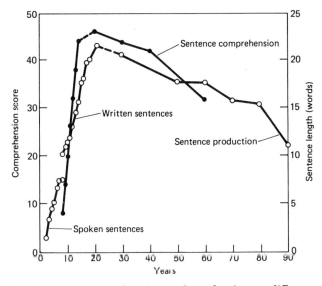

Figure 16.7   Sentence comprehension and production at different age level (Pressey and Pressey, n.d.; Hall and Robinson, 1942; McCarthy, 1954).

lished), and are based on large sets of letters written by three persons during spans of about sixty years. The results on "sentence comprehension" represent the scores on "paragraph meaning" of Pressey's Diagnostic Reading Test. The data on children's performances have been collected by Pressey and Pressey (n.d.), and those on adult Ss by Hall and Robinson (1942).

LIMITATIONS AND FURTHER EXPLICATIONS OF THE MODEL

Since the results of the verbal fluency study seem to limit our statistical interpretation of language development, a reconsideration of the general purpose of the model seems appropriate. Primarily, the model provides estimates of the upper limits of performance for individuals differing in age. Because of additional psychological factors, this upper limit is hardly ever attained in real life situations. In analyzing these interfering factors, we proceeded to explain the observed differences between the data and the model.

Throughout our discussion, the individual has been regarded as an information-handling system that changes over time; that is, with the amount of information handled. The total amount of information provided, and particularly, the total number of different words given, increases with age. But because of intrinsic changes which are not primarily dependent upon environmental conditions, a person's speed of intake, for instance, his reading speed or his ability to comprehend verbal messages, changes as well and thus the relative amount of received (perceived) information is less than the amount of information provided to him. As suggested in Figure 16.8, individuals may be regarded as functioning most efficiently during early adulthood.

For most types of performance, the received information has to be stored temporarily. Since there is a decline in immediate memory span, shown in Figure 16.3, further deficit will occur and the amount of information lastingly stored is less than the amount of information received. Some loss will also accrue when S calls upon the information stored. Because of slowness in expressing himself, such as in writing, and because of difficulties in constructing complex sentences, the amount of information produced is smaller than the amount of information stored. S's failure to formulate complex sentences is, of course, also dependent on his immediate memory span and will lead to the use of short and redundant formulations (Birren, 1955).

Because of these losses in the transmission or handling of information, the overt performance does not match the total information provided. In order to evaluate this loss and the efficiency of the organism,

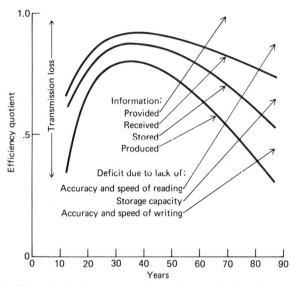

Figure 16.8   Hypothetical representation of transmission losses at different age levels.

we must first obtain an estimate of the information provided. It is here that the model and knowledge from literary statistics is useful and allows us to derive specific hypotheses. What are, for instance, the effects of a poor versus a rich linguistic environment upon the performance of Ss differing in age? How does the environment of a person interact with his verbal abilities? What are the interdependencies of various abilities (perceptual, retentional, retrieval) at different developmental levels? Does the efficiency change with age, and so forth?

In extension to our inquiry we may wish to study the determinants that bring about changes with age in perceptual, storage and retrieval functions. Most likely, these determinants are not exclusively related to the long-term developmental changes we have discussed, namely the overall increase in information provided, but may rather be sought among the changing biological capacities of the organism, that is, among the nonlinguistic, intrinsic conditions of the organism.

Certainly, most of the questions raised are not new. Similar notions are implied in de Saussure's (1931) distinction between *la langue* and *la parole* and in Kurt Lewin's (1954) distinction between a *physical* and *psychological environment*. Although Lewin has not given priority to the study of the physical environment, Barker and Wright (1949) as well as Brunswick (1949) have directed much attention to it. Barker and Wright regard the study of the contingencies in the physical environment rather than the study of particular psychological variables affecting the organism at any given instance as the most appropriate way of analyzing long-term change of the growing organism. Similarly, linguistic contingencies have been analyzed but are still waiting for treatment by psychological ecologists and developmental psychologists. It is here that the proposed model may prove to be useful.

Finally, the present analysis may allow us to redefine the age variable in terms of changes in the response variability. Accordingly, we regard equal intervals on the ordinates of our figures (Figure 16.2 to 16.5) as representing units of "verbal age." Because of the negatively accelerated shape of the curves, corresponding intervals on the abscissa increase toward the right end of the curves. In other words, "verbal age" progresses at a slower pace the larger the total amount of material already accumulated or already emitted.

The possibility of deriving a psychological scale of verbal age brings out an important implication of our discussion: In contrast to most other interpretations in developmental psychology we have remained faithful to an absolute age scale with interval properties instead of regressing to ordinal or partially ordered scales. Piaget's interpretations (see Flavell, 1963), for instance, with their periods of sensory-motor intelligence, preoperational intelligence, concrete and formal operations, constitute an ordinal age scale. Only the sequence of the periods is unambiguously determined, not the distances within or between them, and thus practical as well as theoretical difficulties arise when attempts are being made to map the boundaries of Piaget's periods or stages upon the chronological age scale. Since, we have become accustomed to the notion of developmental periods, however, we must explain how an interpretation of continuous changes would handle these constructs.

First, our previous interpretations were concerned with the continuous growth of the vocabulary as derived from the notion of an accumulating amount of communication. However, we also mentioned the contrasting trend of a continued decline, notably in biological reactions as discussed by Bürger (1953). Both interpretations point to idealized limits. If we were to test any individual in a real life situation various interactions would take place and the observable performances would be like those shown in Figures 16.6 or 16.7. Consider, for instance, age changes in associative reaction time to verbal stimuli. Early in life the speed of reaction is high, but Ss are unable to discriminate and to respond quickly to verbal stimuli. Late in life the verbal habits are developed but the speed of reaction is lowered. Only during the middle period and because of the optimal interaction between two continuous functions, is a maximum level of performance reached. Prior to that period, speed in performance grows; afterwards it declines.

Second, the linguistic environments impinging upon the growing individual differ in quality and type. Undoubtedly the language spoken to a baby differs markedly from that used among playmates, in elementary school, high school, in church, at the job, among friends, and so forth. Thus, we may need to describe language development as a series of shifts between subsystems of the language. Most of these subsystems (*les paroles* of de Saussure, 1931) are preselected for the growing individual in a more or less determined order and the confrontation with a new system is obligatory, as for instance, when a person enters one or another school, is drafted, retires, or such. Any new linguistic environment will shape the individual in a particular manner, and will make behavior periods sufficiently distinct even though it may share many features with the preceding and following ones. The problem raised is identical to the general question of cultural expectations that influence behavior periods like the periods of "nay-say," adolescence, post retirement, and so on.

Third, and most important, the continuous growth, such as of the vocabulary, consists itself of small distinct steps, any one of which may lead to a burst of information and thus provoke qualitative growth. Here, one is reminded of the implications which the acquisition of particularly mathematical and logical concepts, such as *Function, Integral,* as well as concepts like *Property* will have upon the behavior of an individual. More precisely, this sudden increase in the available operations on the basis of additions of single elements has been recently discussed by McLaughlin (1963).

This author attempts to explain the differences in the "logics" of children at the various developmental periods of Piaget, by differences in the immediate memory span. Even though the obtained developmental curves on the immediate memory span, such as for digits or words (see Figure 16.3 and Van De Moortel, 1965), are perfectly smooth, a child in a concrete testing situation recalls either two or three or four, but never 3.6 items. The increase in the number of items retained implies a marked increase in the number of conceptual, logical operation which the retainer can perform. A person who can retain two items at a time is able to perform simple categorizations according to the presence or absence of certain attributes or according to membership or nonmembership in a group. If four items can be retained, items can be simultaneously categorized in two dimensions, and thus, the possibilities of analogical inferences, and syllogisms and many other operations are open to the child.

McLaughlin's interpretation is important because it reduces the typology of a stage model to an interpretation of development as a continuous process. Developmental periods in the logical operations of children are reflections of their memory capacities which attain only discrete values in concrete testing situations. Even though average values may be computed and will yield perfectly smooth developmental curves of immediate memory span, it is more reasonable to separate discrete developmental levels corresponding to ages at which children are likely to retain two, three, four, five items, and so on, respectively.

The significance of McLaughlin's argument is further revealed because it holds equally well in the area of grammar learning. Grammar, and generative grammar in particular, allow us to construct (project) an infinite set of linguistic expressions on the basis of finite sets of elements, classes, and class-relations. In the present discussion we have been primarily concerned with the acquisition of elements (words), which are the prerequisite for the acquisition of classes and class-relations (Riegel and Riegel, 1963). As soon as classes and class relations are acquired, a burst of generative possibilities occurs, and, the child will be able to produce many new formulations which, formerly, he could not generate without matching input information. Depending on the number of grammatical classes and class-relations which a child has incorporated into his language, linguists may be inclined to separate distinct periods in the development of grammar. Again, the underlying learning process is a continuous one but the products to be utilized for grammatical performances, the classes and class-relations, are qualitatively distinct and in turn may create the impression of qualitatively distinct periods in the grammatical development.

SUMMARY

Language is viewed as an objective, quantifiable system confronting an individual throughout life. The sum total of verbal input and output increases with age. As the total amount of verbal input or output increases, the total number of different words perceived or produced has been demonstrated to increase as well. The active verbal repertoire, as tested in word counts of letters, journals, or responses to association tests, increases in a similar manner. Thus, older persons have available a larger passive, as well as active, vocabulary than young Ss.

The verbal fluency test asking Ss to name all the words occurring to them during consecutive time intervals and the equation of Bousfield and Sedgewick (1944) are adapted for the analysis of the data.

Since older Ss were found to have larger passive as well as active vocabularies their performance on verbal fluency tests would be expected to be superior to young Ss. This prediction was not confirmed. The verbal fluency test is a speed task, and time limitations are known to effect differentially the performance of old Ss. In the verbal fluency test Ss have to avoid repetitions of words. This causes interference and might differentially effect old Ss, particularly, since comparisons of any intended response with those previously emitted require a good immediate memory span which old Ss are known to be lacking.

Throughout our discussion, the human organism has been regarded as a system that changes with the total amount of input it receives and/or the output it produces. Analysis of the verbal fluency data has also revealed that external contingencies account for an upper limit of performance only. Of greater importance for many psychologists might be the analysis of factors that prevent the individual from making the most efficient use of the information provided. Deficiencies in perceptual, storage, and retrieval functions exemplify such limiting factors. An explanation for the developmental changes of these functions may need to be based on other than the physical and linguistic conditions in the environment of the individual, and may need to be reduced to developmental changes in intrinsic, biological components.

REFERENCES

Anderson, J. E. *The psychology of development and personal adjustment.* New York: Holt, Rinehart and Winston, 1949.

Barker, R. G., and H. F. Wright. "Psychological ecology and the problem of psychosocial development," *Child Develop.* 20: 131–143, 1949.

Birren, J. E., and J. Botwinick. "The relation of writing speed to age and to the senile psychoses," *J. Cons. Psychol.* 15: 243–249, 1951.

Birren, J. E. "Age changes in speed of simple responses and perception and their significance for complex behavior." In: *Old age in the modern world,* pp. 235–247. Edinburgh: Livingstone, 1955.

Bousfield, W. A., and C. H. W. Sedgewick. "An analysis of sequences of restricted associative responses," *J. Genet. Psychol.* 30: 149–165, 1944.

Brunswik, E. *Systematic and representative design of psychological experiments.* Berkeley: University Calif. Press, 1949.

Bürger, M. "Merkmale des biologischen Alterns. Lebensversicher," *Med.* 5: 41, 1953.

Carroll, J. B. "Diversity of vocabulary and the harmonic series law of word-frequency distribution," *Psychol. Rec.* 2: 379–386, 1938.

Chotlos, J. W. "Studies in language behavior VI: A statistical and comparative analysis of individual written language samples," *Psychol. Monogr.* 56: 75–111, 1944.

Curtis, H. J. "Biological mechanisms underlying the aging process," *Science 141:* 686–694, 1963.

Flavell, J. H. *The developmental psychology of Jean Piaget.* New York: Van Nostrand, 1963.

Hall, W. E., and F. P. Robinson. "The role of reading and life activity in a rural community," *J. Appl. Psychol.* 26: 530–542, 1942.

Herdan, G. *Type-token mathematics.* The Hague: Mouton, 1960.

Lewin, K. "Behavior and development as a function of total situation." In: Carmichael, L. (ed.), *Manual of Child Psychology* (2d ed.), pp. 492–630. New York: Wiley, 1954.

Link, H. C., and H. A. Hopf. *People and books.* New York: Book Manuf. Inst., 1946.

McCarthy, Dorothea. "Language development in children." In: Carmichael, L. (ed.), *Manual of Child Psychology* (2d ed.,), pp. 492–630. New York: Wiley, 1954.

McLaughlin, G. H. "Psychologic: A possible alternative to Piaget's formulation," *Brit. J. Educ. Psychol.* 33: 61–67, 1963.

Pressey, S. L., and Luella C. Pressey. *Pressey diagnostic reading test, grades 3–9, Form A.* Bloomington, Ill.: Public School Publ. Co. (n.d.).

Rader, W. T., and J. O'Connor. *Percentiles for men in ideaphoria, Worksample 161, Based on 33,007 cases, Tech. Rep. No. 625.* Boston: Human Engineering Lab., Inc., 1957.

Rader, W. T., and J. O'Connor. *Percentiles for women in ideaphoria, Worksample 161, Based on 10,127 cases, Tech. Rep. No. 635.* Boston: Human Engineering Lab., Inc., 1959.

Reichenbach, Maria, and Ruth A. Mathern. "The place of time and aging in the natural sciences and scientific philosophy." In: Birren, J. E. (ed.), *Handbook of aging and the individual,* pp. 43–80. Chicago: University of Chicago Press, 1959.

Riegel, K. F. "Speed of verbal performance as a function of age and set: A review of issues and data." In: Welford, A. T., and J. E. Birren, (eds.), *Behavior, aging and the nervous system,* pp. 150–190. Springfield, Ill.: Thomas, 1965.

Riegel, K. F., and J. E. Birren. "Age differences in verbal association," *J. Genet. Psychol.* 106: 1966.

Riegel, K. F., and Ruth M. Riegel. "An investigation into denotative aspects of word meaning," *Lang. Speech 6:* 5–21, 1963.

Riegel, K. F., and Ruth M. Riegel. "Changes in associative behavior during later years of life: A cross-sectional analysis," *Vita Humana* 7: 1–32, 1964.

Riegel, K. F., and Ruth M. Riegel. "Vorschläge zu einer statischen Interpretation von Altersveränderungen sprachlicher Leistungen." In: Hardesty, F. P., und K. Eyferth (eds.), *Forderungen an die Psychologie,* pp. 87–105. Bern: Huber, 1965.

Riegel, K. F., Ruth M. Riegel, and G. Meyer. "A study of the drop-out rates in longitudinal research on aging and the prediction of death," *J. Pers. Soc. Psychol.* (in preparation).

Saussure, F. de. *Cours de linguistique générale* (3d ed.). Paris: Payot, 1931.

Smith, Madorah E. "Linguistic constancy in individuals when long periods of time are covered and different types of material are sampled," *J. Gen. Psychol.* 53: 109–143, 1955.

Smith, Madorah E. "The application of some measures of language behavior and tension to the letters written by a woman at each decade of her life from 49 to 89 years of age," *J. Gen. Psychol.* 57: 289–295, 1957.

Strehler, B. L. *Time, cells and aging.* New York: Academic Press, 1962.

Van De Moortel, R. C. *Immediate memory span in children: A review of the literature. Rep. No. 1,* USPHS Grant No. HD 01368. Ann Arbor: Center Hum. Growth Dev., University of Michigan, 1965.

Wallace, M., and A. I. Rabin. "Temporal experience," *Psych. Bull.* 57: 213–236, 1960.

Wechsler, D. *Manual for the Wechsler adult intelligence scale.* New York: Psychol. Corp., 1955.

Wechsler, D. *Wechsler intelligence scale for children.* New York: Psychol. Corp., 1949.

Wills, Annie R. "An investigation of the relationship between rate and quality of handwriting in primary school," *Brit. J. Educ. Psychol.* 8: 229–236, 1938.

Yule, G. U. *The statistical study of literary vocabulary.* London: Cambridge University Press, 1944.

Zipf, G. K. *Human behavior and the principle of least effort.* Cambridge, Mass.: Addison-Wesley, 1949.

Seventeen

A Field-Theory Approach to Age Changes
in Cognitive Behavior

K. Warner Schaie

Until comparatively recent years, the psychologist interested in the area of human development would have been expected to concern himself primarily with the periods of infancy, childhood, and adolescence. Interest in the problems of old age have led to some theoretical conceptualizations of the final portion of the lifespan (Riegel, 1959), but while there is a noticeable move towards a consideration of the total life cycle it seems still valid to note that developmental theorists usually engage in a meticulous analysis of early childhood, place considerable emphasis on the problems of adolescence, but give only limited recognition to the further developmental crises and behavioral changes in the adult individual. Noteworthy exceptions would be some tentative neo-analytic formulations provided by Fromm (1957) and Erikson (1950) as well as the sociopsychological theory of disengagement (Cumming and Henry, 1961). There is a dearth of comprehensive discussions, however, concerned with the phenomena of age changes in the complex behavior of adults from the period of full maturity to senescent decline.

It is often remarked that the study of adult age changes has but a short history and that one should therefore expect emphasis to be placed upon descriptive studies as a means of building the necessary foundation of empirical data. At least twenty years have passed, however, since the first important cross-sectional studies of the adult age span reached publication, but research emphasis has changed very little. It is timely, therefore, to look for theoretical models which will help integrate the wealth of research data now available.

From *Vita Humana*, 1962, 5, pp. 129–141. Reprinted with permission of author and publisher, S. Karger, Basel/New York.

A preliminary version of this paper was presented as part of a symposium on "Theories of Aging in Adulthood" sponsored by Divisions 7 and 20 at the 1959 meeting of the American Psychological Association. Its completion was facilitated by a faculty summer research fellowship from the Research Council, University of Nebraska, whose support is gratefully acknowledged.

One of the by-products of the earlier emphases of developmental psychology has been the concern with early infancy as a determiner of adult personality development. In revolt against the psychoanalytic insistence upon this importance of early development there has been equally strong preoccupation with the impact of social and cultural factors on human development. It is of concern, however, to find that some individuals develop into well-functioning adults in spite of presumably pathological early experiences; others with benevolent background develop adult pathology and some are sustained by the identical cultural and social factors which overwhelm others. It might therefore appear more useful to study the individual's cognitive response in the here and now as the primary source of information, taking note of, but not being unduly concerned with, his early history or the socio-cultural factors which may often help us to classify but rarely to explain his behavior. To prevent anyone from misunderstanding this to be a thoroughly ahistorical view of behavior, let it be said also that such cognitive responses should be examined at many "here-and-now's" throughout the lifespan; otherwise one is departing from the role of the developmental psychologist.

Recent thinking in the biological sciences suggests that the normal aging process might be described as a gradual decrease in the ability of the organism to adjust to the changes in reaction potential required by a given life situation (Shock, 1951; Linden and Courtney, 1953). A number of interesting physiological hypotheses for such a point of view could be considered at the level of molecular analysis. The present paper, however, is concerned with phenomena occurring at the molar level and requires explanation which involves the realm of complex psychological processes. The controversy over the peripheral as against the central aspects of age changes will therefore be ignored to be left to the physiologically minded. Instead, a frame of reference will be considered which is based on the observation of those complex processes which seem to affect alternation in successful cognitive functioning.

It is here proposed to consider a theoretical model for age changes in cognitive behavior (or what is sometimes referred to as the higher mental functions, including such variables as: abilities, attitudes, motives whether expressed by verbal or nonverbal modes of behavior). Some implications of the model will be indicated and research evidence which may bear upon it will be examined.

AGING AS A "FIELD PHENOMENON"

Lewin has given a comprehensive description of methods for the dynamic analysis of an individual's behavior in terms of his concept of the "lifespace" which he defines as the multiple interaction of the person and his environment (1935). The lifespace is conceived as a force-field consisting of a variety of energy vectors of fluctuating and shifting valence. An analysis of the systematic changes in the nature of the psychological field as they relate to aging was tentatively begun in the context of Lewin's studies of feeblemindedness as well as in the work of some of Lewin's students (notably Kounin, 1943). These writers interpret age changes primarily in terms of the degree of differentiation of the psychological field into well-defined regions. The young child's field is seen to be vague and diffuse. As a more well-defined separation of function occurs, increase in the child's cognitive skills and abilities is witnessed. Lewin briefly suggests that in the case of *senile dementia* a breakdown of barriers between the psychological regions will return the individual to a quite diffuse state. Concomitant with the increasing differentiation of psychological regions, there is an increase in the individual's degree of flexibility, that is his ability to deal with the changing demand of the behavioral field (Lewin, 1946).

Lewin's conceptual scheme for the description of behavior is sometimes characterized as being ahistorical. The alert reader, however, will note important concern with developmental phenomena and suspect that a more formal exposition was forestalled by the theorist's untimely demise. Doerken (1954) derives some developmental implications of a field-theoretical nature. He suggests that the child starts out in a rather constricted environment, that during the period of maturation an expansion of the child's restricted world occurs, and that as a result of increasing opportunity of choice and an increasingly permissive environment the individual emerges at maturity as a rather flexible person. From then on, the individual's lifespace again tends to constrict progressively until in old age a situation somewhat similar to early childhood experience is reached.

For the purpose of a more systematic analysis, this process will be described as a sequence of changes in the individual's capacity to deal with alterations in the phenomenal field delimiting his lifespace. These changes in the individual constitute the process of differentiation of experience as well as the reversal of this process in later life. While the process of differentiation and de-differentiation is presumably subject to all the laws of learning and extinction, it does not appear opportune or necessary to deal here with the problem of describing how the differentiating process occurs.[1] It will simply be assumed that one can describe certain complex behaviors which are considered to be a function of the differentiating process, in order to proceed to the analysis of its maturational sequence.

It is usually considered to be a sign of good adjustment when the individual has available to him techniques (or responses) which enable him to deal with the restraints imposed upon his psychological field with a minimum of conflict. These responses, however, will be effective or will fail not only as a function of the relative adequacy of the cognitive process involved, but will depend also upon more complex traits of behavior which must be viewed in terms of the properties of the phenomenal field. One of the most important properties of the individual's field is the fact that the force-vectors delimiting his lifespace are rarely composed of constant sets of stimuli. The well-functioning individual must therefore be expected to meet without difficulty not only the limiting aspects, but also the constant changes in the patterns of cognitive and social stimuli which compose his phenomenal field.

Experience has shown that it is a formidable task to identify the ever-changing forces which act upon the individual at any given moment as well as to identify his capacity to deal with any particular stimulus. Some more general formulation is possible, however, if consideration is limited to the individual's ability to overcome or deal with the various types of restraints imposed upon his freedom of action which determine the success or lack of success of a given response.

THE BARRIERS OF THE PSYCHOLOGICAL FIELD

Logical analysis suggests several dimensions of restraints to be of significant importance. They may be pictured as a set of concentric lattice-work of field forces surrounding the individual and delimiting his freedom of action. While all these restraints exist at the time the individual leaves the womb and persist until the moment of his death, their impact upon the individual and his awareness, response and adaptation to them show considerable modification at various life periods. Figure 17.1 presents a graphic illustration of the developmental schema involved portraying an oversimplification of the relative importance of the various restraining forces.

Three major dimensions of restraining forces will be postulated each

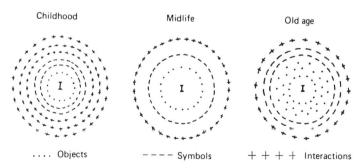

Figure 17.1   The barriers of the psychological field.

of which is met by different modes of adaptation. The first type of restraint will be called the *object barrier*. It consists of rather undifferentiated physical stimuli which may have pleasant as well as noxious attributes. It is the one which faces the infant as soon as self and nonself have become differentiated and is the first barrier to which reaction occurs at a conscious level. Since the components of its latticework of forces consist of the physical universe surrounding the infant, it must be overcome by physical manipulation. Relatively soon the child develops techniques designed to detour this type of restraint. In the adult, facility in coping with familiar situations and problems requiring some positive action will be a reflection of the soundness of his techniques in dealing with the object barrier.

Just as the object barrier is the first set of restraints the infant learns to handle, it is also the first which comes to interfere with the individual's freedom of action in old age. Here the object barrier can no longer be quickly ignored through the use of well-learned and efficient techniques. It becomes increasingly restrictive; more and more energy must be spent on ever clumsier methods of dealing with the obstacles of the physical universe. When the object barrier imposes severe restriction, compensations may be achieved by remaining function at other levels; when it finally closes, death is at hand.

The second set of restraints may be denoted as the *semantic barrier*. It is the restraint imposed by symbolic representations and limitation of the psychological field whether conveyed by verbal or other means. Once the child has learned to ambulate in and handle the forces of the physical world, he begins to note that symbols may be used to restrict his freedom of action. His mother no longer restrains him physically, but a simple "No, Johnny!" may be far more effective restraint than the bars of the play pen have ever been. Soon, however, the child begins to be more comfortable also in the world of symbols and the wall of the semantic barrier opens into a latticework where detours will provide solution of conflicts. In old age, once the decline of mental abilities has begun, loss of memory, decrease in effectiveness of judgment, and so on, will reimpose the semantic barrier as an increasingly effective bar to conflict-free functioning.

The third and final set of restraints shall be called the *pragmatic barrier*. It consists of the restraints imposed upon the individual by the value systems of his society and the implicit limitations of his freedom of action imposed by a variety of interpersonal relations. While the child begins to cope with this barrier, it is usually only in late adolescence or early maturity that the individual acquires proficiency in handling these limitations and only then has full use of techniques for adjusting his behavior so as to achieve maximum freedom of action with minimal social and interpersonal conflict. The pragmatic barrier is the last to become of importance in limiting the freedom of action of the older individual. Even after physical limitations become important and mental abilities have begun to decline, the old individual remains a member of his social group and retains freedom of action based upon respect for his past achievement rather than his current status. In advanced age, however, this barrier also shows increasing restriction as old personal relationships are broken by death and new ones are only rarely established.

OPERATIONS FOR COPING WITH THE BARRIERS

Motivated behavior shall now be assumed to be goal-directed and "learning" will be defined in this context as the process required to develop techniques to cope with the successive barriers. The individual in this situation is therefore faced with a succession of *umweg* type problems. A random response, such as leaving the field, will merely produce a head-on collision with the restraining conditions. As other responses designed to detour the barrier are emitted, the correct response will eventually occur and will be learned. When the learned response, due to changes in the barrier or the organism, is no longer successful or possible, the individual will tend to resort to a response which was previously successful (even though not now appropriate), a phenomenon generally termed regression. When no response at all is possible, cognitive behavior ceases to exist. If such latter situation is generalized and prolonged, panic, intense disorganization, coma, and eventual death will ensue.

Conflict may now simply be described as the psychological situation where motivated goal-seeking behavior is interfered with by one or more sets of barriers, and where the required *umweg* solution for the specific response-sequence has not yet been learned (or has been extinguished). Conversely, conflict is resolved by learning the appropriate solution which will permit resumption of the response-sequence, or by retreating in a nonadjustive manner behind a more primitive barrier level. This latter behavior would include what is often erroneously called "leaving the field," but what should rather be considered as abandoning the on-going response sequence.

It is not enough, however, to specify that an individual's skill in coping with these barriers may be a function of the adequacy of his learning, perhaps reflecting the status of his mental abilities. The results of cross-sectional studies reported elsewhere show that one may differentiate socially positive characteristics of population samples adequately by means of measures of intellectual ability during the period of early maturity, but that such differentiation in later adulthood could best be accomplished by measures of other cognitive variables (Schaie, 1958a; Schaie,

1958b). What seems involved here are the kind of intellective variables which appear in the studies of attitudes and motives and which may represent habit strength, drive strength, and energizing components (see also Herbst, 1957). These, then, are the force-vectors in the individual's lifespace which determine the level of differentiation and consequently of individual adjustment.

OPERATIONS FOR THE MEASUREMENT
OF RESTRAINT-COPING BEHAVIOR

If the postulated field forces are assumed to limit the behavior repertory available to the individual it becomes essential to measure the capacity of the individual to cope with the restraints of the field. Many theoretical constructs have been proposed to describe such a dimension, the most prominent of which may perhaps be the psychoanalyst's construct of ego strength. As suggested earlier, however, what is required in the present context is an intervening variable or variables which can be tied with some sense of security to observable responses to the immediate psychological situation.

It is difficult to predict what the nature of the restraining stimuli will be for a given individual in any particular situation. "Coping" with the restraints, however, would seem to involve the general property of responding to a sequence of *changing* stimulus patterns. The underlying behavior dimension then could be one where responses are characterized to vary from most adaptive flexibility to debilitating rigidity.

It has been suggested that several dimensions of restraints are required to yield a meaningful description of the forces in the psychological field. It is likewise necessary to postulate corresponding dimensions for the individual's efficacy in coping with and acquiring adaptive techniques for dealing with these restraints. This assumption fits well with the experimental literature on rigidity-flexibility (Levine, 1955; Chown, 1959). A variety of researches in this area have failed to produce an unitary factor measuring the central concept, but several clear-cut dimensions have emerged. The writer's work with measures of rigid and flexible behavior, using factor-analytic techniques, points to a three-dimensional system which would fit well with the above-postulated set of restraints (Schaie, 1955; 1960).

The first factor in this system was called "psychomotor speed." It was considered to be a measure of the individual's functional efficiency in coping with familiar situations requiring quick thinking and rapid response. It is in this context considered to be a possible measure of the adult's ability to cope with the object barrier. It is not a good measure, however, as it is based on the emission of verbally mediated responses, while the object barrier is felt to be at the presymbolic level. Perhaps nonverbal material may prove to be more appropriate. The second factor, called "motor-cognitive rigidity," was deemed to be a measure of effective adjustment to shifts in familiar patterns and changing situational demands containing cognitive elements. It is offered as a measure of the individual's capacity to deal with the semantic barrier. The third factor was called "personality-perceptual rigidity," and was considered to be a measure of the individual's capacity to cope with changes in inter-personal and environmental patterns as well as his ability to adjust to new and unfamiliar situations. This latter factor might yield a measure of the individual's ability to cope with the pragmatic barrier.[2]

The widening and constriction of the individual's lifespace over his lifespan might then be explained in terms of increasing and decreasing flexibility as measured along these three dimensions. Individual differences would be explained in terms of the interaction of different rates of maturation with the social situation determining the specific nature of the restraining forces. The pattern of restraint-coping behavior over the lifespan, however, will conform to the general schema outlined above. Thus at different phases in the life cycle the individual's ability to cope with the three postulated restraint dimensions can be described schematically by a pattern of overlapping ellipses as shown in Figure 17.2. Some experimental evidence has been accumulated which would back this schematic (Schaie, 1958a). In a cross-sectional study covering the range twenty to seventy years, peak flexibility was reached for motor-cognitive rigidity at ages twenty to twenty-five, for personality-perceptual rigidity at ages thirty-one to thirty-five, and for psychomotor speed (out of phase for reasons already indicated) at ages twenty-six to thirty. A significant increment in the rigid direction appeared for these variables at ages sixty-one to sixty-five for psychomotor speed, at sixty-six to seventy for motor-cog-

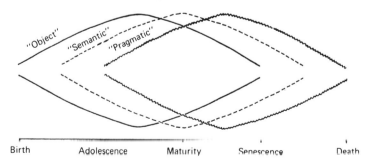

Figure 17.2   Developmental sequence of the expansion and contraction of the restraints bounding the "lifespace."

nitive rigidity, and a gradual slope appeared for personality-perceptual rigidity with a projected significant increment in the seventies.

The psychological field of the child widens as his coping behavior expands and as he learns an increasing range of *umweg* solutions. Experimental evidence for such differentiation of experience is amply available from the fields of ability, personality, and problem solving behavior. The increasing flexibility involved in the child's behavior is suggested also by such evidence as that in current research on color-arranging preferences in children through the school years, where simple and stereotyped patterns are found in the younger children, while more complex and differentiated patterns appear in the older children (Schaie, 1962).

Beyond maturation (that is, that point at which the individual has attained maximum differentiation and which most recent studies would probably place anywhere from twenty to sixty years depending on the specific variable being investigated, but whose average peak lies probably in the late twenties and early thirties) age changes appear to be a function of regressive behavior. Regression, here is defined in the field-theory context as a response which was once the correct solution to an *umweg* problem but is not now appropriate to the developmental level of the individual (see Lewin, 1941).

As differentiation is a prolonged process, proceeding at different rates for different psychological variables, so regression is no all-or-none phenomenon. It follows the same rules as differentiation has. The individual whose restraint-coping behavior has not been well developed at maturity, however, would therefore also be expected to show rapid rate of regression as soon as his restricted behavior repertory becomes inadequate. Thus the well-endowed individual would tend to regress much later than one with little resources, and in fact may expect the object-barrier and physical extinction to enclose him long before significant constriction occurs in other areas. This view of regression does not require recourse to early life experience, but permits evaluation of the current life situation and its immediate antecedents. It permits a more general treatment of age changes by permitting regressive phenomena to occur from one adaptive to another adaptive level. For example, the sexual behavior of a middle-aged individual upon losing his spouse might "regress" to the level of young adulthood rather than to the auto-erotic level.

Toman (1959) in a recent discussion of the general problem of quantifying human motivation, presents evidence that the average interval between the successive satisfactions of motives increases and that their intensity increment decreases as the individual reaches maturity. He suggests that a reversal of this trend would be incumbent if there is such a thing as decline with age. Is this not a necessary corollary of the proposed schema? As the individual's response repertory proliferates greater intervals must be expected between the occurrence of any individual response. Likewise, factorial studies of personality seem to require a larger number of factors in middle adulthood than at any other time in the life cycle, again a necessary consequence of these proposals (Cattell, 1957).

How does restraint-coping behavior relate to strength of motives, attitudes, and abilities at any given point in the life cycle? The approach from the interactional properties of the lifespace is one of optimal strategy since it permits begging many practical questions. These questions, however, cannot be begged forever and need to be worked out in greater detail than the above theoretical formulations so far permit.

SUMMARY

A view of aging in cognitive behavior is offered, which traces the process as a function of developmental changes in the restraint-coping ability of the individual to the widening and constriction of his lifespace. Three principal restraint dimensions were suggested and operations from the study of rigidity-flexibility were offered as possible measures of the individual's ability to cope with the specified restraints. The concepts of differentiation and regression were also examined in the context of a field-theory oriented developmental framework.

NOTES

[1] It does not appear too essential to specify what is meant by learning, or which type of learning-theory would be most useful. It may be suggested, however, that a straightforward contiguity view seems quite compatible with the above analysis.

[2] Alternate measures of the coping behavior at the pragmatic level are also offered by studies concerned with changes in attitudes and interests over the adult lifespan. See, for example Riegel and Riegel (1960) for a study of this kind which would seem to fit the model suggested in this paper.

REFERENCES

Cattell, R. B. *Personality and motivation structure and measurement.* New York; World Book, 1957

Chown, Sh. M. "Rigidity—a flexible concept," *Psychol. Bull.* 56: 195 223, 1959.

Cumming, E., and W. E. Henry. *Growing old: the process of disengagement.* New York: Basic Books, 1961.

Doerken, H., Jr. "Psychometric differences between senile dementia and normal senescent decline," *Canad. J. Psychol.* 8: 187–196, 1945.
Eriksen, E. H. *Childhood and society.* New York: Norton, 1950.
Fromm, E. *Man for himself.* New York: Holt, Rinehart and Winston, 1947.
Herbst, P. G. "Situation dynamics and the theory of behavior systems," *Behav. Sci.* 2: 13–29, 1957.
Kounin, J. S. "Intellectual development and rigidity." In: Barker, R. G., J. S. Kounin, and H. F. Wright (eds.), *Child behavior and development.* New York: McGraw-Hill, 1943.
Levine, A. S. " 'Perseveration' or 'the central factor,' " *Psychol. Rep.* 1: 247–265, 1955 (Monogr. Suppl. No. 6).
Lewin, K. *Dynamic theory of personality.* New York: McGraw-Hill, 1935.
Lewin, K. "Behavior and development as a function of the total situation." In: Carmichael, L. (ed.), *Manual of child psychology.* New York: Wiley, 1946.
Lewin, K. "Regression, re-regression and development." In: Barker, A., T. Dembo, and K. Lewin, *Frustration and regression.* University of Iowa, Studies in Child Welfare 18: 1–43, 1941.
Linden, M. E., and D. Courtney. "The human life cycle and its interruptions: a psychological hypothesis: studies in gerontologic human relations," *Amer. J. Psychiat.* 109: 906–915, 1953.
Riegel, K. F., and R. M. Riegel. "A study on changes of attitudes and interests during later years of life," *Vita Humana* 3: 177–206, 1960.
Riegel, K. F. "Personality theory and aging." In: Birren, J. E. (ed.), *Handbook of aging and the individual,* Chap. 23, pp. 797–851. Chicago: University of Chicago Press, 1959.
Schaie, K. W. "A test of behavioral rigidity," *J. Abnorm. Soc. Psychol.* 51: 604–610, 1955. (a) "Rigidity-flexibility and intelligence: a cross-sectional study of the adult life span from 20 to 70," *Psychol. Monogr.* 1958: 72, No. 9 (Whole No. 462). (b) "Differences in some personal characteristics of 'rigid' and 'flexible' individuals," *J. Clin. Psychol.* 14: 11–14, 1958. *Manual for the Test of Behavioral Rigidity.* Palo Alto: Consulting Psychologists Press, 1960. "Ontogenetic changes in response differentiation on a color arrangement task" (in preparation).
Shock, N. W. "Biology of aging." In: Smith, T. L. (ed.), *Problems of America's aging population.* Gainesville, Fla.: University of Florida Press, 1951.
Toman, W. "A general formula for the quantitative treatment of human motivation," *J. Abnorm. Soc. Psychol.* 58: 91–99, 1959.

Part Four

Ability and Achievement

Having just completed a section devoted in large part to "cognition," the reader may wonder why there is a separate section about ability and intelligence. After all, cognition would seem to be concerned with intellect. This is true of course; but as is usually the case in psychology and other disciplines, a content area reflects its historical origins, and the origins of the concepts of cognition and intelligence are somewhat different.

To oversimplify a bit, the idea of "intelligence" developed rather directly in response to a need to assess and predict performance in school or in school-related tasks. The tests developed to satisfy this need have determined quite specifically the concept of "intelligence" in American psychology. Readers will of course be familiar with Binet's test, developed early in this century in response to the need of the Paris schools to identify children who could not learn the required content of the public schools. Actually, attempts to measure, describe, and predict such performance had been occurring for a number of years before Binet's effort was successful. By World War I, a number of individual tests, derived from Binet's model, had been standardized in the United States and were beginning to be used in clinics and a few schools. With the military draft of World War I came development of the Army Alpha, the first paper-and-pencil or group intelligence test. After the war, a host of group intelligence tests, derived from the Alpha, were produced in America. In succeeding decades, few children escaped being tested by one or several of these devices during their school experience.

The popularity of intelligence tests and the concept of intelligence derived from them has been reduced somewhat in recent years, and in some quarters it has become unfashionable to talk about "intelligence" at all. Then too, as is apparent from Part Three, work in cognition research has provided a great analytical yield (yield in terms of explanation rather than description) and thus has attracted much of the energy that earlier went into intelligence research. And fashion too must be a factor: Fads in developmental psychology (in fact, all of psychology) come and go as they do in dress or music.

But on the other hand, there is an accumulation of more than a half century of data on intelligence and its development, research that has enabled psychologists to provide guidance for both individuals and institutions in matters of ability. This guidance has sometimes, unfortunately, gone beyond the data; but the total effect has been, it is hoped, beneficial. The topic has certainly engaged the interest and efforts of a significant number of able psychologists over the years.

More to the point, here is a quality or characteristic of persons, described and quantified within rather sharply defined limits, related to nearly every imaginable facet of individual origin and experience, and charted through much of the lifespan for both groups and individuals. One reason, perhaps, that there has been so much criticism of the concept of intelligence is simply that so much is known about it!

In this section we present studies that delineate the origin and development of ability, suggest some influences, and describe outcomes in terms of life activities and achievement.

Genetic inheritance has been seen, at various times throughout the twentieth century, as contributing from everything to nothing to the individual's intellectual makeup. Early students tended to ascribe major influence to heredity. Then for a long period a percentage relationship was hypothesized. Behaviorists and their successors threw out the whole notion of heredity as a matter of consequence. One may, of course, find a proponent of almost any view writing today, but most authorities hypothesize a complex, interactive relationship between genetic inheritance and the influence of experience.

A survey of genetic mechanisms and influence is provided by J. C. DeFries of the Institute for Behavioral Genetics, University of Colorado, Boulder, in his article entitled "The Genetics of Intelligence: An Overview." Dr. DeFries provides a brief history, an outline of genetic operation, a discussion of the manifestation of genetic effect on intelligence, and some implications for society. It might be noted that Dr. DeFries wrote this essay before the controversy over the so-called Jensen Report erupted.[1] (The controversy concerned the relationship between race and intelligence.) The article printed here provides useful background for understanding this complex topic.

Experiential influence is obviously multitudinous in source, but of special interest to students of development is the social and cultural milieu in which the child grows up. Alex Inkeles, a sociologist at

Harvard, discusses the influence of these factors in his piece entitled "Social Structure and the Socialization of Competence." Inkeles concerns himself with the concept of competence—a much broader concept than "intelligence"—and relates this to the development of total personality in a cultural context. The dramatic contrast between this article and the previous one by DeFries should be apparent.

Two large-scale longitudinal research projects have commanded attention because data from both have challenged old notions of age in relation to ability. Both are presented here, abbreviated from monographs and extensive research reports. One is William Owens' report on retesting of one-time college students after three and four decades, entitled here "Age and Mental Abilities: A Longitudinal Study." Owens is director of the Institute for Behavioral Research at the University of Georgia. The other is from the most recent monograph (1967) in a series dating back to the 1930s, evaluating life experiences and ability change in a group of subjects described as mentally deficient in childhood. The title of the current report is "Mid-life Attainment of the Mentally Retarded: A Longitudinal Study." Authors of this study are W. R. Baller (United States International University), Don C. Charles (Iowa State University), and E. L. Miller (Ball State University). All were at one time at the University of Nebraska, in whose geographical area the subjects originally lived.

Longitudinal and cross-sectional studies, virtually without exception, produce conflicting data concerning changes of ability in maturity and older age. The cross-sectional studies show marked decline with age, while longitudinal research provides evidence of improvement over time. The late Raymond Kuhlen explored one major influence on these disparate findings in his "Age and Intelligence: The Significance of Cultural Change in Longitudinal versus Cross-sectional Findings."[2]

A researcher interested in age and productive achievement is Wayne Dennis of Brooklyn College. In the article reprinted here, Dennis compares the age-related productivity of scholars and scientists with that of creative persons in the arts, and speculates about the differences he finds.

NOTES

[1] Jensen, A. R. "Social class, race and genetics: Implications for education," *American Educational Research Journal* 5: 1–42, 1968; "How much can we boost I.Q. and scholastic achievement?" *Harvard Educational Review* 39: 1–123, 1969.

[2] Recently new designs have been developed that enable the researcher to separate ontogenetic and generational changes. The reader

is referred to Baltes, P. B. "Longitudinal and cross-sectional sequences in the study of age and generation effects," *Human Development 11*: 145–171, 1968; Schaie, K. W. "A general model for the study of developmental problems," *Psychological Bulletin 64*: 92–107, 1965.

Eighteen

The Genetics of Intelligence: An Overview

J. C. DeFries

A considerable resurgence of interest in the inheritance of behavioral characters, including intelligence, is now taking place. Results from recent research clearly indicate that gene effects are important causes of individual differences in behavior, a finding which has far-reaching implications for our educational, social, and legal systems. The purpose of this essay is twofold: (1) to summarize some of the current evidence concerning the genetics of intelligence; and (2) to discuss briefly some implications for public policy.

HISTORICAL BACKGROUND

Francis Galton, the younger half-cousin of the English naturalist Charles Darwin, conducted the first systematic investigation of the inheritance of mental ability. The accomplishments of eminent men from various walks of life (judges, statesmen, peers, military commanders, artists, scholars, religious leaders, and so forth) and their relatives were evaluated. The most eminent man in each family (as judged by Galton) was taken as a reference point and relatives who attained eminence in one profession or another were then tabulated according to the closeness of relationship to this individual. Although this approach may appear crude by present standards, the results were striking: A far greater number of outstanding men were found among the relatives of eminent men than would be expected to occur by chance alone; the closer the degree of relationship, the higher the incidence of eminence.

The results of this investigation were published in 1869 in a volume entitled *Hereditary Genius, an Inquiry into its Laws and Consequences*, ten years after the publication of Darwin's *The Origin of Species*. These results, plus those of a later study of similarities and differences between

This paper has not previously appeared in print. It was written in 1968, prior to the appearance of the paper by Arthur R. Jensen ("How much can we boost I.Q. and scholastic achievement?" *Harvard Educational Review* 39. 1–123, 1969).

members of fraternal and identical twin pairs, demonstrated to Galton that "nature prevails enormously over nurture" (1883, p. 241). Darwin, who had evidently previously accepted the Victorian middle-class view that you can accomplish anything if you only work hard enough (Darlington, 1962), was also converted by these results to the view that human intelligence was hereditary.[1]

Interest in the inheritance of behavioral characters continued until the 1930s, when a sharp decline occurred, principally due to the advent of "behaviorism" in psychology. The extreme environmentalism characteristic of this behavioristic point of view is well illustrated by the familiar challenge of J. B. Watson:

> Give me a dozen healthy infants, well formed, and my own specified world to bring them up in and I'll guarantee to take any one at random and train him to become any type of specialist I might select—doctor, lawyer, artist, merchant-chief and, yes, even beggarman and thief, regardless of his talents, penchants, tendencies, abilities, vocations, and race of his ancestors. (1930, p. 104)

However, a reawakening of interest in the genetics of behavior began to take place in the 1950s and research activity in this area is now occurring at an accelerated rate. This is the subject matter of the relatively new, interdisciplinary field of behavioral genetics, a focal interest of which is the inheritance of mental characters.

GENES AS CAUSES OF INDIVIDUAL DIFFERENCES IN INTELLIGENCE

The Phenotype

Before a character may be subjected to genetic analysis of any sort, a satisfactory yardstick or means of describing it must be available. Sometimes, individuals may differ to such an extent that they may be easily classified without measurement. Examples of such "qualitative" differences are normal intelligence versus mental retardation, normal pigmentation versus albinism, tall versus short, and so on. Upon closer examination, however, more subtle differences become apparent, even among normal individuals. Among people of "normal" intelligence, "normal" pigmentation, or "normal" height, some "normal" range of variation is expected. In fact, when the means of measurement is precise, most characters are found to vary "quantitatively."

Some considerable disagreement exists among psychologists as to whether or not I.Q. tests are valid measures of intelligence. Although the question of what intelligence tests really measure has been a lively issue in psychology, a full discussion of this point is not within the scope of this paper. However, these measures of intelligence are reliable (both repeatable and internally consistent) and have predictive validity. For example, these tests are useful indicators of success in school and, when the range is large enough, success in later life. Thus, performance on such tests ("measured intelligence") is a satisfactory character or "phenotype" for genetic analysis.

THE GENOTYPE

Early in the history of genetics, the distinction between the phenotype of an individual and its genetic constitution or "genotype" was made clear. Chromosomes are carried in pairs, one member of each pair being derived from the father and the other from the mother. Thus, genes which are located on chromosomes are also carried in pairs. However, a gene which occurs at a particular locus or point on a chromosome may assume different states. These different forms of a gene are referred to as alleles. For example, assume that normal development of an individual requires the presence of an allele, symbolized A_n, whereas another allele (A_d) is deleterious and may result in abnormal development. If individuals which are heterozygous, that is, of genotype A_nA_d manifest abnormal development, then A_d is dominant to A_n. However, if heterozygotes are phenotypically normal and indistinguishable from individuals which are homozygous A_nA_n, then A_d is recessive to A_n.

SINGLE GENE EFFECTS

Several of the clearest examples of gene effects in man are those associated with some forms of mental retardation. Huntington's chorea, characterized by loss of motor control, progressive dementia and, when the age of onset is early, mental deficiency, has been traced through many generations. A consistent pattern of transmission is observed: (1) afflicted patients usually have a parent who was also afflicted, and (2) approximately one-half of the children of an afflicted parent eventually develop the disease. Since age of onset is variable, death occasionally precedes the onset of the disease in individuals genetically predisposed to develop it. Nevertheless, this pattern of transmission conforms very closely to that expected of a condition caused by a dominant gene effect. Individuals which develop the disease received this dominant gene from a parent who was also afflicted. Since afflicted individuals are heterozygous, one-half of their children would be expected to receive the dominant gene and thus develop the disease, while the others would receive the normal allele and develop normally. How can this dominant gene which results in death be maintained in the population? This persistence appears to be due to the fact that the disease does not usually develop until after the childbearing years.[2]

Recessive gene inheritance results in a quite different pattern of transmission. Individuals which are homozygous recessive for some undesirable gene usually have parents which are phenotypically normal. In addition, since inbreeding increases the likelihood of homozygosity, the incidence of genetic diseases due to recessive inheritance is increased in children of consanguineous marriages, for example, cousin marriages.

A condition which follows this pattern of transmission is phenylketonuria or PKU, characterized by severe mental retardation and other behavioral and physiological effects. PKU is due to a deficiency in a liver enzyme which is required for the metabolism of an amino acid, phenylalanine, a common constituent of proteins. Although the precise etiology of the mental defect is still unknown, the information which is available has nonetheless led to the development of a successful treatment. It is now well established that commercially prepared diets which are extremely low in phenylalanine will prevent the development of severe mental retardation, provided that the treatment is begun at an early age. In untreated cases, a rapid decline in intelligence begins at four to six months of age. If the special diet is not administered until after six months, the prognosis becomes increasingly unfavorable.

Although there is a possibility that some small deficit may remain even in patients treated early, treatment effects are nonetheless dramatic. Since these effects are age dependent, it is imperative that children be tested for the disease during the first few weeks of life, or better still, prior to discharge from the hospital. Practical methods for the screening of new-born infants are now available. In addition, tests are also available which permit the detection of carriers (heterozygotes) of this recessive allele. By such testing of relatives of PKU patients, the recessive mode of inheritance of this disease has been confirmed.

Phenylketonuria has provided a model for research concerning the genetics of mental retardation. Some 30 other defects in amino acid metabolism have since been described in which mental retardation is a frequent symptom. Many other examples of mental deficiency due to single gene effects are also known (see Penrose, 1963, and McClearn, in press). The cases discussed, however, sufficiently illustrate the point that development of both mental and physical characters is a function of the genotype. It will be shown below that development is not only dependent upon the kind of genes present, but also on genic balance.

CHROMOSOMAL ANOMALIES

The normal number of chromosomes in man is 46, that is, 23 pairs. Two chromosomes (one pair) are associated with sex determination and are thus referred to as the sex chromosomes. The other 22 pairs are called autosomes. Members of the sex-chromosome pair may be distinguished on the basis of their size: One is relatively large and is designated as the X chromosome; the other is small and is called the Y chromosome. Normal females possess two X chromosomes, whereas normal males have one X and one Y, symbolized XX and XY, respectively.

Approximately twenty years ago it was discovered that nondividing cells from males could be distinguished from those of females under the microscope. In a proportion of cells from normal females, a dark staining mass called a chromatin body is observed at the membrane of the cell nucleus. It is now believed that this chromatin body is an X chromosome which is genetically inactive. In normal males, no such chromatin body is observed. For this reason, normal females are referred to as chromatin positive, whereas males are chromatin negative.

These techniques were soon applied to the study of two forms of abnormal sexual development, characterized by sexual infantilism, plus various other abnormalities. In one condition, known as Turner's syndrome, individuals are apparent females. When cells of such females were examined, they were found to be chromatin negative. Shortly thereafter, cells of apparent males who failed to mature sexually (Klinefelter's syndrome) were found to be chromatin positive. From subsequent work it was confirmed that females with Turner's syndrome have only one X chromosome (XO) and that males with Klinefelter's syndrome have two X chromosomes and one Y (XXY). Many other sex-chromosome anomalies have since been reported. In general, when extra X chromosomes are present, some mental retardation occurs; the greater the number of extra X chromosomes, the more severe the deficit (Moor, 1967). However, when one Y chromosome is present, the individual is phenotypically male, regardless of the number of extra X chromosomes.[4]

A well-known example of an autosomal anomaly which results in mental retardation is Down's syndrome or Mongolism. For years, Down's syndrome was one of those baffling conditions which defied explanation. Although it was occasionally found to be familial, it was clearly not due to simple dominant or recessive inheritance. Its higher incidence among children of older mothers gave rise to many environmental theories of causation (reproductive exhaustion and such) which have since been laid to rest.

With improved methodological techniques, more detailed examination of human chromosomes became possible. As a result, in 1956, it was shown that the number of chromosomes in man was 46, not 48 as was previously thought. With these improved methods it then became feasible to study the chromosomes of individuals with aberrant conditions. It was soon found that individuals with Down's syndrome had 47 chromosomes; instead of having a pair of small chromosomes (designated 21), such individuals were found to have three such chromosomes. Although this condition may arise in several different ways, it is usually due to a failure of the chromosome pair of the mother to divide properly during germ cell formation.

Strange as it may seem, the degree of mental retardation is apparently more severe in individuals with an extra chromosome 21 (which is relatively small) than in those with an extra X chromosome. This is apparently due to the inactivation of X chromosomes when the number is greater than one. Recall that normal females (XX) have one sex chromatin body which is thought to be an inactive X chromosome. Females with three X chromosomes have two such chromatin bodies, and so on. Thus, although individuals who have chromosome 21 in triplicate are carrying less extra genetic material, the genic imbalance is greater, leading to more severe effects (Parson, 1967).[5]

INTELLIGENCE: A QUANTITATIVE CHARACTER

The examples cited in the previous sections demonstrate that at least some individual differences in intelligence are due to gene effects. But these gene effects were associated with abnormalities. What about the differences among individuals within the "normal" range of intelligence?

There is wide agreement, at least among geneticists, that measured intelligence is a quantitative or polygenic character. Quantitative characters are influenced by genes at many loci, each with relatively small effect, and by environmental factors. (Of course, major gene effects on quantitative characters, such as those already described, are also found.)[6] Because of their relatively small individual effects and because these effects may be masked by environmental factors, effects of individual "polygenes" are not readily discernible. For these reasons the techniques of classical genetics are not appropriate for the study of quantitative characters. However, various statistical techniques are available which provide a means for their analysis.

One important concept in quantitative genetic theory is that of heritability. The heritability of a character may be defined (somewhat imprecisely) as the proportion of the observed phenotypic variation which is due to gene effects. Thus, heritability is a measure of the extent to which individual differences in a character are due to hereditary differences. The heritability of a character is a function of the population on which it is measured. Therefore, it may vary from population to population, time to time, and so on. However, it is clear from the results of much research on the heritability of quantitative characters in domestic animals that at least some generality is possible.

For a given character, the observed correlation between relatives is a function of its heritability. A correlation is a statistical index of the relationship between measurements; a correlation of zero indicates no relationship, whereas a correlation of one indicates perfect correspondence between the measures. If all gene effects were completely additive (no dominance or other form of gene interaction) and if no environmental factors influenced the character, that is, if heritability were one, the correlation between relatives of the same degree in a large random mating population should be as follows: identical twins, 1.00; full sibs, fraternal twins, and parent-child, 0.50; half-sibs and grandparent-grandchild, 0.25; unrelated individuals, 0.00. Thus, these are the correlations one would expect if only genetic factors contributed to the similarity between relatives.

Although such a model sounds incredibly naïve for a character as complicated as measured intelligence, the data agree remarkably well with this expectation. In 1963, Erlenmeyer-Kimling and Jarvik published a comprehensive review of 52 previous studies which reported correlations between the measured intelligence of various relatives. These data were collected in eight different countries, on four different continents, and during a time span of over two generations. Individual investigators had different orientations regarding the importance of heredity and not all used the same measures of intelligence. Nevertheless, an impressive orderliness of results was obtained. The median correlations on the tests were as follows: identical twins reared together, 0.87; identical twins reared apart, 0.75; parent-child, 0.50; siblings reared together, 0.49; fraternal twins, both opposite-sex and like-sex pairs, 0.53; unrelated children reared together, 0.23; unrelated persons reared apart, −0.01.

Erlenmeyer-Kimling and Jarvik (1963) concluded that these results were compatible with the hypothesis that mental ability is a polygenic character. However, the authors were careful to point out that they did not mean to imply that the environment was without effect on intellectual functioning.

When taken at face value, these results suggest that individual differences in measured intelligence are largely due to gene effects; more specifically the correlation between members of identical twin pairs which were reared apart indicates that the heritability may be as large as 75 percent. Environmentally oriented investigators and even some geneticists (myself included) were quick to pooh-pooh this report. A consistent pattern of correlations is not in itself conclusive proof of a sizable genetic determination of intelligence. Human relatives develop in similar environments; thus, any observed correlation between human relatives must be a function of the common environment, as well as the common genes. This is even true for twins reared apart; we all know that agencies look for homes for children where the foster parents will match the natural parents in many respects.

In spite of its comfortable fit, this well-worn argument may have to be discarded. In 1966, Sir Cyril Burt, University College, London, whose active research interest in the inheritance of mental characters has covered a span of over 50 years, reported the most convincing evidence to date concerning the heritability of intelligence. In Burt's study, three different measures of intelligence were used and data were obtained on 95 identical twin pairs reared together and 53 identical pairs reared apart. The correlations between twin pairs reared together for a group test, an individual test, and a final assessment were 0.94, 0.92, and 0.93, respectively. The correlations between twins reared apart for the same measures were 0.77, 0.86, and 0.87.

At face value, these correlations indicate that intelligence, as measured by the individual test, has a heritability of over 85 percent. But may we not invoke the same argument as before regarding the similarity of environments of separated twins? Additional data presented by Burt indicate that this criticism is no longer valid. Data were also obtained concerning the occupational class of the breadwinner in the home in which the separated twins were reared. In most cases, one of the twins was brought up by his own parents, whereas the other twin was reared by foster parents. If the above argument were valid, the correlation between the occupational classes of parents and foster parents which reared the separated twin pairs should be at least moderately large. From Burt's data, this correlation is essentially zero.

Burt has also correlated differences in material and cultural conditions between homes in which separated twin pairs were reared with differences in their intelligence test scores. These correlations indicate that material conditions in the home are not significantly related to measured intelligence. However, a significant correlation (0.37) between differences in material condition and differences in school attainment was found. Differences in cultural conditions were significantly correlated with differences in school attainment (0.74) and performance on the group intelligence test (0.43), but were not significantly correlated with performance on the individual test or on final assessment. These results indicate that environmental factors are important causes of individual differences in school attainment, but that they are relatively unimportant with regard to intelligence as measured by performance on an individual test.

As stated previously, heritability may vary from character to character, and from population to population.[7] Therefore, it is hoped that other investigators will attempt to assess the environment, as well as the intelligence, of human relatives in other populations. Only in this way will the generality of Burt's results become known. However, until some serious fault is found with Burt's measures of home environment or until contradictory evidence is obtained, the following conclusion seems inescapable: Galton was right—nature does prevail enormously over nurture, at least with regard to measured intelligence.

INTELLIGENCE AND FERTILITY

On an evolutionary time scale, intelligence in man has increased enormously; thus, at least in man's evolutionary past, higher intelligence must have conveyed some advantage regarding the likelihood of leaving descendants. But what is happening today? If intelligence and family size are now inversely related as is generally assumed, intelligence must be declining.

Very few studies of change in intelligence level in a population over a span of years have been attempted. However, when such attempts have been made, no decline in intelligence has been observed. If anything, a slight increase is evident. But if individuals with low I.Q. have more children than average, this should not occur. Higgins, Reed, and Reed (1962) first reported data which lead to the resolution of this "paradox." In earlier studies of the relationship between intelligence and fertility, I.Q. scores of children were correlated with the number of children in the family. In this manner, sibships of size zero were never included. Reed and his co-workers, however, collected data on parents, their children, and also on nonreproducing sibs of the parents. Although parents with I.Q. scores somewhat below average had relatively high fertility, sibs which were severely retarded did not reproduce at all. In addition, brighter individuals also were found to have families which were larger than average. Thus, by including childless members of the family, a small, but *positive*, correlation between measured intelligence and family size was observed. This important finding has been subsequently confirmed by Bajema (1963). Both studies demonstrate that the relationship between measured intelligence and fertility is not a simple one. The distribution of mean number of children, plotted as a function of I.Q., is definitely bimodal, that is, bright and moderately dull individuals both have more children than average.[8] This "balanced" form of selection, coupled with some degree of assortative mating (tendency of like to mate with like), may be factors in the maintenance of a relatively large amount of genetic variation in measured intelligence.

IMPLICATIONS FOR PUBLIC POLICY

In the preceding sections I have attempted to summarize the current evidence concerning the genetics of intelligence. The evidence, in my view, clearly indicates that a substantial proportion of the variation in measured intelligence is due to heritable differences. In this section I shall briefly discuss one of the implications of this finding for educational, social, and populational policy. In addition, I shall also discuss some of the implications of behavioral genetics for our legal system. It should be emphasized, however, that the speculations which follow are those of a concerned geneticist, not those of a professional educator, social scientist, or lawyer.

EDUCATION

The educational implications of genetic and environmental determinants of individual differences in intelligence have recently been discussed in some detail by Jensen (1968). According to Jensen, present methods of education do not work for perhaps as many as 20 percent of our population. For these children, school is an experience of frustration and lack of accomplishment. Existing inequalities of educational opportunities and facilities account for only a small fraction of the variation in educational attainment; biological and social environmental factors account for most of this variance. Jensen suggests that the solution to this problem is not in literal equality of opportunity, meaning uniform treatment, but in equal opportunity for diversity of educational experiences, including diversity in our methods of instruction.

To a geneticist, the above argument has much appeal. The literature of genetics contains many examples which demonstrate the principle that different genotypes require different environments for optimal development. This is the concept of "genotype-environment interaction." What is true for the production of corn is probably also true for education: Children with low I.Q. scores are unlikely to succeed without special education; with special education, they may succeed.

Unfortunately, very little is known concerning the feasibility of individualizing educational experiences. Nevertheless, the process of individualizing education should be continued, at least as it involves special education for exceptional children at both ends of the distribution. Only further research can indicate the desirability of also individualizing education for students within the normal range of ability.

In spite of the best efforts to overcome the biological and social handicaps of our children, it is highly unlikely that anything approaching equality of performance will ever be achieved. Someday all children may have an equal opportunity to pursue an education; however, they will not be equally likely to obtain a college degree. As a consequence we should make every effort to keep the aspirations of children and their parents realistic. We should caution the more expansive educators and politicians against raising false hopes: To suggest that all children are potential candidates for a college degree not only raises false hopes, it is irresponsible; it can only lead to frustration, resentment, and reaction.

Social Class and Race

The implications of behavioral genetics for the social sciences have only recently been considered (Eckland 1967, 1968). Most theories of social stratification accept the premise that everyone has the same capacity to learn. However, heritable differences in capacity to learn do exist and such differences are important bases for social stratification. Therefore, efforts to provide an equal opportunity are not likely to eliminate the tendency toward social stratification. This argument becomes even more compelling when one considers the fact that assortative mating is occurring. Since the family is the child-rearing unit, the biological and the social transmission of class structure are both facilitated (Eckland, 1968).

It is an unfortunate fact that social class differences are confounded with racial differences. Would equal opportunity for members of the white and Afro-American populations eliminate the correlation between race and social class? We simply do not know. It is also an unfortunate fact that although there is a considerable overlap in the distributions, a difference of about 10 points separates the mean I.Q. score of the white and Afro-American populations. Is this difference between the means of the two populations due in part to gene differences or is it entirely due to environmental factors? Again, we simply do not know. Although perhaps as much as 80 percent of the variation in I.Q. scores within the white population (and presumably also within the black) is due to gene defects, we have no unambiguous evidence concerning what proportion of the variance *between* the populations is genetic. It is possible that this difference is entirely environmental. In fact, it is theoretically possible that the mean scores of the Afro-American population could exceed that of the white population if both populations were reared in the same environment. However, we must also recognize the possibility that at least some of the observed difference between the populations may be due to gene effects. The concept of genotype-environment interaction may be useful in this context: If one changed the environment in which the populations developed, the rank order of their mean scores might also change.

This is obviously an emotionally charged area and one in which ignorance abounds.[9] It is also an area which competent geneticists have tended to avoid. Nevertheless, an expansion of research concerning the relative effects of heredity and environment on intelligence and performance, especially with regard to racial differences, has been vigorously advocated by the Nobel laureate William Shockley. In response to this call, the National Academy of Sciences has issued a statement (*Science*, 1967) prepared with the assistance of three eminent geneticists; their task was a difficult one. After a discussion of the problems inherent in such research, they state the following:

> . . . we question the social urgency of a crash program to measure genetic differences in intellectual and emotional traits between racial groups. In the first place, if the traits are at all complex, the results of such research are almost certain to be inconclusive. In the second place, it is not clear that major social decisions depend on such information; we would hope that persons would be considered as individuals and not as members of groups (p. 893).

Although I do not advocate a crash program to study the biological bases of racial differences, neither can I agree with the conclusions expressed above. First, the results of such research would not necessarily be inconclusive. By incorporating environmental assessments into such a study, some information concerning the relative importance of environmental and genetic variables might be obtained. Secondly, major social decisions may depend on such information. Many programs which have been developed to improve the conditions of the disadvantaged are based upon the assumption that what works for the middle-class white will work for everyone. Perhaps additional information concerning the biological bases of racial differences would facilitate a more rational approach to solution of this problem. Finally, I question the assumption that if each person were only considered as an individual and not as a member of a group, the racial problem would disappear. Racial identity would still exist and the correlation between race and social class would still be glaringly obvious. In my view, the racial problem will continue to exist until all races have equal access to all social strata.

POPULATION CONTROL

The importance of heritable differences in intelligence has implications for our population beyond that of class structure. It also has implications for an issue which should be of great public concern, namely population control. The issue of population control may seem somewhat out of context in an essay on the genetics of intelligence; however, the subject of population control is not limited to considerations of quantity. Once the subject of population control is contemplated, the question of the nature of the control must automatically follow.

The case for population control is so well documented (see Archer, 1968; Davis, 1967; Ehrlich, 1968; Paddock and Paddock, 1967) that I shall not attempt any detailed justification here. It is a simple fact that continued population growth cannot be tolerated indefinitely. Further growth of the human population will inevitably be halted, either by a deliberate self-imposed policy of population control or by environmental exigencies such as famine and disease. Assuming that self-imposed population control is more desirable than leaving it up to the painful balancing forces of nature, why should we evade the responsibility of taking action now? Further delay will only make the problem increasingly burdensome and will accelerate the continuing deterioration of our environment. As Paddock and Paddock (1967) suggest, it may already be too late to avert widespread famine and suffering due to overpopulation in some countries.

But what sort of population control (ignoring problems of enforcement) should be advocated? Should individuals with different abilities reproduce differentially or should each couple contribute more or less equally to the next generation? Without doubt, deliberate selection for higher intelligence would be successful. However, we are almost totally ignorant of the genetic correlates of intelligence and, hence, cannot begin to predict the correlated changes which may accompany such a selection response. Our population is the product of millions of years of natural selection; thus, it seems somewhat presumptuous to assume that we can improve upon it using only the meager store of information which is currently at our disposal. In my view, any conscious attempt to alter the genetic constitution of our population should be deferred until considerably more is known about the possible consequences of such action. On the other hand, the time for *quantitative* population control is now at hand.

Law

A jury in Melbourne, Australia, recently acquitted a man charged with murder on the grounds of legal insanity after a defense witness testified that the man had an extra Y chromosome. This decision was based upon evidence collected within the past few years which suggests that XYY males may be more predisposed to commit violent acts than individuals with a normal chromosome complement. Several similar cases and appeals are now pending elsewhere. The self-evidence truth that all men are created equal is no longer apparently true, even before the law!

The evidence that individuals with an extra Y chromosome are more predisposed toward criminality is far from conclusive. Nevertheless, this precedent raises many interesting legal questions: For example, may other genetic anomalies lead to a predisposition toward criminality? If so, should such conditions always be expected to be associated with an abnormal number of chromosomes? If not, how could such individuals be identified?

Evidence from laboratory and domestic animals clearly demonstrates that aggressive behavior is heritable, although subject to experiential factors. Therefore, it is likely that heritable differences in aggressive behavior (and possibly criminal tendency) also exist in man. But aggressive behavior is almost certainly polygenic and, thus, only rarely associated with chromosomal anomalies.

If it were found that some forms of criminality had a biological basis, the possibility of prevention or rehabilitation would not be precluded. On the contrary, environmental prevention and therapy for various genetic diseases are well known, for example, PKU diabetes. However, considering criminality as a genetic disease could explain some of the difficulties which are frequently encountered in rehabilitation.

It would clearly be far simpler to ignore the possibility of heritable differences in criminal behavior and to judge everyone by one standard. But if heritable differences in criminal behavior do exist, an enlightened society should take cognizance of them and at least attempt to employ differential treatment and punishment.

NOTES

[1] For a discussion of the pioneering contributions of Galton to such diverse fields as genetics, psychology, and statistics and of his place in the history of behavioral genetics, see McClearn (1963).

[2] See Reed (1966) for a discussion of the dilemma faced by a genetic counselor when giving advice to young people who have a 50 percent risk of being heterozygous for this insidious gene.

[3] See Hsia (1967) for a more detailed presentation of the clinical features, inheritance, and treatment of this hereditary metabolic disease.

[4] For a more detailed discussion of chromosomal anomalies and behavior, see McClearn (1963) and McClearn and Meredith (1966).

[5] Recall, however, that severity of retardation is a function of the number of extra X chromosomes.

[6] See Morton (1967) for a discussion of the relative merits of quantitative genetic analysis versus the characterization of major genes in the study of mental disorders.

[7] See Vandenberg (1968) for evidence concerning the differential heritability of different components of cognitive functioning.

[8] For a more detailed discussion of the relationship between fertility and intelligence, see Carter (1966).

[9] See, however, Spuhler and Lindzey (1967) for an excellent dis-

cussion of the genetic basis of racial differences and McClearn (in press) for a discussion of the mistaken notion that performance on a one-dimensional scale is an index of generalized superiority.

REFERENCES

Archer, E. J. "Can we prepare for famine?" *BioScience* 18: 685–690, 1968.

Bajema, C. J. "Estimation of the direction and intensity of natural selection in relation to human intelligence by means of the intrinsic rate of natural increase," *Eugenics Quarterly* 10: 175–187, 1963.

Burt, C. "The genetic determination of differences in intelligence: A study of monozygotic twins reared together and apart," *British Journal of Psychology* 57: 137–153, 1966.

Carter, C. O. "Differential fertility by intelligence." In: Meade, J. E., and A. S. Parkes (eds.), *Genetic and environmental factors in human ability*. Edinburgh: Oliver & Boyd, Ltd., pp. 185–200, 1966.

Darlington, C. D. "Introduction." In: a reprint of *Hereditary Genius, an Inquiry into its Laws and Consequences*, by Francis Galton. Cleveland: The World Publishing Co., 1962, pp. 9–21.

Eckland, B. K. "Genetics and sociology: A reconsideration," *American Sociological Review* 32: 173–194, 1967.

Eckland, B. K. *Genetic variability and cultural deprivation in education: Equal opportunity versus the family*. Paper presented at the Annual Meetings of the American Sociological Association, Boston, Mass., August 1968.

Ehrlich, P. R. *The Population Bomb*. New York: Ballantine, 1968.

Erlenmeyer-Kimling, L., and L. F. Jarvik. "Genetics and intelligence: A review," *Science* 142: 1477–1479, 1963.

Galton, F. *Hereditary genius, an inquiry into its laws and consequences*. London: Macmillan, 1869.

Galton, F. *Inquiries into human faculty and its development*. London: Macmillan, 1883.

Higgins, J., E. Reed, and S. Reed. "Intelligence and family size: A paradox resolved," *Eugenics Quarterly* 9: 84–90, 1962.

Hsai, D. Y. "The hereditary metabolic diseases." In: Hirsch, J. (ed.), *Behavior-genetic analysis*. New York: McGraw-Hill, 1967, pp. 176–193.

Jensen, A. R. "Social class, race and genetics: Implications for education," *American Educational Research Journal* 5: 1–42, 1968.

McClearn, G. E. "The inheritance of behavior." In: Postman, L. (ed.), *Psychology in the making*. New York: Alfred A. Knopf, 1963, pp. 144–252.

McClearn, G. E. "Behavioral genetics," *Proceedings of XIIth International Congress Genetics, Tokyo, III*, in press.

McClearn, G. E., and W. Meredith. "Behavioral genetics," *Annual Review of Psychology* 17: 515–550, 1966.

Moor, L. "Niveau intellectual et polygonosomie: controntation du caryotype et du niveau mental de 374 malades dont le caryotype comporte un exces de chromosomes X ou Y," *Revue de Neuropsychiat. Infantile* 15: 325–348, 1967.

Morton, N. E. "Population genetics of mental illness," *Eugenics Quarterly* 14: 181–184, 1967.

Paddock, W., and P. Paddock. *Famine—1975! America's decision: Who will survive?* Boston: Little, Brown, 1967.

Parsons, P. A. *The genetic analysis of behaviour.* London: Methuen, 1967.

Penrose, L. S. *The biology of mental defect.* New York: Grune & Stratton, 1963.

"Racial Studies: Academy states position on call for new research," *Science* 158: 892–893, 1967.

Reed, S. C. "The normal process of genetic changes in a stable physical environment." In: Roslansky, J. D. (ed.), *Genetics and the future of man.* New York: Appleton-Century-Crofts, 1966, pp. 9–21.

Spuhler, J. N., and G. Lindzey, "Racial differences in behavior." In: Hirsch, J. (ed.), *Behavior-genetic analysis.* New York: McGraw-Hill, 1967, pp. 366–414.

Vandenberg, S. G. "The nature and nurture of intelligence." In: Glass, D. C. (ed.), *Genetics.* New York: Rockefeller University Press and Russell Sage Foundation, 1968, pp. 3–58.

Watson, J. B. *Behaviorism.* New York: Norton, 1930.

Nineteen

Social Structure and the Socialization of Competence

Alex Inkeles

I will define competence as the ability effectively to attain and perform in three sets of statuses: those which one's society will normally assign one, those in the repertoire of one's social system one may appropriately aspire to, and those which one might reasonably invent or elaborate for oneself. In contrast to socialization, then, the concept of competence stresses the end-product, the person as he is *after* he has been socialized, rather than the formative process itself. This conception is also broader than that of socialization in that the latter usually is defined with reference to a fixed repertoire of roles provided by a given socio-cultural system, whereas competence is here defined to include an individual's capacity to move to *new* statuses and to *elaborate* new roles. Despite these differences, however, the concepts socialization and competence are intimately linked. In general, the objective of socialization is to produce competent people, as competence is defined in any given society. It aims to develop a person who can take care of himself, support others, conceive and raise children, hunt boar or grow vegetables, vote, fill out an application form, drive an auto, and what have you.

As soon as we specify some of these qualities, it becomes evident that the research on socialization in our scientific literature has little to say about these matters. Research on socialization addresses itself predominantly to understanding how the child learns to manage his own body

Alex Inkeles, "Social Structure and the Socialization of Competence," *Harvard Educational Review*, 36, Summer 1966, 265–283. Copyright © 1966 by President and Fellows of Harvard College. Reprinted by permission from the author and the publisher.

This paper was initially prepared for a Conference on the Socialization of Competence sponsored by the Committee on Socialization, Social Science Research Council, meeting in Puerto Rico, April 29–May 1, 1965. M. Brewster Smith summarized the results of the Conference in *Items*, published by the Social Research Council, Vol. XIX, (June, 1965). The work of the Committee on Socialization is supported by the National Institutes of Mental Health (Grant #MH4160), whose aid is gratefully acknowledged.

and his primary needs. It inquires mainly how the child is guided in learning to manage the intake of food, the discharge of waste, and the control of sexual and aggressive impulses. Except for the rather isolated, even if highly interesting, forays in the direction of studying modes of moral functioning,[1] little is done in socialization research to study the acquisition of a broad array of qualities, skills, habits, and motives, which are essential to the adequate social functioning of any man or woman and in fact occupy the great bulk of the time of all socializing agents. A discussion of competence provides an opportunity to correct that imbalance in some small degree. I propose, therefore, to emphasize some of the qualities of individuals which are of most interest to society (and, incidentally, the focus of a good part of its socialization effort after infancy) but which seem largely to have escaped systematic study by students of socialization.

Two paths are open to me. One would be to list demands on the individual typically made by society, followed by specification of the requisite personal qualities these demands assume and of the socialization patterns presumed to engender these "socially demanded" personality dispositions. While this might be manageable if I were dealing with a particular stratum of a single society, it is otherwise too large and diffuse a perspective for the limited space at my disposal. I have therefore chosen to follow a second path, that of presenting a model of the personal system, essentially an accounting scheme of the elements of personality, broadly conceived, which I have found highly serviceable in all my efforts systematically to relate personality to social structure. The elements of this scheme will then serve me as indicators, pointing to more specific personal attributes I consider relevant to a discussion of competence.

A MODEL OF PERSONALITY
AND ITS IMPLICATIONS FOR RESEARCH
ON THE SOCIALIZATION OF COMPETENCE

In this section, I will present a model of the personality, one which represents essentially an accounting scheme rather than a theory.[2] I make no claim for originality in this scheme, but rather emphasize its practical usefulness as a framework for the discussion of personality in those researches in which interest is centered not in intra-personal dynamics but rather on interaction between the person and the socio-cultural system. I have used the scheme here, as elsewhere, mainly as a way of organizing the discussion of a social issue which we assume cannot be dealt with adequately unless we take systematic account of personality.[3] In this case, the theme is competence to perform social roles. Each element in the scheme will serve as an opportunity to point up an aspect on the "social demand" side of the equation which requires certain personal system attributes in the incumbents of social statuses. Further, to keep the discussion focussed, I approach competence mainly as a requirement for participation in contemporary and "modern" urban industrial settings. And rather than attempt any thorough coverage of the relevant issues, I have concentrated my energies on pointing to what seem to me important neglected opportunities in socialization research.

TABLE 19.1   An Accounting Scheme for Personality Study

Psychomotor System	{	Temperament *Aptitudes *Skills
Idea System	{	*Information Opinions and Attitudes
Motivational System	{	Values *Motives and Needs
Relational System	{	Orientation to Authority Figures Orientation to Intimates and Peers Orientation to Collectivities
Self System	{	Conceptions of Self Modes of Defense Modes of Moral Functioning
Modes of Functioning	{	*Cognitive Modes Affective Modes Conative Modes

* Indicates an element discussed in the text.

One of the virtues of the model is that it encompasses, in a decidedly limited set of some twelve to fifteen major headings, a great deal of what is ordinarily included in "personality." Even within that restricted frame, space limitations require that I forego altogether discussion of most elements of the model. My objective here is not to be exhaustive, but rather, by suggestion and illustration, to open up a discussion. The following list (Table 19.1) presents the scheme in full, and asterisks indicate those elements of the model which are actually taken up in the subsequent presentation.

The scheme, I repeat, is arbitrary, meant to serve as an accounting device. This applies not only to the main entries but also to the subsystems into which I have suggested they may be grouped. But I hope that in the discussion which follows, the utility of the scheme will nevertheless be manifest. As already indicated, because of space limitations I have in the section which follows undertaken to discuss only five of the sixteen main entries.

By an aptitude I mean an innate capacity or potential capacity to perform exceptionally well specific and difficult acts of sensing, muscular coordination, or the like. I include this heading more for the sake of completeness than out of a conviction that it has a definite relevance to the study of social competence. At the present time we are not at all sure which, if any, qualities meet the test of biological distinctiveness as special rather than general aptitudes characterizing all men more or less equally. Musical ability very likely does, manual dexterity and coordination may. Intelligence seems to be the clearest case, although some will question all the evidence yet available. Of course, at the extremes of the distribution, there may be nothing very problematical here. A child with impaired brain functioning is clearly not competent to learn more than the most rudimentary forms of social behavior and can never become a full participant in his society. Such a person may be defined legally as incompetent and therefore may be barred from exercising any of the formal rights allowed to most individuals in our society, such as rights to hold property, to vote, to conduct vehicles on the public thoroughfares, or to choose one's place of residence or occupation. Yet an adult male moron or imbecile may be perfectly competent to impregnate some female. So the *physical* capacity, drive, and coordination aspect of competence—the aptitude if you wish—must be clearly discriminated from the social and legal definition of competence.

Large numbers of our citizens have been and are defined, by the powers that be, as legally incompetent to participate in society on the basis of their alleged lack of aptitude. I have in mind not only the 600,000 who occupy our mental hospitals, but also much of the Negro population of the American South and a great many of the American Indians. Not unlike many tribal and colonially dominated people in other parts of the world, they have often been deemed by authorities as incompetent to exercise the rights of citizenship and in other ways conduct their own affairs. This legal definition of incompetence has frequently been justified on the grounds that mentally and psychologically, the Negroes affected are incapable of managing their own affairs, presumably because endowed by our maker with insufficient innate intelligence or social maturity. But in our Northern cities, the argument takes a rather different form. There we can encounter many social workers, often motivated by the highest ideals, who believe that the disadvantaged minorities are basically equipped by nature as are other men. Yet large numbers of these social workers in Harlem and Brooklyn have come to the conclusion that many of their clients are not able to manage their own lives and will likely be more or less permanent wards of the State or of private welfare agencies.

If we hold fast to the belief that this is not due to *innate* defects, that is, to lack of aptitude, then we must ask: What is lacking in their clients and what produced this condition? Presumably we may find the clue in some other aspect of the personal system.

SKILLS

Skills are aptitudes which have been trained or developed in accord with some cultural pattern. In other words, a skill is a socialized aptitude. Skills may therefore be based on the special aptitudes such as musical or mathematical ability, or on the more general aptitudes for muscular coordination, or for learning and using language, which virtually everyone possesses. The essential condition of the social groups and individuals who perform at a disadvantage in our society is their lack of the primary social skills. Our disadvantaged minorities are disadvantaged not only in winning so small a share of the available goods, services, and psychic rewards, but precisely because they so often lack the specific skills which could enable them to win a larger, or at least more adequate, share.

The main facts are today so well known that they have become almost clichés. While the end results, as reflected in low I.Q. scores, abysmal performance on aptitude tests, and consistent failure in the classroom, are well known, we must admit that students of socialization have done little to study the process which presumably yields this low capacity, especially as the process unfolds in the home, on the street, and in the primary grades of the classroom.

Of course, as we begin to explore these antecedent conditions, we may be led to re-examine our impressions of the actual pattern of skill deficiencies. The Harlem boy and girl may have an extremely meager vocabulary and very little ability to manipulate concepts—but are they also less well coordinated muscularly? Is the little Indian girl less able to cook or care for a younger sibling? And is the little Puerto Rican boy less able to bat a ball, to put it through a basket, or to sew on its torn cover? These questions, which unhappily we cannot pursue here, point to the important distinction between the social desirability of skills and their intrinsic difficulty, rarity, or aesthetic value. Some of the skills one must possess in minimum degree to participate in one society might be totally irrelevant or even outright disadvantageous in another.

Yet it does not do much good to mourn the fact that the other skills men *do* have are fine, even exquisite. The problem is painfully evident in the developing countries. There, the demands of the industrial order and urban living insistently undermine the relevance of venerable and often exquisitely developed skills which were highly important in the past; and they elevate to great importance *new* skills which, to the citizens of more developed countries, seem most elementary, yet which may seem difficult to master, even occult, to those in the developing nations.[4] It is notable, at least to me, that many of these skills are precisely those which *also* seem to be seriously underdeveloped in those groups in American society which are most disadvantaged, that is, our ethnic minorities, including those in our Indian population and in our pockets of Protestant white poverty such as Appalachia and the South. Among the more obvious skills which are relevant to adaptation to modern life and insufficiently mastered by these disadvantaged groups I note, for illustrative purposes:

The Telling and Management of Time   I know of no simpler or better indicator of a man's desire to show himself as modern than the acquiring and *demonstrative* wearing of a wrist watch. Of course, what is involved here is a value as well as a skill. However we may *feel* about the coercion of the clock, we must recognize the ability to tell time and to order one's affairs in relation to the clock as a critical skill for participation in the modern world. From social workers in the more deteriorated slums, and researchers who have worked with juvenile delinquents, one hears the frequent complaint that "they *never* keep appointments." Although my search is probably not exhaustive, I have not come upon a single article comparing children of different backgrounds in their ability to tell time and to meet the exigencies of the clock. This is all the more striking since the literature on developing countries gives the problem so much attention, and long ago Lewis Mumford argued that the industrial revolution and the modern world were ushered in not by the steam engine but by the elaborate organization of time in Christian monasteries.

The Command of Language, Especially in Its Written Form   The point certainly needs no elaboration, but perhaps I may be permitted to inquire what light our numerous studies of socialization can throw on the persistent failure of our Negro slum children to master language in a way appropriate to adequate performance in our schools? We cannot be satisfied with the usual reference to "poverty," not only because it is so global but also because it seems hardly to explain the extraordinary richness of the language and the elaborateness of the spoken style of Oscar Lewis' poor Mexican families.[5] And poverty can hardly explain the late readers and language cripples who bloom so profusely in some of our most favored suburban communities. Here again, then, we would like to see fuller study of those practices of child rearing, those patterns of interaction between parent, environment, and child, which lead to greater or lesser degree of command over spoken and written language.

The research cupboard is not so bare in the case of studies of the acquisition of language skills as it is of those involving management of time, but neither are we presented with a bursting granary. We may eagerly await the second volume of *The Review of Child Development*, which is to give us a chapter on "Language Development and its Social Context." If we are to judge from advance reports on the studies of Martin Deutsch and Irving Taylor at the Institute for Developmental Studies, we may yet meet some surprises in discovering that it is not the number but the use of words that distinguishes the underprivileged child.[6] For them, apparently, words stand for objects and objects for action. One word does not elicit another word, but rather elicits the *image* of action. This would seem to be highly congruent with the findings of Miller and Swanson.[7] It could also serve somewhat to explain not only the disadvantaged minority groups' performances on our typical word tests, but also the problems of later incorporating these groups into environments where one word leads to another word and not to an action.

Arithmetic   I suppose I cannot mention the language of words without also mentioning the language of numerical and mathematical symbols. On a four-item arithmetic test we gave to thirty-seven miners and ex-miners in Appalachia whose education averaged three to four years, we found that two "occupational" groups—the physically disabled and those now employed on relief jobs—could collectively answer only 6 percent and 17 percent of the problems, respectively. In the physically disabled group, for example, only three of the twenty men could correctly subtract 9 from 23! Two of these three could also multiply 8 x 9. So two men got half the problems right, one man got a quarter right, and the remaining seventeen men missed all four. We must note not only the appalling average lack of skill, but the individual variation. The group employed on relief did better, collectively, but the difference is due largely to individual variation. Two of the seventeen could answer all four questions, one could do three, another man one, and the remaining thirteen could answer none!

We had a very similar experience with tests of verbal ability. Using a larger sample from the same region of Appalachia, now augmented by groups with an average of seven or eight years of school, we asked the respondent to give the opposite of each of fifteen words. Before the fifteen words were presented, the interviewee was given an example such as "black" and "white." Many of the men simply could not grasp the task, that is, they could not dominate the *concept* of a word's opposite. By contrast, there were some who could give twelve correct answers, including words as difficult as "intelligent," "modest," "corpulent," and "affluent." Although education played a large role in this outcome, the variation within groups of the same education was very wide.[8] Again we are led to wonder what are the precise qualities of intelligence, of home environment, or of later experience which yield these differences.

Getzels and Jackson have presented some highly suggestive research relevant to these issues as they apply at a quite different educational and cultural level.[9] In their research they distinguished between adolescents with a high I.Q. on standard tests and those with the special skill to use ideas and information creatively. Perhaps this is a matter not merely of basic skill, but rather of cognitive style, which will be discussed below.

INFORMATION

The cognitive element of the personality may be divided into two broad categories levels of information and styles of thinking. I refer to the former here. Almost every public opinion poll ever conducted has shown great differences in the sheer quantity of fact known by middle and upper class individuals as against those in the lower classes. Such differences are not restricted to news events and public figures, but includes as well all sorts of practical and useful information, such as how to get your gas turned on or your garbage collected, how to get permission to conduct publicly controlled activities, how to organize a meeting, and so on. We may assume that this kind of knowledge about getting along in the everyday world is an important pre-condition to effective and independent participation in the modern social order.

The difference in information among various social class groups is generally attributed to the differences in their average education. While acknowledging the importance of education, I am rather of the opinion that these differences are already quite marked by the time children enter grade school, and that a great deal of the variance is to be explained as a result of the different early socialization experiences which the children of the several groups receive.

I am further strengthened in this belief by the clear evidence of great individual variation in information levels among persons with equal education. Thus, when we asked former Soviet citizens what they could do if some bureaucrat were taking an action injurious to them, the most common response among workers was "nothing," whereas most members of the intelligentsia cited at least two sources to which they might turn for help.[10] But there were ordinary workers who named four, five, six, and more agencies they would write to—the Communist party, the trade union committee, the factory manager, the newspapers, and finally Stalin himself! These different levels of information undoubtedly reflect the influence of motivation to know and perhaps also of a sense of efficacy— but that does not make it any less interesting to inquire what are the mechanisms of socialization which lead to these differences both in the desire to know and the resultant knowing.

The topic of language skills and information provides an opportunity to highlight the conceptual problems of relating major social-structural

factors to a socialization problem such as "the development of competence." Again I emphasize the necessarily schematic form which must
characterize my discussion. The points I can make within these limits are
rather obvious, but perhaps they will serve to provide a starting point for
future discussion.

Let us consider the command of language, including the size and
content of vocabulary and the capacity to form sentences and larger units
of speech in grammatical and culturally acceptable ways, thus enabling
one to bring his language capacity effectively to bear on situations of
action.[11] Assume we deal with a Negro boy in Harlem with normal
intelligence and no more than the average number of situational disabili-
ties which affect residents of that area. How would the hypothetical
"self-sustaining vicious circle" we so often hear about be apparent in his
case? In the home, we may identify the following inputs:

Low Capacity Models   From a sheer learning point of view, the
total vocabulary available to be learned in the home is likely to be quite
small because of the limited education and experience of the parents or
others in the household. Each incumbent of the home probably adds very
little that is not already in the vocabulary of others, so that the total word
pool is likely to be small and restricted. Since many homes are broken
and one parent absent, the available pool of models is further reduced.
Inevitably, fewer words will be learned than actually are available in the
limited total potential vocabulary of even this group. The prognosis is for
a very limited vocabulary.

Interaction Effects   Whatever the pool of potential words to be
learned, they cannot be learned if those who know them do not use them
in situations in which the child can learn them. If mother and father are
gone most of the day and/or leave the child alone a great deal, effective
learning is greatly reduced. Mother and father may be there, but not
communicate much with each other. Thus, in my current cross-national
research I found in all countries marked social group and individual
differences in the frequency with which men indicate any interest in communicating with their wives on a variety of themes, such as work, educating the children, running the house, and sex.

Content and Tone   Interaction effects are felt not only in that they
make words "available" for learning. The quality of the interaction clearly
will affect the relative frequency of words and, beyond that, the emotional
tone associated with verbal exchange in general. If the parents talk to
each other mainly to complain, grumble, or quarrel, the words in that
realm will obviously be learned sooner and more fully. Perhaps more
important, the unconscious conception of what language is mainly

"about" or "for" will be affected. With parental communication mostly demanding or quarrelling, the likelihood is greater that, unconsciously, language and verbal expression will be associated mainly with unpleasant experiences and hence be something more or less avoided.

Social Valuation   The interest or motivation for involvement in language will be affected not only by unconscious association with situation and content, but also by the more explicit cultural valuation put on relative degrees of skill and interest in the use of language. If the most common evaluations are generally negatively toned, as I believe they are in Harlem, the effect will most likely be to induce the child to view language skills as relatively undesirable qualities to cultivate in oneself. I cannot establish the point as a matter of fact, but I feel fairly certain that in Negro Harlem, the most common evaluations of language facility are negative: "big talker" and "talking big" are clearly negative. "Loudmouth," "shooting off the mouth" speak for themselves. "Preaching" is not too good. "Nagging" is, after all, mainly verbal behavior, and "talking foolish like a woman" reflects a similar feeling. There may be some positive associations to the facile use of language, but they seem fewer and less strongly toned than the negative associations.

The brief sketch of the Harlem boy's start in language in the home could be extended, but I trust I have made my point. The same mode of analysis could be applied as well to later stages in his language training, but I had better forbear. Let me merely note a few salient points. In the peer group, again, the pool of words collectively shared will be small and each boy will add little new. New vocabulary may be amply introduced, but mainly in areas relating to sex, aggression, and the law, and of such nature that the words cannot be carried over for use in polite society. On the streets, the skills valued and encouraged will be mainly physical, and indeed, the verbal may be actively disvalued.

When the boy later arrives at school, interesting new elements are introduced. In contrast to the rewards the school offers the middle class child for what he already *knows*, it is likely to greet our Harlem boy with horror for what he does *not* know and *cannot* do with language. The result, on his part, will be more avoidance of words and language. Much of the language he does know will be unacceptable, and if expressed will produce more negative reactions. The mode of expression of which he is most capable, the physical, will likely find no valued outlet at school, or may even be punished.

This analysis could be carried further in the life cycle, to the first job and beyond, but I suspect everyone can tell it for himself. Whether the story is accurate or not is unfortunately not as well documented as it might be, but I doubt that it qualifies as a "just-so" story.

MOTIVES

The preceding discussion points to the importance of motives as an underpinning for the acquisition of information, and thus alerts us to the relevance of motivation for a discussion of competence. Certainly at first blush, motivation would seem to have little to do with competence, since competence refers to the *ability* to do something whereas motivation deals with the *desire* or wish to do it. But if we recognize skills and information as contributing to competence to perform social roles, then we must recognize the *motive* to attain socially valued status positions as a necessary, even if not sufficient, requirement for competence in social action.

It might be argued that the point is trivial, on the grounds that little if anything socially desirable, yet relatively scarce, will be acquired without motivation to do so. The more challenging question is whether there are any motives which in and of themselves can be seen as more adaptive, and in this sense contributing more to one's competence to attain and perform in available and respected social roles. I am rather of the opinion that this question cannot be answered if it is put in general terms, as applying to any sociocultural system. Every sort of motivation, including aggression and hostility or extreme dependency and passivity, has been found to be relevant and adaptive in some culture somewhere. If we specify performance in a particular type of social system, however, then we can identify more or less precisely motives which are differentially adaptive. For example, cross-cultural anthropological research indicates that hunting and gathering societies are more likely to train youngsters for autonomy and independence, while pastoral societies give more emphasis to inculcating compliance and dependence.[12] In the context of modern industrial society, we might expect the need for achievement to be more adaptive than the need for affiliation, the need for autonomy more productive than the need for dependence, at least for those competing for middle-class positions. Studies of the adult population of the United States indicate that there is some such pattern in the distribution of motives,[13] and studies of child rearing in the different class and ethnic groups suggest that these adult differences most likely rest on differences in socialization practices.[14]

COGNITIVE MODES OF FUNCTIONING

When we considered the theme of information, we were concerned mainly with the amount and type of knowledge possessed by the individual. The cognitive modes refer to the forms of thinking and to the "style" characterizing the individual's mental processes. We ask: Is thinking abstract, concrete, or both? Is it slow and deliberate, or quick and mercurial? Is interest focussed or diffuse? Is the language of emotion more elaborated than is the conceptual apparatus for dealing with objects or material relations?

Cognitive functioning, at least in the realm of concept formation, has certainly been of interest to students of child socialization, but the interest has been mainly the usual developmental one of fixing the ages at which different conceptual skills emerge, with little systematic attention paid to individual, and even less to group, differences.[15] The work of Miller and Swanson and their students in establishing the stronger tendency to conceptual expression in the middle-class child and of motoric expression in the working-class child points the way. Unhappily, this path has been little followed by other workers in the field.

The modes of cognitive functioning clearly influence the child's initial performance in meeting the demands of the school and other social agencies; his later preferences for academic work, trade school, or practical apprenticeship; and his eventual choice and performance in his occupational and other adult roles. The boy who feels inadequate or is made uncomfortable in an environment which gives much emphasis to manipulating symbols instead of things will soon be drifting out of school to a world in which experience is more immediate and concrete. And this applies not only at the level of primary school and in disadvantaged neighborhoods. The problem is also very real for those who are already in college and choosing their professional careers. Thus, Stern, Stein, and Bloom report how great a role cognitive styles played in the adjustment of freshmen at the College of the University of Chicago. The College program stressed and rewarded "abstract analysis and relativity of values and judgment rather than fixed standards" (p. 191). Teachers introduced a good deal of ambiguity and often departed from conventional standards of judgment. It was precisely those students whose cognitive style inclined them to concrete thinking, to an insistence on one "correct" answer, who made up the bulk of the academic casualties at the end of the year. And this was true despite close matching of the students on measures of intelligence and scholastic aptitude.[16]

Unfortunately, Stern, Stein, and Bloom did not systematically explore the home environments which produced these different types. Indeed, we have very little knowledge about the socialization experiences from which stem one or another style of cognitive functioning. Rokeach, for example, tells us almost nothing of the home environment of those with "open" and "closed" minds. He does suggest, however, that ambivalence toward parents which is not permitted expression generates anxiety and narrowed possibilities for identification with persons outside the family. Rokeach sees both of these conditions, in turn, leading to the development of closed belief systems.[17] The point seems closely paralleled by the observation of Getzels and Jackson with regard to the home environments producing more creative adolescents.[18] The creative family, they concluded, "is one in which individual divergence is permitted and risks accepted." Those which produced a mere high I.Q. without "creativity" seemed more conventional, with the mothers stressing cleanliness, good manners, and studiousness. Miller and Swanson help us to see that physical punishment is more intimately tied to motoric than to conceptual expression, whereas psychological discipline, such as the threatened withdrawal of love, more often yields conceptual expression.[19] Hoffman's research also contributes to isolating the style of training he calls "inductive discipline," referring to efforts to explain to the child the effects of his action on others. The outcome which interests Hoffman, however, is less a mode of cognitive and more a mode of moral functioning.[20] We should, of course, here recall the explorations into the origins of the authoritarian personality, one component—some would argue the main component—of which is a certain cognitive style. Adorno and others suggest a series of antecedents in the home environment of those who display prejudice and authoritarianism, but their evidence is mainly clinical and has not been more systematically tested on large samples.[21]

Cognitive styles emerge as an extremely important component of the individual's equipment for coping with the demands of society and a critical element in determining what kinds of roles he may seek out and successfully play. Cognitive style will evidently play an early role in school performance; it will channel—and limit—the choice of occupations, and will affect the nature of one's political participation. The evidence seems unmistakable that the observed adult differences in cognitive style have their origins in childhood experience. Unfortunately, very little has been done by specialists on socialization to follow these insights in programs of systematic research. The implications for future research seem clear.

CONCLUSIONS

This presentation of a comprehensive conception of the elements of personality may have led the reader astray, and may leave me exposed to the charge of neglecting my announced subject. Admittedly, I have not discussed as fully as I perhaps should have social structure and the socialization of competence. If this is so, I beg indulgence on the ground that it seemed to me the main topic could not be properly understood unless we first dealt with matters more fundamental. In my discussion of the elements of the personal system, I have in effect sought to establish the basis on which—and in a sense the language *in* which—a more meaningful discussion can be presented. But it has taken so long to compile the vocabulary and explain the grammar that little space remains for telling the tale I meant to recite.

The message is very simple. Like so many obvious things, it is not only fundamental but also much overlooked. The main business of socialization is the training of infants, children, adolescents (and sometimes adults) so that they can ultimately fulfill the social obligations that their society and culture will place on them. Implicit in this statement is the expectation that, in meeting these societal demands, the individual will not be placed under so much strain as to fall apart psychologically. And not excluded is the thought that the term "social obligations" includes elaborating and acting effectively in roles not commonly assigned by the given sociocultural system. Indeed, we do not by any means exclude the possibility that the most creative way of meeting the demands of a given social situation may be to reject that situation as it presents itself, to insist on a new deal, and to forge new roles and new styles of life.

I am firmly convinced that concern about the ultimate playing of social roles is the decisive element in the child-rearing behavior of most parents or chief parent-surrogates. This is not meant to deny that at some periods in a person's life—especially in early infancy and perhaps again at puberty—the problems of sheer management of physical need or the facts of physiological change may not come briefly to dominate the concerns of the socializers. Yet, I believe that even in dealing with such ultimately physiological needs as the hunger drive, socializers never lose sight of the long-run adaptive significance, both of the sheer mastery of this drive and of the *way* in which it is mastered. Evidence for this can be found in the common speech of every mother. Nor do I mean to deny that there may be periods in the child's development when what is done to him more expresses the psychic needs and desires of the parent for giving or withholding, for restraining or indulging, than it represents any conscious or subconscious thoughts about the social roles the child will ultimately play in society. The issue is clearly one of relative emphasis. My chief point is that the degree to which, and the ways in which, socialization is a relatively conscious process of training in anticipation of future social roles, have been neglected relative to the conception of socialization as mainly a process in which adults cope with the challenge of the infant and child as *organism*. The same criticism applies to those who approach the study of socialization as mainly expressive of the parents' needs and dispositions.

So far as competence is concerned, some children face a situation in which almost everything conspires to insure that most of the more favored positions in society will be closed to them. They will grow up ill-equipped to compete for entrance into the more advantaged roles, and those desirable positions they may acquire they will be unable to hold successfully.

This is what we mean by competence—the ability to attain and perform in valued social roles. In our society this means, above all, the ability to work at gainful and reasonably remunerative employment, to meet the competition of those who would undo us while yet observing the rules for such competition set down by society, to manage one's own affairs, to achieve some significant and effective participation in community and political life, and to establish and maintain a reasonably stable home and family life. We should not for a moment forget the massive and cruel formal obstacles our society has devised to prevent the disadvantaged minorities from sharing equally in the opportunities inherent in the level of wealth and civilization we have attained. But we must also recognize that these obstacles—such as overt discrimination, segregated schools and communities, color-bar hiring practices, and even legal disfranchisement—are not the *only* barriers to effective functioning on the part of disadvantaged minorities. The most cruel aspect of discrimination and disadvantage lies in its ability to deprive the individual of that competence which is essential to effective functioning once the formal barriers to free competition have been breached. Lack of competence effectively to take advantage of new opportunities in a competitive system can make the attainment of nominal legal equality a hollow victory, and make a self-fulfilling prophesy of the bigots' claim that minority members are unable to perform effectively even when not formally discriminated against. To deny people the means for attaining competence while yet granting them technical equality under the law is the contemporary equivalent of saying that the majesty of the law confers on the rich as on the poor alike the right to sleep under bridges.

To perform effectively in contemporary society, one must acquire a series of qualities I believe to be developed mainly in the socialization process. Effective participation in a modern industrial and urban society requires certain levels of skill in the manipulation of language and other symbol systems, such as arithmetic and time; the ability to comprehend and complete forms; information as to when and where to go for what; skills in interpersonal relations which permit negotiation, insure protection of one's interests, and provide maintenance of stable and satisfying relations with intimates, peers, and authorities; motives to achieve, to master, to persevere; defenses to control and channel acceptably the impulses to aggression, to sexual expression, to extreme dependency; a cognitive style which permits thinking in concrete terms while still permitting reasonable handling of abstractions and general concepts; a mind which does not insist on excessively premature closure, is tolerant of diversity, and has some components of flexibility; a conative style which facilitates reasonably regular, steady, and persistent effort, relieved by rest and relaxation but not requiring long periods of total withdrawal or depressive psychic slump; and a style of expressing affect which encourages stable and enduring relationships without excessive narcissistic dependence or explosive aggression in the face of petty frustration.

This is already a long list, and surely much more could be added. My purpose here is not to strive for an exhaustive list. I want simply to indicate the *kinds* of personal attributes which I feel a modern industrial society requires in significant quantity of substantial numbers of its citizens. Without most of this array, one is not competently prepared for life in our society, and must sink into some form of dependency or deviance. There is no great difficulty in demonstrating that these qualities are very unevenly distributed in the several strata of our society—educational, occupational, ethnic, and regional. The challenge for the students of child-rearing is to show whether, and explain how, these differences came about as a result of differential socialization practices and experiences. My assessment of recent work in the field of socialization research is that very few of these issues have been the object of much systematic study on a significant scale. More than that, I incline to the conclusion that this situation is not now rapidly changing. The cause, I believe, lies in the scientific "culture" of those doing socialization research; they are beginning at the wrong end.

The master key to understanding socialization, in my opinion, lies not in further deepening our involvement in the innate propensities of the child and the situation of action this defines for the parent. The key lies rather in a redefinition of the problems of socialization research which starts with a clear statement of what are the massively evident observed *differences* among adults which appear socially important enough to be worth the trouble of explaining. I have tried to show through a brief and limited discussion of competence what a few such differences may be. But any other social issue—mobility, political participation, delinquency and crime, or occupational performance—could have served the same purpose. I applaud the fact that the six culture study sponsored by the Laboratory of Human Development at Harvard included, in the nine standard behavior systems to be observed, not only the old and tried themes of nurturance, succorance, aggression, and obedience, but also responsibility, sociability, achievement, and self-reliance.[22] Even in this case one may ask: If you have not first defined what is the quality of the adult you wish to understand, how do you know what to look for in the disciplining of the child when you study "responsibility" and "achievement"? And beyond responsibility, sociability, and achievement, we still want to know about information, values, motives, skills, moral functioning, self-conceptions, cognitive, conative, and affective modes; about the ability to trust others and enter into enduring relationships of cooperation or undestructive competition; about images of and relations to authority figures, and the sense of membership in, and feeling of obligation to, the community.

Before concluding, I should clarify some issues not necessarily important to the student of socialization, but nevertheless fundamental to the functionalist perspective on social structure, a perspective which my analysis in this paper represents.

The first issue concerns the appropriateness of a *general* model of competence, such as I have presented, for the analysis of performance in what is inevitably a highly differentiated social structure. Another, and blunter, way of making the point is to claim that I have presented a model of competence as defined by the middle class in American society. The charge is correct. The aspects of competence I have sketched above are precisely those which one requires either to continue as part of, or to attain to a position in, middle-class America. Every model of competence is, in large measure, specific to some culture, and often even to some stratum of a particular society. The elements of competence, as I have sketched them above, would not necessarily loom equally important for a man who was hoping to be the world's heavyweight champion. And they might be quite beside the point for a Trobriand Island fisherman or an Arctic Eskimo. The point I mean to make in this way, again, is that socialization research generally begins from the wrong end. In my opinion, the starting point of every socialization study should be a set of qualities "required" by, that is, maximally adaptive in, a given sociocultural system and/or manifested in a given population. The task of students of socialization should be to explain how these qualities came to be manifested by individuals, thus rendering them competent, or why individuals failed to manifest these same qualities, thus being rendered less competent to perform in the given social setting.

The very form of my last sentence raises the second and last issue on which I wish to touch briefly—the issue of "competence for what?" or "adjustment for whom?" A functional perspective always runs the risk of leading one to assume that what is good for society is good for the individual, and vice versa. Those interested in encouraging competence, or excellence, or whatever desirable quality, run the same risk. If we define competence as the capacity to organize one's life and to strive so as to achieve some degree of social stability or desired mobility, it means that many individuals, in seeking to meet the competence requirements of their society, may in that very act also be inviting more or less certain frustration. The Negro in Harlem who is quite comfortably able to accept his dependence on welfare authorities, to be passive in the face of middle-class society's expectation of constant effort and striving, and to find release from his tensions through extreme physical and vocal expression in his store-front church, may be making a more appropriate adjustment to the realities of his situation, and to that degree be more *competently* managing his life than is his neighbor who has all the white middle-class virtues, which will in turn increase the probability that he will run up against a solid wall of frustration and futility.

Everywhere today—by continent, by nation, by region, by class—there is a vast process of social change exerting its force. To manage their lives in a satisfying way, men need new information, skills, motives. New problems and situations everywhere constantly challenge their competence. Tragically, men find that the skills and talents which formerly made them models of competence in their community are of no value or are even demeaned and degraded in the new scheme of things. In this turbulent sea, we often glimpse some remarkable people who seem especially equipped to navigate freely and easily through conditions which are tumbling most people overboard. Are there then some qualities of man which give him a general competence useful in all places and times, qualities especially suited to adapt a man to all waters no matter how fast the current or sudden its changes? What are these qualities and what are the special forms of socialization which bring them into being? Here is a challenge to the student of socialization worthy of *his* competence.

NOTES

[1] See the review by Lawrence Kohlberg, "Development of Moral Character and Moral Ideology" in Martin L. and Lois W. Hoffman, *Review of Child Development Research*, Vol. I. New York: Russell Sage Foundation, 1964.

[2] The model is as much Daniel J. Levinson's as it is mine. Although we have used it extensively in our work, we have never published a full and systematic account of the scheme. We sketched some of its elements in the article "National Character" in Gardner Lindzey (ed.), *Handbook of Social Psychology*, Vol. II, 1954, and later in "The Personal System and the Sociocultural System in Large Scale Organizations," *Sociometry*, Vol. XXVI (June 1963). I used the scheme systematically in empirical research reported in A. Inkeles, E. Hanfmann, and H. Beier, "Modal Personality and Adjustment to the Soviet Socio-Political System," *Human Relations*, Vol. XI (1958). The conception is outlined in schematic form most fully in A. Inkeles, "Sociology and Psychology" in Sigmund Koch (ed.), *Psychology: A Study of a Science*, Vol. VI. New York: McGraw-Hill, Inc., 1963.

[3] For a fuller statement of this issue see: A. Inkeles, "Personality and Social Structure," in Robert Merton, et al. (eds), *Sociology Today*, New York: Basic Books, 1959.

[4] One of the most sensitive accounts is offered by Erik Erikson in his remarks on the Dakota Indians, in *Childhood and Society* (New York: W. W. Norton & Company, 1950), whom he describes as no longer

having a socially satisfying mode of using either their skill as riders and hunters or the character traits of cruelty and generosity, which were apparently meaningful, rewarding, and encouraged when their culture was whole.

⁵ The richness of the vocabulary and the fluency of expression of his Mexican subjects has so struck readers of Lewis' accounts that many wonder to what extent these qualities are the product of translation and editing. But those who have heard the tapes in Spanish say they are often quite poetic. And Lewis says: "Despite their lack of formal training, these young people express themselves remarkably well. . . ." [Oscar Lewis, *The Children of Sanchez* (New York: Random House, 1961), p. xii.] Of course, one might well retort: "So do many Harlem Negroes." Has Lewis been unusually selective? In any case, we may ask how two environments so much alike in their poverty and related conditions can produce groups of individuals so different in their command of language. Surely socialization practices played some role.

⁶ These have been briefly summarized by Frank Riessman in *The Culturally Deprived Child*. New York: Harper & Row, 1962.

⁷ I have in mind their finding, reported in *Inner Conflict and Defense* (Holt-Dryden, 1960), that middle class boys more often adopted a conceptual mode of expression, whereas the motoric mode was more typical for working class boys.

⁸ These results are presented in an unpublished senior honors thesis by William W. Lawrence, Department of Social Relations, Harvard College, April 1965.

⁹ Jacob W. Getzels and Philip W. Jackson, "Family Environment and Cognitive Style: A Study of the Sources of Highly Intelligent and of Highly Creative Adolescents," *American Sociological Review*, XXVI (1961), 351–359.

¹⁰ See Alex Inkeles and Raymond Bauer, *The Soviet Citizen*. Cambridge: Harvard University Press, 1959.

¹¹ Any discussant of this topic is necessarily heavily indebted to Basil Bernstein's pioneering work in the development of socio-linguistics. See "Some Sociological Determinants of Perception: An Inquiry into Sub-Cultural Differences," *British Journal of Sociology*, Vol. IX (1958) and "Language and Social Class," *British Journal of Sociology*, Vol. XI (September 1960).

¹² Barry Herbert, A. Irvin Child, and Margaret Bacon, "Relations of Child Training to Subsistence Economy," *American Anthropologist*, LXI (1959), 51–63. Also see David Aberle, "Culture and Socialization," in F. S. Hsu (ed.), *Psychological Anthropology*. Homewood, Ill.: Dorsey, 1961.

¹³ See Joseph Veroff, et al., "The Use of the Thematic Apperception Test to Assess Motivation in a Nationwide Interview Study," *Psychological Monographs*, Vol. LXXIV, (1960).

¹⁴ See B. C. Rosen, "The Achievement Syndrome," *American Sociological Review*, XXI (1956), 203–211; F. L. Strodtbeck, "Family Interaction, Values, and Achievement," in D. C. McClelland, et al. (eds.), *Talent and Society*. New York: Van Nostrand, 1958; D. R. Miller and G. E. Swanson, *op. cit.*

¹⁵ This should be readily apparent to anyone who consults the review

by Irving I. Sigel, "The Attainment of Concepts," in *Review of Child Development Research*, Vol. I, M. L. Hoffman and L. W. Hoffman (eds.). New York: Russell Sage Foundation, 1964.

[16] G. Stern, et al., *Methods in Personality Assessment*. Glencoe: Free Press, 1956.

[17] M. Rokeach, *The Open and Closed Mind*. New York: Basic Books, 1960.

[18] Getzels and Jackson, *op. cit.*

[19] Miller and Swanson, *op. cit.*

[20] Martin L. Hoffman, "Report of Research Sponsored by N.I.M.H." (Merrill-Palmer Institute, October 1964). Mimeographed.

[21] Adorno, et al., *The Authoritarian Personality*. New York: Harper & Row, 1950. Also see Else Frenkel-Brunswik and J. Havel, "Authoritarianism in the Interviews of Children," *Journal of General Psychology*, LXXXII (1953), 91–136.

[22] Beatrice B. Whiting (ed.), *Six Cultures*. New York: John Wiley & Sons, 1963.

Twenty

Age and Mental Abilities: A Longitudinal Study

William A. Owens, Jr.

This is a report on the development of intelligence in a group of men tested in 1919, 1950, and in 1961. The subjects were 127 male freshmen of 1919, who had originally taken the Army Alpha Test, Form 6, as an entrance examination at what was then Iowa State College. They were retested during 1949–1950, with the identical examination accompanied by a Personal Information Sheet composed of twenty "open-ended" items. This initial study covered the subjects' mean ages nineteen to fifty. A follow-up was undertaken after approximately another decade, at the subjects' mean age sixty-one. In addition to retesting with the Alpha, a scored life history blank was also administered in an attempt to reveal characteristics or patterns of living which are typical of those whose mental abilities age has treated more kindly versus less kindly.

I. 1951 STUDY

A. PROBLEM AND METHOD

The problem of course was to determine the effect of a thirty-year age increment upon the mental functions measured by the eight subtest scores and by the total score of Army Alpha Form 6. Of concern also were (1) the effect of the given age increment upon the relative magnitudes of individual differences and trait differences and (2) the relationship of certain personal-social variables to the observed temporal shifts in subtest scores and total Alpha score.

The study reported here has been summarized and abbreviated from three publications. First, "Age and Mental Abilities: A Longitudinal Study," *Genetic Psychology Monographs 48:* 3–54, 1953. Second, "Age and Mental Abilities: A Second Adult Follow-Up," *Journal of Educational Psychology 57:* No. 6, 311–325, 1966. Third, "Life History Correlates of Age Changes in Mental Abilities," Cooperative Research Project No. 1052, Purdue University, Lafayette, Indiana, 1963.

These studies have been summarized by Don C. Charles with the permission of the original author, William A. Owens, and of the publishers.

The method employed involved the retesting during 1949 and 1950, of 127 males to whom the Army Alpha Form 6 had been administered as an entrance test at the Iowa State University during January of 1919. In addition to repeating the test each subject was required to complete a twenty-item Personal Information Sheet covering many of the presumed personal-social correlates of age changes. The subjects retested represented approximately 65 percent of the total number potentially available. Of these, approximately one-half were still resident in Iowa and were retested by representatives of the college; the other half were scattered throughout the United States and its territorial possessions and were retested by examiners resident in their vicinities.

All test scores were reported in normalized standard score form from norms for a thousand comparable cases in order to obtain reasonably comparable units of measurement and to provide a stable referent for the evaluation of shifts in performance levels. The significance of the effect of the given age increment upon the magnitude of individual differences in a given function was appraised by obtaining an initial (1919) and a final (1950) variance and by testing for the significance of the difference between the correlated variances. The comparable effect of age upon the magnitude of trait differences was evaluated by computing two "V" scores for each individual, one for each testing, equal to the variance of his eight standard scores on the given occasion. A distribution of the differences between V scores was then obtained and its mean tested for the significance of its departure from that of a distribution with mean zero. The analyses of the personal-social correlates of the thirty-year shifts in mental abilities were made by attempting to relate responses on each item of the Personal Information Sheet successively to each of the nine series of D-scores; that is, 1950–1919 scores for each of eight subtests and the total score.

B. RESULTS

The results of the analyses to determine the effects of the thirty-year age increment upon subtest scores and total score appear in Table 20.1. The column headed $Mn_{50}-Mn_{19}$ contain the differences between the 1950 mean scores and the 1919 mean scores in units of the norm sigma. It will be noted that scores on the retesting were significantly *higher* for four subtests and the total, and that the apparent direction of the change was *positive* on all save one subtest, arithmetical problems. The magnitudes of the shifts range from approximately zero to over nine-tenths of a sigma, at the mean.

Table 20.2 contains the results of the analyses to determine the effects of the given age increment upon the magnitudes of individual differences in the several functions. It should particularly be noted that there was a

TABLE 20.1   Temporal Shifts in Subtest and Total Alpha Scores

Content	$Mn_{50} - Mn_{19}$	t
1. Following Directions	0.042	0.43
2. Arithmetical Problems	−0.101	1.53
3. Practical Judgment	0.542	8.22**
4. Synonym-Antonym	0.549	8.26**
5. Disarranged Sentences	0.633	7.29**
6. Number Series Completion	0.001	0.01
7. Analogies	0.143	2.02*
8. Information	0.931	17.44**
Total score	0.550	11.17**

Norm Mn = 5 & σ = 1.
** = P < .01; * = P < .05.

significant decrease in individual differences on subtest five (Disarranged Sentences), a significant increase on subtest seven (Analogies), and "probably significant" (p > .05) increase on the total score.

Analysis of the twenty personal-social correlates revealed significant relationships between D-scores and present chronological age, years of education and field of specialization; pre-college residence and rural-to-urban migration; number of hobbies and recreations; and earned income.

C. Conclusions

Within the limitations of the methods and circumstances of the present investigation, the following conclusions seem warranted.

1. With respect to the effects of the thirty-year age increment upon subtest and total Alpha scores:

TABLE 20.2   The Effects of Age upon Individual Differences

Content	$1919a^2$	$1950a^2$	t
. Following Directions	0.7972	0.8433	0.329
. Arithmetical Problems	0.8071	0.9211	1.027
. Practical Judgment	0.6483	0.5976	−0.550
. Synonym-Antonym	0.7900	0.7530	−0.349
. Disarranged Sentences	1.1155	0.6827	−3.164**
. Number Series Completion	0.8559	0.6986	−1.451
. Analogies	0.5149	0.8697	3.569**
. Information	0.5177	0.4479	−1.040
Total score	0.5890	0.7431	2.055*

* = P < .01; * = P < .05.

a. There were significant *increases* in score on the Practical Judgment, Synonym-Antonym, Disarranged Sentences and, Information subtests which were of the order of magnitude of one-half to nine-tenths of a sigma from the original mean.

b. There was a significant increase in the total Alpha score which was of the order of magnitude of one-half of a sigma.

c. There was an increase in score, significant at the 5 percent level, on the Analogies subtest, but this increase was of the order of magnitude of only one-tenth of a sigma.

d. There were insignificant changes in score on the Following Directions, Arithmetical Problems, and Number Series Completion subtests.

e. There was no significant decrease in score on any subtest.

2. With respect to the effects of the given age increment upon individual differences and trait differences:

a. There was a significant decrease in the magnitude of individual differences as measured by the Disarranged Sentences subtest.

b. There was a significant increase in the magnitude of individual differences as measured by the Analogies subtest.

c. There was an increase, significant at the 5 percent level, in the magnitude of individual differences as measured by the total Alpha score.

d. Trait differences remained remarkably constant.

3. With respect to the personal-social correlates of the observed temporal shifts in test scores (1950–1919):

a. The younger subjects tended to make the higher total D-scores

b. On the Analogies subtest, and on the total, subjects with over five years of college had significantly higher mean D-scores than those with less than four years of college.

c. On the Synonym-Antonym subtest, there was significant variation among the mean D-scores of subjects who had specialized in agriculture, engineering, or "other" curricula.

d. On the Disarranged Sentences subtest, subjects who had come to college from rural areas had significantly higher mean D-scores than those who had come to college from urban areas.

e. Also on the Disarranged Sentences subtest, subjects who had migrated from rural areas (pre-college) to urban areas (post-college) had significantly higher mean D-scores than those who had not so migrated.

f. On the Synonym-Antonym and Information subtests, there was significant variation among the D-score means of those subjects having 0 or 1, 2 or 3, or 4 or more hobbies and/or recreational activities. On the Information subtest, those with the largest number of hobbies and recreations had significantly higher mean D-scores than did those with the smallest number. This difference closely approached significance on the Synonym-Antonym subtest as well ($t = 1.91$).

g. D-scores on the Number Series and Analogies subtests were significantly positively correlated with reported earned income for 1948.

II. THE SECOND ADULT FOLLOW-UP

The second adult follow-up extended the coverage approximately another decade to mean age sixty-one. The problems to be investigated were three in number: (1) to describe the impact of an additional ten-year age increment on the mental abilities measured by the Army Alpha; (2) to appraise the effects of the given age increment on the magnitudes of individual differences and trait differences as measured; and (3) to reveal the scored life-history characteristics or patterns of living which are typical of those whose mental abilities age has treated more kindly versus less kindly.

A. METHOD

The problems to be attacked were (a) noting shifts over time in the levels of mean verbal (V), numerical (N), and reasoning (R) components of Alpha or its subtests; (b) comparing earlier and later variances across tests and across people; and (c) item-analyzing an extensive scored life history against criterion changes in the Alpha component scores from 1919 to 1950, and from 1950 to 1961, with subsequent factoring of the "valid" items.

Subjects Subjects were 96 males from the earlier study of mean age sixty-one in 1961. This subject group represents a 75 percent sample of the 127 first retested during 1950. Of the 31 lost, 13 were deceased, 5 were disabled, 5 could not be located, and 8 refused to participate.

Measuring Instruments Two measuring devices were utilized in the study. The first is the Army Alpha examination Form 6 as used in the first study. The second device was a so-called life experience inventory especially constructed for project purposes. It contained 228 objective items of which 115 appear to have their options on a continuum and 113 do not.

Data Collection All subjects were first invited to attend a class reunion and banquet held in conjunction with the project. Of the 96 ultimate participants, 28 attended this reunion at which one-half day of their time was used for test and questionnaire administration. Subjects unable to attend the reunion were retested by examiners resident in their vicinity.

Component Scores   In the interest of parsimony, shifts in Alpha scores are reported in terms of the verbal (V), numerical (N), and reasoning (R) components reported by Guilford (1956) wherever this is appropriate. As a consequence, life history items potentially diagnostic of shifts in intellective abilities were selected for their demonstrated relationship to a temporal shift in one or more of these component scores or the total.

Statistical Analysis   As previously, all computations involving the test scores were performed on the normalized standard scores for the subtests and total or on the component scores derived from them (norm mean $= 5$ & $\sigma = 1$). Shifts in mean score are thus reported in units of the norm standard deviation. Significance of these shifts was evaluated through application of the T-test to a given distribution of different scores to determine whether or not its mean departed significantly from that of a distribution with mean zero.

An analysis of the life history correlates of the shifts in Alpha scores over time was made by relating responses on each item of the life experience inventory successively to each of the eight criterion segments (V, N, R, or T for either 1919 versus 1950, or 1950 versus 1961). In the case of the 113 noncontinuum items each option was regarded as a dichotomy, checked or unchecked; the given criterion was also dichotomized and a direct binominal test of significance was applied.

In the case of continuum items, product-moment correlations were run between item score and each of the eight criterion scores. Those items which did not correlate with at least two criteria above the 20 percent probability level, or with one criterion above the 10 percent level, were eliminated. A test was then made to determine the probability of obtaining the given number of significant items by chance.

B. RESULTS

Level of Abilities   The first major finding is that the Alpha scores of these 96 subjects remained remarkably constant over the period from 1950 to 1961. Table 20.3 contains the mean differences and associated t-values. At the subtest level none of the differences is significant.

Of the apparent "gains" five are negative and three positive. In the case of the components, the largest and third largest subtest losses have combined to produce the significant downward shift in the N component. On the total score however, the t-value does not closely approach significance and the apparent loss is of the order of magnitude of only

TABLE 20.3   Temporal Shifts in Subtest, Component, and Total Alpha Scores

Subtest or component	$\bar{x}\ 1961-$ $\bar{x}\ 1950$	t
1. Following Directions	.001	.012
2. Arithmetical Problems	−.237	−.436
3. Practical Judgment	− 118	− .192
4. Synonym-Antonym	−.049	−.072
5. Disarranged Sentences	−.036	−.052
6. Number Series Completion	−.084	−.116
7. Analogies	.013	.025
8. Information	.043	.089
Verbal Component	−.032	−. 90
Numerical Component	−.203	−4.38**
Reasoning Component	.015	.28
Total	−.047	−.138

Note. −M = 5; SD = 1.
** = p < .01.

one-tenth of a standard deviation. Figure 20.1 shows this trend in component scores and the total over the forty-two-year period of the three
testings. Since the true functions cannot be determined from a three-point
estimate it seems most parsimonious to connect these points with straight
lines. However, the argument from the data is clear: *Accelerating declines
in these abilities begun thirty to forty years ago should be more apparent
in the last decade than these declines are.* The implication is that any

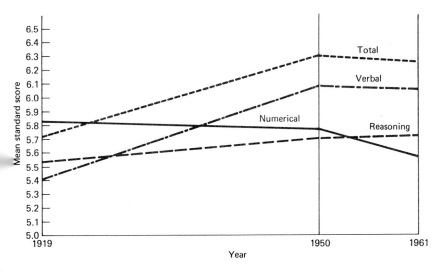

Figure 20.1   Test performance trends for components and total.

declines which have begun, have done so more recently. A possible exception is the case of the N component.

Individual and Trait Differences In examining these differences, a comparison of the variances for 1950 and 1961 reveals that four differences are apparently negative and four positive and that none is statistically significant. It is therefore implied that no significant shift in the magnitude of individual differences took place over the eleven-year period in question. An examination of the effects of the given age increment upon trait differences provides comparable evidence; significance is not approached.

Life Experience Inventory Correlates Eighty-two of the 115 continuum type life history items utilized were found to be related to shifts in test score over time. The probability of obtaining 82 significant items from a pool of 115 by pure chance is far less than .0001.

Factor Analysis Eighty-two life history items plus the eight criteria constituted the entries in a 90 x 90 correlation matrix which was factored by the method of principal axes (Harman, 1960) with adjusted communalities on the main diagonal. Twenty factors were extracted from this matrix: five rotated to a biquartimin criterion of oblique simple structure. Since orthogonal and oblique solutions were virtually identical, percents of variance accounted for *are* quoted.

A concensus of five judges who reviewed these factorial results was employed in the interpretation and naming of the factors.

Factor I This factor, which accounts for 12 percent of the common factor variance, was designated as Socioeconomic Success. Responses of high scorers on the factor to the seventeen items with loadings above 0.30 imply a high level of occupational and financial success. There are understandable overtones of what might be called "executive attitude" exemplified in a desire for minimal external control, acceptance of responsibility, and a high drive level at work.

Factor II This factor, which accounts for approximately 8 percent of the common factor variance, was designated as Sensitive Intelligence. Responses of a high scorer to the thirteen items with loadings above 0.30 evidences above-average intelligence in his devotion to reading, his superior memory, and his production of ideas judged worth recording. Coupled with these is a syndrome of emotionality and sensitivity which may be inferred from a stated youthful tragedy, some tenseness and nervousness, some dissatisfaction with self, some perseveration with problems

Factor III The judges described and labeled this factor as Physical Vigor. It accounts for a little less than 8 percent of the common factor variance. Responses of the high scorers on the factor to the fourteen items with loadings above 0.30 relate to physical vigor, work attitudes, self-perceptions, and health history. All suggest the appropriateness of the assigned name.

Factor IV Introversion was the label assigned to this fourth factor, which accounted for about 6 percent of the common factor variance. Responses of high scorers on the factor to the fifteen items with loadings above 0.30 seem clearly relevant to the implied lack of social enjoyment and activity, somewhat negative attitude towards others, possibly overlong association with mother, and suggestions of poor health, low energy level, and social reticence.

Factor V The fifth and final factor was called Egocentric Independence, and it accounted for a bit less than 6 percent of the common factor variance. A high scorer on the factor to the eleven items with loadings above 0.30 would appear to be characterized by some selfishness and solitariness, by a preference for working alone, by an absence of religious belief, and by a feeling that he has done well, probably better than his father and better than he himself expected.

Factor Criterion Relationships Table 20.4 contains critical evidence regarding relationships of Alpha component gains, representing intellective changes, to life-history factors, representing demographic characteristics and experiences. From the data available the following general statements can be made:

Factor I The subjects who stand high on Factor I, Socioeconomic Success, tended to show above average increases in Total score and Reasoning score from 1919 to 1950.

Factor II The second factor, Sensitive Intelligence, is the most important factor in the sense that it contains four criterion loadings greater than .20. Individuals scoring highly on Sensitive Intelligence tended to

TABLE 20.4 **Loadings .20[a] of Alpha Component Gains on Life-History Factors**

	Verbal component		Numerical component		Reasoning component		Total score	
Life-history factors	P_1	P_2	P_1	P_2	P_1	P_2	P_1	P_2
I. Socioeconomic Success	.52				.28		.23	
II. Sensitive Intelligence					.31	.22	.69	
III. Physical Vigor			.35					
IV. Introversion								.29
V. Egocentric Independence		−.50	.36	.25				

Note. — $P_1 =$ 1919–1950; $P_2 =$ 1950–1961.
[a]This level was chosen because of the priorly noted unreliability of the component gains.

show substantial increases in Total score, Verbal score, and Reasoning score from 1919 to 1950, and Reasoning score from 1950 to 1961.

Factor III   People described by Factor III, Physical Vigor, showed a substantial shift on only one criterion—an increase in Numerical score from 1919 to 1950.

Factor IV   People characterized by Factor IV, Introversion, also showed a substantial shift on only one criterion—an increase in Total score from 1950 to 1961.

Factor V   Individuals described by the last factor, Egocentric Independence tended to show a better than average increase in Numerical score for both time periods and also a substantial decrease in Verbal score from 1950 to 1961.

Separating Cultural Change from Maturational Change   Some years ago Kuhlen (1940) pointed out that cultural changes might well be mistaken for age changes in the context of a cross-sectional study. Clearly, the longitudinal study may also be distorted by cultural factors. Projecting intellective changes found in this group from 1950–1960 into those expected of comparable age groups in future decades must be conditioned on the assumption of comparable cultural evolution over the later time period; this is of course unrealistic. Therefore, it seems reasonable to ask what a given age increment may be expected to do to the mental abilities of individuals in a *constant culture*. To accomplish this requires that one "baseline" the apparent age changes noted by measuring a sample of present nineteen-year-olds comparable to the Ss of this study at their age nineteen. Assuming comparable individuals, one may then attribute group differences observed to cultural change. If these differences are subtracted from apparent or observed changes in the same individuals over time, the residuals should represent "true" or intrinsic changes in these individuals. To fulfill the indicated purpose Alpha, Form 6 was administered during 1961–1962 to a random sample of 101 male freshmen at the same institution attended by Ss of the longitudinal study, that is, Iowa State University.

Table 20.5 contains the relevant data. Column 2 shows total "apparent" gains from 1919 to 1961. Column 3 contains as "cultural changes" the scores of the 1961–1962 freshmen minus those of the freshmen of 1919 Under "true gain," by subtest and component, appear the differences between columns 3 and 4. It is in the case of R that the correction introduces the most dramatic change. The explanation may well be fairly simple. The Analogies test, for example, was probably quite unfamiliar to freshmen of 1919, as were all group tests of intelligence. In 1961 it was not unfamiliar to either nineteen-year-olds or sixty-year-olds. Thus, what appeared as a small intrinsic gain in R for the sixty-one-year-olds may be better accounted for in terms of a vastly increased general level of sophistication with this type of test, in the midst of which the senior Ss really lost ground in the relative sense.

The corrected components seem to be in better accord with both theory and practice than the uncorrected. For example, deterioration is commonly measured in terms of a discrepancy between verbal and reasoning functions.

C. IMPLICATIONS

1. Since Total Alpha score did not change significantly from the subjects' ages fifty to sixty-one, it is implied that programs like adult education, posited on the continued intellectual competence of persons through this age range, are on firm ground.

TABLE 20.5 Estimated "True" or Intrinsic Changes in Alpha Score, 1919 to 1961

Subtest	Apparent gain 1919–1961	Cultural change ±	True gain[a] Subtest	Component
Practical Judgment	.42	.18*	.24	
Synonym-Antonym	.50	.00	.50	.37**(V)
Disarranged Sentences	.60	.88**	− .28	
Information	.97	− .03	1.00	
Arithmetical Problems	− .34	− .31**	− .03	
Number Series				− .18*(N)
Completion	− .08	.63**	− .71	
Following Directions	.05	.03	.02	
Analogies	.16	.90**	− .74	− .61**(R)
Total	.50	.44**		.06

[a] M of 101 male freshmen of 1961 minus M of 96 male freshmen of 1919.
* = $p < .05$.
** = $p < .01$.

2. Several implications follow from the fact that age had a differential affect upon Verbal, Reasoning, and Numerical component scores which showed an increase, constancy, and a decrease, respectively, from 1919 to 1961. (See Figure 20.1.)

a. If an "age-free" test for the adult years were desired, some sampling of reasoning functions would be most likely to provide it.
b. If one were concerned with diagnosing "psychometric age," a discrepancy between verbal and numerical abilities would provide the best index to it.

c. In training adults a cognitive appeal through verbal channels would utilize their most age-favored ability.

d. Insofar as mean shifts in abilities are concerned, the most appropriate jobs for adults would be those which place increasing premium upon verbal abilities and a decreasing premium upon numerical.

3. Since individual differences do not appear to increase with age, it is *not* to be expected that age will treat the initially more able any more kindly than the initially less able.

4. Since trait differences do not shift with age, it must *not* be assumed that age will treat more kindly those functions which are practiced, in an occupation or elsewhere, such as the numerical skills of an accountant.

5. Overall, since there *are* many more significant life history correlates of age changes in mental abilities than could reasonably be attributed to chance, it is implied that experiential patterns of living *do* play a significant role in at least this aspect of the aging process.

a. In terms of the saturation of discriminating items in the totals of this type, it is implied that the life history areas of Education, Occupational Endeavors, and Interests are particularly important.

b. In terms of the factorial analysis of discriminating life history items the following seem to be the clearest and most important implications:

(1) Intelligence, accompanied by some emotional-motivational component involving youthful trauma, tenseness, perseveration, and self-dissatisfaction argues for an increase in Total, Verbal, and Reasoning scores to age fifty and for a further increase in Reasoning through age sixty-one.

(2) Physical Vigor, accompanied by Egocentric Independence, argues for an increase in Numerical score through age sixty-one. It may well be that vigorous manipulation of the problem constitutes a linkage between physical health and problem-solving ability.

REFERENCES

Guilford, J. P. *Psychometric methods.* New York: McGraw-Hill, 1954, pp. 523–525.

Harman, H. H. *Modern factor analysis.* Chicago: University of Chicago Press, 1960.

Kuhlen, R. G. "Social change: a neglected factor in psychological studies of the life span," *School and Society* 52: 14–16, 1940.

Twenty-one

Mid-life Attainment of the Mentally Retarded: A Longitudinal Study

Warren R. Baller, Don C. Charles, and Ebert L. Miller

One of developmental psychology's interesting questions is, "What becomes of persons who, during childhood and youth, are identified as mentally retarded?" This question prompted a series of studies beginning in 1935 and continuing to the present. The major endeavor in the earlier studies was largely that of securing information about location, economic status, and social adjustment of a group of persons so described. The present research not only was planned to update information about the conditions and factors reported in the early studies but to go beyond these kinds of descriptions to the question of the antecedents of the changes apparent in the lives of the members of the several groups.

I. BACKGROUND OF THE STUDY

The present research originated in a study conducted by Baller in 1935. The subjects were 206 individuals who had been in the "Opportunity Rooms" of the Lincoln, Nebraska, Public Schools from the time of classification until the termination of their public school education. Each individual had been classified as mentally deficient on the basis of an individual intelligence test score of 70 I.Q. or below, and failure to do acceptable work in the regular school classes. For more than three-fourths of the individuals, there were at least two individual mental test ratings on file with no rating above 70 I.Q. These persons were matched on the basis of sex, race, ethnic origin, and age with control individuals whose I.Q.s ranged from 100 to 120 on the Terman Group Test of Mental Ability. In this first study, Baller completed family and individual subject interviews as well as investigations of court and social agency records. In

W. R. Baller, D. C. Charles, and E. L. Miller, "Mid-life attainment of the mentally retarded: A longitudinal study," *Genetic Psychology Monographs* 75: 235–329, 1967. The report of this study has been abbreviated for inclusion with these readings by the second author, with the permission of the other authors and publisher.

1953, Charles reported a follow-up study on 151 of the subnormal group identified in Baller's research. The subjects were again studied in 1961–1962 by Miller.

A different group of individuals whose I.Q.s ranged from 5 to 85 were studied by Baller in 1937. There were 307 persons who comprised this "dull" group.

By the early 1960s, the subjects of the studies just mentioned were well into mid-life, with average age in the fifties. Presumably they had become what they were to be as mature adults.

Purpose and Rationale of the Study

Most of the literature pertaining to mental deficiency is concerned with diagnosis, etiology, and situational problems. To the question asked earlier, "What becomes of retarded children" there have been few answers supported by evidence. These efforts have been primarily descriptive and have not been much concerned with cause. It was the purpose of the current research to bring up to date the records covering very nearly thirty years of the lifespan of the persons comprising these Nebraska studies and to attempt to answer some of the "why" questions, as well as the "what happened" questions.

II. PROCEDURE OF THE STUDY

The first requirement of the study was to locate the subjects of the Baller studies and also persons of a comparison group. Then their present social and intellectual status was to be determined. Finally, through interviews and examinations of the life records, antecedents of good and poor adjustment were to be inferred.

A. Subjects

1. *Low Group Subjects* These were originally 206 persons identified as mentally deficient in 1935, and described earlier in this report. A total of 61 males and 48 females were found with a mean age of fifty-six years. The sex ratio was approximately the same in 1935, 1951, and 1964. Race and ethnic origin of subjects suggests that no marked selection took place in the follow-up studies and the mean of original I.Q. scores varied little in the three study periods. On the basis of available evidence subjects of the present study seemed to be representative of the original population.

2. *Middle Group Subjects* These were originally 206 persons described as "dull" (I.Q. 70 to 80) in 1937, and matched with the above group. A total of 152 Middle Group subjects were located. Again the sex ratio remained identical to the early study. The racial and ethnic distribution remained quite similar to the original and little difference was observed in the mean intelligence of the present group as compared to the original group.

3. *High Group Subjects* This was originally selected as an "average" population in 1935, and used as a control for the study of the Low Group. One-hundred-twenty High Group students were located at a mean age of fifty-three years and again the sex ratio and mean intelligence remained about the same as the original. The ethnic distribution differed in that there was a deficiency of "German-Russian" subjects. On the whole it would be safe to say that the present studied subjects were not unlike the original in most ways.

B. Procedure

Subjects of all three groups were sought out and their present social status was evaluated. Some subjects from each group were given an intelligence test. All subjects available, that is, living in the area and willing to cooperate, were interviewed or filled out life-history blanks. Social status of each group was reported in tabular form and comparisons were made between groups. Test scores were reported and ability estimates were made on the Low Group from accumulated evidence. Life-history and interview data were presented for each group separately and compared. Some "most successful" and "least successful" males and females were selected out of the Low Group, and all available evidence was examined for clues to good and poor adjustment. Special reports were made on these individuals from the Low Group.

III. FINDINGS OF THE STUDY

A. Life Adjustment

1. *Location* More than 65 percent of Low Group subjects had remained in or returned to the home community, probably thereby easing or avoiding some adjustment problems. About 40 percent of the other two groups had remained.

2. *Institutionalization* Very few subjects from either of the lower groups were institutionalized (only 9 out of 247; none of these was a recent institutionalization) and none out of the High Group.

3. *Mortality* Both Middle and High Groups had low death rates. The Low Group had a very high rate—nearly a third of "located" subjects were deceased. Death rate, especially accidental death, had been lower since the 1950 study than earlier.

4. *Marital Status* Divorce rate was from two to 10 times above the national average in the two lower groups. Percentage of Middle and High Group subjects married and living with spouses compared favorably with national average, but Low Group subjects had less success than other subjects in getting and keeping mates and were consequently more likely to be living alone.

5. *Economic Status* Although unemployment and dependence on public support was somewhat higher for Low Group subjects than the national average, they had shown steady improvement in self-support over the years. In 1935 only 27 percent were found to be fully self-supporting and 57 percent more partially self-supporting. By 1950 the figures on complete self-support had risen to 36 percent. By the time of the current study, self-support had risen to 67 percent with another 16 percent needing only some relief help to get along. Nearly 80 percent of the group was described as usually employed, and half had been continuously employed at the same job for some years. Most jobs were in the labor and service category, but there were some in the higher occupational ranks.

Middle and High Group subjects were almost all self-supporting. Jobs of Middle Group resembled those of Low Group, while High Group subjects had a much greater proportion of professional and business level jobs with few labor and service listings.

6. *Law Conformity* All three groups were generally law abiding. Less than 10 percent of the two lower groups had convictions on civil offenses and most of these were typically lower class (for example, drunk and disorderly) rather than vicious or truly antisocial. High Group subjects had no convictions.

7. *Social and Recreational Activity* As anticipated, membership and active participation in community activities reflected class, education, and intelligence. "Zero membership" in community clubs and the like ranged in stairstep fashion from High to Low Groups: 23 percent, 48 percent, 60 percent. Participation in these activities reversed that order: Low Group, 4 percent; Middle Group, 7 percent; High Group, 33 percent. It

was noted, however, that even in the Low Group 40 percent belonged to some organizations and a few spent some time in club or social organization activity.

B. Ability and Performance

1. *Low Group* These subjects originally (1935) averaged about 60 I.Q. Upon being retested with the Wechsler Bellevue in 1950, a sample averaged near 80 I.Q. The total group was categorized on test scores and performance: 20 percent clearly defective, 10 percent testing low but managing to get along in society, 65 percent dull normal or average, and about 5 percent physically handicapped.

A sample was retested for the present study. The sample maintained its 1950 status with slight but nonsignificant gains in scores. On the basis of recorded data and personal acquaintance with subjects' lives a "clinical-social" evaluation was made of all subjects with these results: (a) permanently retarded: 14 percent; (b) low test scores but "getting along" in society: 14 percent; (c) below average or borderline (A.A.M.D.—1 level): 46 percent; (d) average or better: 24 percent; and (e) victim of circumstances (multiple handicaps): 2 percent.

2. *Middle Group* A small sample of this group (18) was retested and showed a mean rise from 80 I.Q. in 1935 to 88 I.Q. in the current study. Because both were group tests and the sample retested was so small, there seemed little reason to doubt the accuracy of the original description of "dull" (—1 A.A.M.D. level). Insufficient evidence on lives was available to make clinical evaluations of ability as was done in the Low Group.

3. *High Group* This originally was supposed to be an average "control" population. Mean I.Q. of 107 was secured on these subjects in the 1920s and again in the current study on a small sample (18 cases). General social performance of the total group suggested that the test sample may have been of lower ability than the true mean of the group. However, the test evidence available gives no grounds for doubting the "average" appellation.

C. Life History

Life-history information was secured for subjects of each of the three groups through paper-and-pencil questionnaires or interviews. Life experiences were described for each group and between-group comparisons were made.

Desirable experiences ranged downward in the three groups in stair-step fashion with the Low Group on the bottom. The subjects of this group suffered in comparison with both of the two higher groups on such experiences as having had health and appearance instruction, having a source of advice in general, likelihood of growing up with own parents, having mother at home, going to Sunday School, having had parents interested in life work, amount of education and vocational instruction and the like. The bleakness of the early experiences of this group suggested the term "culturally deprived."

Some useful, "pin-pointed" evidences were derived from the relatively intensive study of the small samples of "most successful" and "least successful" Low Group subjects. Successful as compared to unsuccessful males were likely to have acquired a skill early and worked at it continuously, probably with a large paternalistic employer like a railroad rather than by trying a variety of occupations. They were likely to have stayed in one community rather than having drifted about. Successful females were likely to have learned principles of good grooming and health care early, to have married well, and to be working steadily. Unsuccessful females generally had learned habits of dependence and attachment to mother and home.

D. Case Reports of Relatively Successful and Relatively Unsuccessful Low Group Subjects

As Mary Cover Jones has said in commenting on some individuals in the California Growth Studies, "Neither computers, case numbers, group averages nor theoretical formulations can separately, or combined, submerge the individual *if you know the data as persons*." (Italics added, see Jones, 1964.) The authors have come to know these data as persons and it is their hope that case reports can supply some understanding about Low Group subjects that will effectively supplement the data given elsewhere in the report. Group averages and similar kinds of descriptions tend to obscure the "flavor" of individual life. The purpose of the life-history sketches that follow then, is to identify in them certain characteristics and associated factors that tend with considerable regularity to contrast the relatively successful with the relatively unsuccessful members of the low mentality group. Four different "portraits" will be presented from a large accumulation of the authors. First, there is a description of a relatively successful male followed by one of a relatively unsuccessful male; then, there is a portrait of a relatively successful woman followed again by an unsuccessful woman. While no one individual can represent such a large and diverse group, each of these case reports contains many of the features and conditions descriptive of members of the larger sample from which this was drawn.

Selection as "relatively successful" or "relatively unsuccessful" was made on the basis of the subject's occupation, home and family life, the status of his activity in his community, and the quality of his relationship to life in general. Social data records, interview forms, and the researchers' personal acquaintance with the subjects were used as the basis for selection. At the beginning of each of the four portraits will be found a name (a pseudonym) plus a number of parentheses followed by a date. This number represents the highest intelligence quotient recorded for the subject during the time he was in school. The date that follows the parentheses is the individual's birthdate.

1. A relatively successful low ability male. Jack M. (68) 1903. Jack's life history as mentioned above has a number of features in common with the larger proportion of relatively successful Low Group males. His first work (except for after school odd jobs during his early teens) was for a well-established manufacturing firm. After nearly seven years of employment with this company and largely because of his record of trustworthiness and conscientious effort even in quite unskilled tasks, he secured a steady job in the equipment building shops of a major railway company. He had remained in this work for more than twenty years when he was interviewed in 1963. Securing work with a relatively large and well-established company is the first of the features alluded to above.

A second feature of some apparent importance in Jack's record is that of acceptance of himself as a worthwhile person because of his willingness to "stay by a job" and do his best in it "even if it wasn't much to look up to." The relatively successful males of the Low Group quite consistently exhibit this attitude toward work.

Jack did not marry early ("I guess I was too busy and didn't have the money"). He was twenty-nine when he married. The woman he married was some three years younger, had completed two years of high school, was well trained by her own mother in housekeeping, enjoyed church attendance, and had considerable interest in music. She was described by one interviewer as "very clean and nice appearing."

The fact that Jack's marital life has been characterized by stability and congeniality constitutes a fairly consistent common denominator. He and his wife and their one son (now grown and father of two children) were, and are, regular in church attendance. Jack played in a small band "combo" as a drummer largely for his own enjoyment until quite recently after "teaching myself" nearly thirty years ago. Jack's wife works as an "ironing woman" for a successful and long established laundry; she has remained at this job since leaving school to marry Jack.

Jack and his wife share several recreational and other avocational interests. He mentioned with quite evident pride his membership in the "Moose Lodge"; he had considerable fishing equipment on display to prove his devotion—and that of his wife—to this form of sport; and he emphasized the regularity of his attendance at "The Railroad Veterans' Club."

Neither Jack nor his wife has any history of serious illness or accidents. The remark, "a couple of little operations—that's all," appears to sum up the family medical history.

At the time of this writing Jack and his wife live in what the interviewer described as, "a nice, neat, well-kept-up six-room house that has an attractive back yard." The property is owned by Jack and his wife who occupy three rooms and rent the other three to another married couple. The house would be valued from $9000 to $10,000 on the present Lincoln market. Jack and his wife have lived at this one address for thirty years—practically their entire married life.

The comparatively long residence in one place is still another life history feature that characterizes the more successful of the Low Group males. Remaining in the same city but moving fairly regularly is descriptive—on the other hand—of the less successful males.

In one respect Jack differs from the majority of the relatively successful low ability males. He disavows any feeling that there were certain persons who were particularly important for the way they influenced his life —none, that is, except his wife. Additionally (and unlike most of the successful males) he finds nothing of any significance to attribute to his school experience. He remarked, "I went to school just because I had to."

2. A relatively unsuccessful low ability male. Carl W. (54) 1909. When Clarence W. (Carl's father) talked about his ten children, he was not hesitant about singling out one of them as "no good." That child was Carl. Today Carl has similarly rejective feelings about his parents for he does not view them as making any contribution to his life except for providing the basic essentials until he quit school in the third grade. In Carl's eyes he is a self-made man, and he owes all that he is to himself. He taught himself what he needed to know about work, personal appearance, and life in general; he thinks so little of the guidance of his parents that he can no longer remember what his father's occupation was.

Similarly he thinks so little of the contribution of the schools that he can no longer utilize anything that he was taught in the schools or even remember how he felt about them. School for Carl was one frustrating experience after another, for he could not or did not want to learn anything offered by the school. When the expectations became too demanding in the third grade he simply quit attending with no regrets at the time or now.

He says as a description of his entry into the working world, "I quit school while I was in the third grade and grabbed the first train out of here and I've been on my own ever since." Being on his own has meant a succession of jobs in definitely unskilled labor connected with farming, construction, and mining. These jobs have taken him all over the country. Although he has always returned to his childhood home, he has continued to be highly mobile and has lived in from three to five different cities in the past ten years.

In describing himself as a self-made man, Carl says, "I ain't never asked no one for nothing." He sees no contradiction between his denial of any outside influences or help in his life and his total support today by welfare agencies. As is often the case in such families, Carl continued the pattern characteristic of his parents who relied on welfare support for more than half of their subsistence. Carl claims his right to such support because, "I hurt my back in the mines and two years ago I fell off a ladder and since then I can't do nothing. We're getting $254 a month from the government and my lawyers are fighting for the $4000 I got coming from compensation."

Carl is presently living with his second wife in a rented home. They have five children and "Two hundred fifty dollars a month don't go far with these five kids of mine." Totally dependent upon outside sources for their subsistence, the W. family is living in a precarious economic situation with little anticipation of increased income. Carl's life has followed a pattern characteristic of a large percentage of the relatively unsuccessful low ability subjects. It is a pattern of early withdrawal from school, transient residential status, unskilled labor, and reliance on social welfare for much of life's requirements. His success has been minimal.

General Inferences from the Records of Eight Males   The cases selected as the *most successful* males were men whose lives were characterized by stability. Each had acquired a particular skill early in his life and had worked at it with almost no change for years. Two were self-employed after an early "apprentice" period of working for someone else. The other two were employed by large organizations. There would seem to be significance in the fact that the men with the best records for the Low Group in general—not just this select sample—had in a sizable percentage of cases worked throughout life for one large firm or organization (the Burlington Railroad most frequently). Many, at the time of this writing, were looking forward to company-supported retirement with these kinds of organizations.

The successful males tended to stay in a single (that is, the same) community throughout their adult life. In some instances, however, this was not the community of the individual's childhood. They all married well; they married better educated, personally "well adjusted" women who contributed stability and planning to their lives. Marriage age varied from twenty-two to thirty-two. From the record they could be said to be rather clearly family and occupation oriented.

The characteristics and associated conditions descriptive of the four least successful males contrasted considerably with those of the more successful males. They personified instability rather than stability. Two were described as emotionally unstable. One of these was said to be brain damaged ("kicked by a mule") but there was no medical evidence of such damage. They showed two striking personal-vocational traits: (a) they possessed no well-developed skills (drifting from one odd job to another and almost never being engaged by a large organization), and (b) they moved frequently from one place of residence to another.

Many of the unsuccessful males outside this sample exhibited the same drifting tendency just mentioned. Some established a sort of "circuit" which they followed, showing up eventually at given places again and again. All four of the least successful male subjects lost their fathers while they were in their teens. The married life of all four has contained much discord and other kinds of trouble.

Subjects in the unsuccessful group were less fortunate than those of the successful group relative to help from their parents or other persons in their vocational planning. The same was true regarding advice about personal appearance and personal health. Without exception they professed to think of no one to whom they could look for constructive examples or for beneficial consultation at any time in their life.

1. A relatively successful low ability female. Evelyn J. (70) 1907. Childhood for Evelyn was a time of severe poverty and deprivation. There were eight children in her parental home, and most of Evelyn's out-of-school hours were spent in helping with the care of her brothers and sisters. Her father, an unskilled truck driver, was kept busy providing the money to meet the bare necessities of life for his family. Her mother was fully occupied with the care of the children; there was little time for any motherly attention to individual ones. The parents, however, seemed to instill in the children some hope for a better life. Evelyn says today, "I owe everything to the folks." The "everything" that she owes may be inferred from the progressively more satisfying life that followed her childhood.

Attendance at school was not a pleasant experience for Evelyn. The difficulties that she encountered in academic work are reflected in her reluctance to discuss her school experiences some thirty years after leaving the "Opportunity Room" at the end of "grade 7." She says only that school was necessary and expresses the view, "Kids should take every advantage nowadays. Wish my folks had made me."

Evelyn began her paid working experience at the age of sixteen. During all but a few months of the seven years that she was employed, she worked as a "chocolate dipper" in a candy factory. The record shows that she was considered "a very good worker." She attributes her success to an "old man who ran the candy company who asked me if I would like to learn to dip chocalets (sic), and so I did and I loved it." Work for Evelyn was enjoyed for its own sake; it was not simply a way to earn a living.

Evelyn married at the age of twenty-three. Since her marriage she has spent her time in caring for a home, rearing three children, maintaining friendships, and, more recently, working in such organizations as the PTA. The barriers imposed by a childhood environment of poverty were not insurmountable for Evelyn; her life testifies to the possibilities of social class mobility. Today she lives with her husband, a successful masonry contractor, in an attractive stone home well furnished on the inside and well landscaped. In a city where "average" housing would cost from $16,000 to $18,000, the house that Evelyn and her husband own would have a value between $25,000 and $30,000. Their leisure time is spent in a manner consistent with their level of income. They own a good motor boat and camping unit which are frequently utilized as they pursue their hobbies of camping and fishing.

Evelyn's participation in organizations, her regular appointments at the beauty parlor, her well-managed home, and leisure time activities seem to indicate concern for the maintenance of the higher social status which has been attained in spite of lower class beginnings. Evelyn would appear to have attained top success for a person of her intellectual capabilities as assessed in childhood and again during her middle forties. The first recorded individual mental test I.Q. rating was 65; the later rating was an I.Q. of 70.

Evelyn is quite clearly happy and contented with the way life has treated her. She has utilized her potential to good advantage, and she seems capable of inculcating middle-class habits of thinking, believing, and working into the lives of her three children.

2. A relatively unsuccessful low ability female. Pearl D. (60) 1907. When Pearl secured her first part-time job at age sixteen she dropped out of school and also separated almost completely from her parents and her five siblings—two brothers and three sisters. Her father was a frequently unemployed plasterer. Pearl found part-time work as a maid in various homes. No employment lasted long—never more than two or three months. The result was that she became for several years a sort of perpetual itinerant "helper" going from one housecleaning job or dishwashing job to another (some hotel experience included). This was her roving existence

until age twenty when she married. The person she married was also one of the subjects (low mental ability) of the present study.

When Pearl was contacted in 1935 (age thirty-two) she had been married eight years and had five children. The dwelling in which she, her husband, and the five children lived consisted of two exceedingly filthy rooms with dirt floors. The family was on relief. Pearl's husband was able to secure only part-time employment at a low rate of pay. He worked at that time for an automobile parts firm and his job was to dismantle salvageable parts of wrecked vehicles. This was his kind of employment as long as he lived. He died of lung cancer at age forty-five (in 1953).

After the death of her first husband Pearl remarried and was soon a widow again. Her second husband died of a heart attack in 1959. When she was interviewed in 1961 and in 1963 she was living in a two-room, very dirty apartment located behind a small grocery store. She had one of her children living with her, a mentally retarded son. Pearl had six children born of the two marriages. (Court records indicate considerable sexual promiscuity dating before Pearl's first marriage and continuance of this pattern throughout much of her life.)

When she was seen in 1963, Pearl was desperately trying to "get on welfare." She had very limited earnings "from sewing that people bring in for me to do and from things I make with my sewing and sell at the church." She had some three years previously become a member of a church and intimated in the interview that "some women in the Church Circle" helped her get to meetings of the Circle and to secure different kinds of sewing jobs. During the latest of the interviews Pearl complained of being anemic and of having an injured back. She quite obviously was living at a bare level of subsistence. There was evidence that welfare agencies, which had throughout her life provided various kinds of help, were in touch with her situation and prepared to supply at least limited relief.

The entries on the life-history blank for Pearl spell out more than fifty years of buffeted experience where there appears to have been few, if any, constructive forces with which she could identify. She left school with no truly salable abilities and was described by those who knew her before she left school as "dirty," "unfriendly" (friendless?), "a girl who hated her parents," and "sort of shiftless." In none of the interviews with her from 1935 to 1961 did she ever name any person who strongly and beneficially influenced her life. In the 1963 interview she mentioned the helpfulness of certain "women I know in the Church Circle." Responses to the life-history blank for Pearl (1963) included numerous statements, such as "No one was ever important in my life." "They didn't like me." (Employers with virtually no exception released her after only brief periods of work.) "School I didn't like. I liked one teacher, just some." "I never talked with my parents about anything—especially life work." "I wish school had taught me sewing which is the thing I like."

General Inferences from the Records of Eight Females Four women were selected as fairly representative of the relatively successful females. Certain personal traits and conditions of life not only appear with regularity in their histories but also in the histories of the larger sample of the comparatively successful females. For all four females an interviewer at some time or other (1935, 1950, or currently) commented, with reference to these women, on their "neat, personally quite attractive" appearance and also noted that their housekeeping was good. The relatively successful females (in quite marked contrast to the less successful ones) evidenced concern for their own personal appearance. These are traits that seem to have been developed early in life. Apparently, if a girl cannot be particularly bright, being neat and attractive is a most helpful substitute. The more successful females all married well in the sense of finding mates who were steady workers and good family men.

Most of the relatively successful women (referring to all in this group and not just the few mentioned in the portraits) volunteered favorable recollections of individual persons who gave them encouragement and especially the kind of encouragement that instills a sense of personal pride. Such favorable comments were almost entirely absent from the records of the less successful females.

The element of reinforcement in a work situation (for example, praise from an employer, enjoyment in duties) is quite regularly to be found with the more successful individuals and quite consistently absent from the history of the less successful females. The relatively successful females were able to recall without difficulty persons and incidents contributory to a sense of self-significance.

Of the four females who were categorized as "unsuccessful," two had remained dependent and homebound most of their life. The phrase "stays home and helps with the housework" was a common description in the records of the majority of the less successful females as a group.

There seemed to be generally among the less successful females during their early life a strong attachment to the mother. To what extent this dependent status was necessitated by inadequacy and to what extent encouragement of dependence compounded existing inadequacy is something the life histories do not divulge. Perhaps it is a reasonable hypothesis that there is in this circumstance considerable interaction and that excessive protection of children of low ability encourages life-long inadequacy.

IV. GENERAL OBSERVATIONS

From both objective data and personal acquaintance with subjects, some inferences about their lives can be made. The suggestions which follow are applicable to all retarded children but in many ways can be generalized to children of any intellectual level or background. These suggestions have to do with early and continued socialization and acculturation, with schooling, and with assistance in vocational choice and adjustment.

Socialization and Acculturation

All children require socialization and acculturation. Since so many persons of limited ability come from lower class families they not only need to learn the ways of society but need to modify many of the ways they learned at home or wherever they grew up. These needs were especially apparent in the Low Group of this study. A distressingly large number of low ability subjects reported that as children "nobody" helped them with their dress, their health, or their personal appearance. They could turn to "nobody" for advice and for help in general. A statement made earlier might be repeated here: "Most successful" females were observed by the researchers over a period of thirty years to be neat, clean, and attractive. It is not clear in all instances who helped these women as girls to learn the arts of grooming but the arts have been rewarding for them. For both sexes, it appears that the relatively successful have more middle class attributes of dress, speech, personal habits, and the like than have those of poorer adjustment. Of course cause and effect relationship is obscured here but early acquisition of middle class values and ways would seem to be highly desirable for retarded children.

Schooling

It is obvious that formal education of an appropriate nature is near the top of the list of the needs of the mentally retarded. It seems to the writers that the word "opportunity" in the name of the school which the subjects of the present study attended early in life has special significance. The school is a bridge between home (or institution for some) and the world these persons must be able to adjust to. Whatever its specific design that "bridge" must be long enough and high enough to deliver such persons to the adult world in such a fashion that they can cope with what they meet.

Vocational Assistance

Accompanying "general education" may be assistance in learning an appropriate vocational skill, in finding a job, and in continuing to develop vocationally. Perhaps no other single factor is more important in life adjustment and success in our adult society than this.

Examination of the lives of the best adjusted Low Group subjects of both sexes revealed that they either joined a big paternalistic organization early and stayed with it (usually in the same job or one quite similar) or learned a single skill early and continued without a change or break.

The least successful subjects in adjustment on the other hand had drifted from job to job, usually working for one person or a small firm rather than for large, well established, and continuing organizations. Almost none had a single readily identifiable vocational skill.

These generalizations are of course based on past performance of persons born before World War I. The rate of cultural change is accelerating sharply and stability of any kind will be harder to achieve as time goes by. Ability to adapt to change will be increasingly important to individual "survival" and success. One might speculate that even after early training and preparation, persons of low intelligence will need periodic rehabilitation, literally. "Stability" would not be a lifetime characteristic at least vocationally, but a terminal one to be renewed through a period of retraining. Change of location might also be a necessity; witness the current community and area depressions where an industry has been outmoded or a business has closed or moved away.

Development of Self-concept

One of the strong impressions one has in working with these persons is the necessity of developing an adequate self-concept—the importance of "being somebody."

How does one convince himself that he is somebody instead of nobody? This aspect of personality cannot be explored in depth here with all its connotations. However, three factors influencing the self-concept have been inferred from subjective evaluations of the persons included in this research. They are: (a) early experience; (b) the influence of significant persons; and (c) vocational experience.

Early Experience The negative effects of early cognitive and sensory deprivation are so familiar as to need no explication here. There is no question but that many of these subjects as preschool children lived in circumstances that could only be described as culturally and intellectually impoverished. It would seem self-evident that the affectional and cognitive deprivation of the early lives of the subjects could not lead to the development of a very positive self-concept.

Significant Persons   Every individual needs some other person as model, guide, and support at each stage of life, but this need is especially great in childhood and adolescence. A person cannot seem significant to himself when he is significant to no one else. Parents of the subjects of this study were often quite inadequate or in some cases nonexistent. Unfortunately, teachers sometimes seem to be threatening or too distant from a lower-class child to meet his needs for significance. Such a child is fortunate indeed when he finds an appropriate adult to use as model and support.

Many of the subjects of the present research reported the rewarding experience of finding such a person. Sometimes (but not often enough) it was a teacher—a teacher who "made me want to be my best, to do the right thing . . . to become something." Sometimes it was an employer during adolescence who guided and led and helped the youngster to a feeling of adequacy. Not infrequently it was an older relative or a neighbor who filled the need.

Who this significant person happened to be was not nearly so important as the fact that there was someone. Success or failure in living adequate lives appeared among the subjects to be closely linked in a high percentage of cases to the presence or absence of a "significant person." The schools and society in general would seem to have a responsibility for helping to expose children to a variety of persons who qualify as *appropriate*.

Vocation   When we meet a stranger in our society we are likely very early in our acquaintance to ask, "What do you do?" or "What business are you in?" From the reply we estimate education, social class, income, and a host of other qualities. In childhood and adolescence a frequent question is, "What are you going to be?" asked in an attempt to forecast future status.

The adolescent or adult of limited ability and background has at least as much, if not more, need for a ready and "respectable" answer to the question of occupation as does the normal or bright person. Anyone who lived through the depression of the 1930s (or who has studied the period) can recognize the damage to self-esteem and sense of personal adequacy caused by joblessness in the competent. The less competent are no different in their reactions and needs. Society, through government, is taking greater cognizance of the problem of vocational adequacy. Those who work with the retarded need to recognize the importance of vocational competence to personality as well as to economic self-sufficiency.

ABILITY AND PERFORMANCE

The tests given subjects at school age were not wholly accurate predictors of individual achievement or success, especially for the Low Group. It is certainly no news to hear again that intelligence tests are not entirely reliable predictors of individual behavior, but some school and rehabilitation workers still seem not to have heard the message.

Error in ability measures on the subjects—especially Low Group—of this study could have stemmed in part from cultural and language influences; there remains the possibility that some individuals develop intellectually slowly over a long period of time. It should be remembered also that some persons got along quite well in society despite continued poor ability.

New and better tests might be one answer, but little improvement has occurred in testing and prediction since these subjects were children. It would seem reasonable to infer from the evidence of this study that:

1. Especially with persons of low ability, intelligence test scores should be regarded as quite tentative.
2. Periodic re-evaluations beyond childhood would seem necessary whenever the person is involved in an advising, training, or placement situation.
3. More "whole person" evaluation must be made including cultural, motivational, and personality characteristics. Some of the Low Group subjects of this study were rather obviously disturbed emotionally and a few were physically malfunctioning. Psychotic, psychoneurotic, and physically inadequate states can lead to behavior resembling mental deficiency but which requires quite different treatment.

The upward trend in ability of subjects was discussed in an earlier chapter, but additional comment might be made about the relationship of personality and intelligence.

In other studies a relationship has been found between personality characteristics and intellectual change. It would seem reasonable that the relationship was due to other than chance factors.

From long-term study of the subjects of this research, the writers hypothesize that enhancement of self-concept and personality may be related to improvement of general intellectual functioning. Perhaps "freeing of intellect" would be a better term. Whether improved self-concept is related to intellectual change per se, it is certainly a factor in general adequacy and life adjustment.

REFERENCES

Baller, W. R. "A study of the present social status of a group of adults, who, when they were in elementary school, were classified as mentally deficient," *Genetic Psychology Monograph* 18: 165–244, 1936.
Baller, W. R. "A study of the behavior records of adults who, when they were in school, were judged to be dull in mental ability," *Journal of Genetic Psychology* 55: 365–379, 1939.
Charles, D. C. "Ability and accomplishment of persons earlier judged mentally deficient," *Genetic Psychology Monograph* 47: 3–71, 1953.
Jones, M. C. "Psychological correlates of somatic development," *Newsletter: Division of Developmental Psychology,* Fall, 1–21, 1961.
Miller, E. L. "Ability and social adjustment of mid-life of persons earlier judged mentally deficient," *Genetic Psychology Monograph* 72: 139–148, 1965.

Twenty-two

Age and Intelligence: The Significance of Cultural Change in Longitudinal versus Cross-sectional Findings

Raymond G. Kuhlen

A number of hypotheses may account for observed adult age trends in various psychological traits, and may serve to explain or reconcile the contrasting trends in mental abilities revealed by cross-sectional and longitudinal studies of aging. In a paper published some twenty years ago the writer (Kuhlen, 1940) emphasized the role of cultural change in *cross-sectional* studies, and concluded by stressing the merit of the longitudinal study as a means of minimizing the impact of that variable. In contrast, the present paper argues that cultural change contaminates *both* types of studies, but with opposing effects, such as to promote exactly the contradictions that have been found.[1]

In the comments to follow, this argument is pursued, first, by examining the nature of the data to be "reconciled"; second, by noting the character of some of the cultural changes that have been evident during the last forty or fifty years; third, by suggesting ways in which such changes may have differential impact upon different age groups as they are studied in cross-sectional and longitudinal procedures; and, finally, by suggesting ways in which research designs and procedures of data analysis might provide a better perspective in this matter.

THE NATURE OF THE DATA ON "INTELLIGENCE"

Putting aside for present purpose any detailed consideration of what is meant by "intelligence," the term may be defined in its barest operational sense: Intelligence is what intelligence tests measure. Granting that this definition is a crude one, it is important to recognize that the basic data to be "reconciled" come from culturally based tests. It is well known that performance on such tests is influenced by cultural variables, by

The article originally appeared in *Vita Humana*, 1963, 6, 113–124. Reprinted by permission of the publisher, S. Karger, Basel/New York.

experience variables, and by personality and motivational variables; and that to a considerable degree such tests are broadly based *achievement* tests. This consideration is of first importance in evaluating longitudinal and cross-sectional data.

A second major consideration is that the trends suggested by current longitudinal studies are not yet determined with precision. We cannot be sure what it is we are contrasting when we compare the findings from longitudinal to the findings from cross-sectional data. Existing longitudinal studies of mental abilities suffer from one or more of the following deficiencies:

1. Abilities have usually been measured at only two, often widely-separated ages. In view of the fact that most functions studied are curvilinearly related to adult age, it is apparent that two points, which can establish only a straight line, are insufficient. Thus, Owens (1953) demonstrated that fifty-year-olds did better than they themselves did at age nineteen; but previous studies indicate that performance on the Army Alpha test does not reach a peak until age twenty.

2. The tests employed in some studies are relatively homogeneous, and instead of yielding measures of "general" ability, may provide trends for some single function similar to vocabulary ability. This criticism may be made of the Bayley and Oden (1955) follow-up of Terman's gifted group, though it is to the credit of this study that the data were obtained at a number of age-points in the adult life span.

3. A third criticism applicable to certain longitudinal studies is that they are, in a sense, make-shift studies, capitalizing on existing initial data. Useful as such analyses may be for exploratory purposes, it is extremely likely that influences which operate at one age and not at the other may result in differences which are mistakenly attributed to age. Tests administered to *groups* during freshman week in college, for example, are seldom administered under conditions of high motivation. When these same subjects are retested individually and under vastly different conditions of motivation at a later time, important questions can be raised as to the meaning of the obtained differences.

4. The carefully designed longitudinal studies now in progress have not been pursued long enough to yield definitive data. Probably the most important studies are those under way at the University of California (Berkeley), which have revealed continued gains in intellectual ability up to around thirty years of age, as thus far reported. It is not yet known whether performance begins to decline as these groups move beyond age thirty.

In evaluating the findings from current "two-point" longitudinal studies, it may be worth recalling an earlier semi-longitudinal investigation which is rarely referred to in current papers, but which provided longitudinal data at *several pairs of points* during the adult years. In 1934 Catharine Cox Miles published a brief paper reporting data from the Stanford Later Maturity Studies. She had retested, after two years, different age samples of the basic study group. Scores did not continue to rise, but paralleled rather closely the decrements apparent in the cross-sectional data. McCulloch (1957) used a very similar semi-longitudinal procedure in the study of age trends in 937 retarded adults. His summary curve, covering the age range of sixteen to sixty, showed a high point at thirty, with decline thereafter. These studies suggest that longitudinal data may also reveal a downward turn after a peak in relatively young adulthood, just as do the cross-sectional studies.

It is noteworthy that the trend apparent in the Miles study, although true for the total sample of 190 Ss, was not evident in a subgroup of 135 Ss who had voluntarily returned to be retested because of their interest in the study. In fact gains, instead of losses, were evident for this group, except for the oldest members. Mean losses throughout the age range resulted for the total sample when those additional 55 subjects were included who agreed to be retested only after vigorous recruiting efforts. This is a rather important point since it suggests that those volunteering for longitudinal studies may be different Ss over time from those not volunteering.

In contrast to the relatively tenuous longitudinal data now available, the cross-sectional data inspire more confidence in the curves they demarcate. The four principal cross-sectional studies are the 1933 Jones and Conrad investigation of a rural Vermont population with the Army Alpha test; the Stanford Later Maturity Studies (Miles and Miles, 1932) involving a sample in California tested with a brief form of the Otis; the standardization data of the Wechsler Bellevue Scale (Wechsler, 1939); and the recent data of the Revised Wechsler Adult Intelligence Scale (Wechsler, 1958). In three of these studies performance reached a peak at age twenty and declined shortly thereafter. The 1955 Wechsler data, based on a stratified sample of the population, identified peak performance at around age twenty-five, with linear decline evident shortly thereafter. Although there is a possible effect of cultural change in comparisons of the Wechsler data of the mid-1930s and the mid-1950s all four cross-sectional studies are consistent in showing an early adult peak performance, followed by regular decline.

Thus it may be argued that at the present time fairly adequate cross-sectional data are being compared with extremely tenuous longitudinal data; and the attempts to "reconcile" the findings or to draw firm conclusions may be somewhat premature.

CULTURAL CHANGE AND ITS IMPACT

There is now more "cognitive stimulation" in our culture than was true in earlier times. Older people when they were young not only received less stimulation through formal societal efforts (public schooling), but they were exposed to less informal stimulation through the media of mass communication. As is well known, length of schooling is highly correlated with measured intelligence; thus, "decline" in test performance evidenced in cross-sectional studies may be partially explained by the lower educational level of older individuals. It would seem clear that if intellectual stimulation is most important in younger years, older individuals, those who grew up at a time when verbal stimulation was relatively meager, would be handicapped.

If we consider the broad problem of cultural change as related to studies of aging, rather than considering only studies of intelligence,[2] further complications are encountered. Different aspects of the culture do not change to the same degree or in the same direction over the same period of time. An interesting illustration comes from a study published recently (de Charms and Moeller, 1962). As revealed in an analysis of children's readers, there is a marked curvilinear relationship over time in the pressure put upon children for achievement. The evidence suggests an increase in achievement pressure from 1800 to 1890, a steady decrease thereafter through the 1940s, and other evidence indicates a recent increase again in the last few years.

If such a curvilinear relationship between cultural trends and time characterizes the lifespans of adults now living (instead of encompassing a century and a half, as in the foregoing illustration), then cultural change will have a differential impact upon different age groups in cross-sectional studies. However, cultural change will also produce age "gains" or "losses" throughout the age span studied in the longitudinal studies, depending upon the historical time setting in which the longitudinal measures are taken.

To take another example, some data suggest a turn away from religion during the first three decades of the twentieth century and a shift in the opposite direction during the last two decades. Thus if studies of age changes in religion had been conducted in 1930, one might well have expected, in a cross-sectional study, that older people would be *more* "religious" than those younger. At the same time, 1930, a longitudinal study might well have shown that, with age, people became *less* religious. The reverse might be expected in studies carried out in 1960—in which during the first half of the adult lifespan, a cross-sectional study might well show younger people more religious than middle-aged, whereas a longitudinal study might show that, as persons moved from early to middle adulthood, they became more religious.

It becomes evident from the foregoing that a longitudinal study will, under the cultural circumstances of the last several decades, result generally in data showing that older individuals are "better" in the respect being measured than they were when first tested at a younger age; whereas

a cross-sectional study will put the older individual at a disadvantage compared to the younger person when they are both being tested at the same
time.

That cross-sectional studies are influenced by cultural change has, of
course, long been recognized. It is not so widely recognized that the longitudinal study may yield data spuriously favorable to the older individual
when marked cultural change is occurring in a positive direction. Indeed,
developmental psychologists appear to have become so enamoured of the
longitudinal study that many view it as the design of choice for almost any
problem. Yet a careful comparison of the advantages and disadvantages of
the cross-sectional and longitudinal methods for research on aging will
show the two methods to be fairly well balanced. For some purposes the
longitudinal study is the only method of obtaining the desired data; for
other purposes the cross-sectional study is the method of choice.

Two studies demonstrate the point that what appeared to be age
trends in longitudinal data were instead artifacts of cultural change: Nelson
(1954) tested college students with an array of attitude scales and retested
them fourteen years later. Marked trends in the direction of increased
liberalism were noted. However, at the second testing a new sample of
college students was obtained. The scores of this new group were about
the same as those of the retest scores of the first sample. In another study
by Bender (1958), Dartmouth College students were retested after a
fifteen-year period with the Allport-Vernon Scale of Values. A marked
trend in the direction of greater religious interest was noted. But again, test
results from a new group of Dartmouth students tested at the time the first
group was retested suggested that the change in the first group was likely
due to cultural change, not to age per se.

There seems little question that the tremendous changes in cognitive
stimulation over recent decades can affect performance on intelligence tests
just as dramatically as indicated in these attitude studies. Indeed, Tilton
(1949) showed substantial gains in intellectual performance of World War
II soldiers over World War I soldiers.

It might well be asked why cultural changes should affect different age
groups differentially. Why are not people of all ages influenced to the same
extent by the changing world in which they live? Several reasons suggest
themselves. First, at least in the American culture, there are massive efforts
to transmit the culture to the young through formal education. Thus the
young get the quick advantage of new advances in knowledge. (A dramatic
illustration is the fact that engineers can become rapidly outdated because
of the rapid cultural change designated as "technical advance.") In addition to a very probable decline in learning ability with age, with handicaps
especially evident when faced with new and different learning tasks, older
people, as compared with younger, tend not to experience so directly the

impact of cultural change. This is true, (a) because of reduced need or motivation to learn (reflecting the decreased demand of the culture that they learn),[3] (b) because of pressure of the work-a-day world, which denies the adult opportunities to interact with his broader environment, and (c) because of the tendency of older persons to insulate themselves psychologically from new features of their environment. One notes, in this last connection, that as people become older they tend to live in an increasingly restricted social matrix. This circumstance probably results partly from losses in energy, and possibly from psychological trends such as increasing habits of conservatism, caution, and avoidance tendencies—habits or practices which serve to reduce the anxiety generated by age-related losses. Despite these reasons, however, some of the cultural change reaches the older individual, with the result that compared to himself when younger, he is at a "higher" level.

DESIRABLE CORRECTIVES

While the arguments advanced above may explain why cross-sectional data may yield one type of age trend whereas longitudinal data may yield another, the basic question remains: What is the pattern of intellectual development and decline during the adult lifespan? As pointed out earlier, an answer to this question must necessarily await better data. If it should turn out that longitudinal data tend to show a peak at about age thirty, the findings will be reasonably consistent with the cross-sectional data now available, and no reconciliation between the two types of studies will be necessary. If, however, longitudinal data should show that the peak is not reached until age forty, will this finding be due to practice effect, or to cultural change? Or do people actually continue to mature in the abilities measured by the tests? These questions cannot be answered in the absence of the data; yet when the data are in, they are likely to be subject to varied interpretations. Accordingly various procedures should be explored for checking on the effect of cultural change and other influences.

One approach is the systematic replication of studies across time and across cultures. Data presently available from both types of "replication," though admittedly sparse, tend to emphasize the influence of culture. Probably the major replication study across time is to be found in the Wechsler data of the 1930s and the 1950s, both sets of data based on substantial samples. In one case the peak of intellectual ability is shown at age twenty; in the other, at age twenty-five. Has the peak of the growth curve of intelligence shifted in approximately one generation? Or is there reflected here the greater verbal character of the culture, the increased levels of education (the larger proportions of the population in high school; the widespread training programs available to young people in industry, in adult education programs, as well as in college)? Perhaps in the United States the important changes in cognitive stimulation have occurred in the last two or three decades, with the onset of commercial radio in the early 1920s to the saturation of the American public with television in the 1960s. Hereafter such cultural changes may be less dramatic. Replication studies across time may thus provide intelligence test data of considerable relevance.

A second approach to the problem is to examine trends on individual test items, items which are judged to be especially reflective of cultural change. While there are important advantages in generalizing across groups of items, as subtests conventionally do, a detailed inspection of single *item* trends would yield useful insights.

Third, an analysis of the cultural, dietary, "life style," and other characteristics of subgroups of subjects showing differential age trends might prove to be instructive. Although test-retest correlations are high, even over thirty-year intervals, they are not so high as to preclude quite different trends for different individuals, especially when longitudinal studies suggest larger variance in distribution of scores at older ages. There may be some advantage at this stage of our research in relatively exploratory types of studies, in this respect. It is becoming increasingly evident that we need a better conceptualization of the aging process than is presently available. Age itself is not a significant theoretical variable, but is mainly an index of time—time in which other things of psychological importance happen. An empirical approach to the identification of age-related variables which may prove to be the important independent variables should not be neglected.

A fourth suggestion concerns the general design of studies. Cross-sectional and longitudinal studies are generally considered to represent different basic designs. The importance of cultural change in the outcomes of both types of studies, however, makes obvious the necessity of combining the two approaches. Given the substantial amount of time and money required for a longitudinal study, it would be wise if the investigators added the small amounts necessary to obtain cross-sectional data at the same time, as a check on the influence of cultural change. In this connection, it may also be noted that the design of many current studies does not provide for control over practice effects. The recent study by Jarvik, Kallmann, and Falk (1962) and some observations of William A. Owens (1962), as yet unpublished, suggest that practice effect on test-retest may be an exceedingly important influence even over ten-year intervals.

Finally, it may be that we need some new conception and definition of intelligence. Many psychologists still take a somewhat biological view of intelligence, and interpret cultural influences upon test performance primarily as a source of error. Perhaps studies of the central nervous system will eventually provide "answers" regarding the nature of intelligence. Be that as it may, current intelligence tests are fundamentally based in the culture; and the growth curves of intelligence obtained by means of the tests are to a considerable degree a function of the particular measuring instruments used. For these reasons we shall probably never know the inherent nature of the intellectual growth curve through the use of such tests. Catharine Miles (1934) and Demming and Pressey (1957), among others, have pointed out that tests standardized on middle-aged adults rather than on children and adolescents would yield quite different "growth" curves.

New conceptions of intelligence may be emerging which place greater emphasis upon perception, thought processes, and reasoning; and which accord a greater role to experience and thus to cultural influences. For example, the distinction between "fluid" and "crystallized" intelligence, recently given elaboration by R. B. Cattell (1963), leads to specific predictions regarding adult age trends. Cultural and personality variables are assigned an important role in the determination of crystallized ability.

To take another example, Piaget's work has received much attention in recent years. That American psychologists may be ready for a new look at the nature of intelligence is suggested not only by Cattell's research, but also by J. McV. Hunt's recent volume (1961) on intelligence and experience, and by the recent monograph (Kesson and Kuhlman, 1962) reporting a conference which focused on Piaget's views of cognitive development. Commenting briefly upon deficit, Hunt (1961, p. 297) notes that one of the contributions of Piaget's procedure "may be to help to clarify the nature of various kinds of psychological deficit. . . . The fact that logical operations clarify the variations in thought that occur with the development of thought suggests that they might help clarify the variations that occur with the dissolution of thought." New approaches to the definition of intelligence and to the study of cognitive functioning during the lifespan may serve to reconcile the findings of longitudinal and cross-sectional studies and thus to advance our understandings.

SUMMARY

Contradictory trends between cross-sectional and longitudinal studies of aging are especially evident in studies of mental abilities. This paper argues that cultural change contaminates both types of studies, but with opposing effects which promote the contradictions apparent in current findings. Attempts to reconcile contrary findings must take cognizance of the cultural base of "intelligence" tests, the tenuous character of present longitudinal data, the cultural change in cognitive stimulation which will differentially affect different age groups. Remedies include more complete longitudinal data, replication of studies across time and cultures, analysis of trends for test items presumed to be differentially affected by culture, identification of variables related to differential age trends among individuals, and better research designs to check on cultural change and practice effects. Perhaps most promising with respect to a better understanding of adult age trends in mental abilities are new conceptions of "intelligence" which seem to be emerging.

NOTES

[1] The problem is, of course, intrinsic to aging studies generally, and in no sense peculiar to investigations of mental abilities. For example, Kelly (1955) noted in his presidential address to the APA some years ago that he found a trend toward increasing masculinity among women in a longitudinal study covering the age range from the twenties to the forties, whereas Terman and Miles (1936) had found a trend toward increased femininity in a cross-sectional study. Similarly, young adults were found in a longitudinal study to become more liberal (Nelson, 1954; Bender, 1958) over a fourteen- or fifteen-year period whereas cross-sectional studies show trends toward conservatism.

[2] Toch (1953) has argued that cultural change is not a factor in certain types of psychological age data. Though he presents evidence from public opinion polls supporting his position, the evidence relates to a short span of years.

[3] In the instance of "intelligence" tests, what is learned and practiced in adult years is often *not* the type of learning tested.

REFERENCES

Bayley, N., and M. H. Oden. "The maintenance of intellectual ability in gifted adults," *Journal of Gerontology* 10: 91–107, 1955.

Bender, I. E. "Changes in religious interest: A retest after 15 years," *Journal of Abnormal and Social Psychology* 57: 41–46, 1958.

Cattell, R. B. "The theory of fluid and crystallized intelligence: A critical experiment," *Journal of Educational Psychology* 54: 1–22, 1963.

de Charms, R., and G. H. Moeller. "Values expressed in American children's readers: 1800–1950," *Journal of Abnormal and Social Psychology* 64: 136–142, 1962.

Demming, J. A., and S. L. Pressey. "Tests 'indigenous' to the adult and older years," *Journal of Consulting Psychology* 4: 144–148, 1957.

Hunt, J. McV. *Intelligence and experience.* New York: Ronald Press, 1961.

Jarvik, L. F., F. J. Kallmann, and A. Falek. "Intellectual changes in aged twins," *Journal of Gerontology* 17: 289–294, 1962.

Jones, H. E., and H. S. Conrad. "The growth and decline of intelligence:

A study of a homogeneous group between the ages of 10 and 60," *Genetic Psychology Monographs* 13: No. 3, 1933.

Kelly, E. L. "Consistency of the adult personality," *American Psychologist* 10: 659–681, 1955.

Kesson, W., and C. Kuhlman. "Thought in the young child," *Monographs of the Society for Research in Child Development* 17: No. 2, 1962.

Kuhlen, R. G. "Social change: A neglected factor in psychological studies of the life span," *School and Society* 52: 14–16, 1940.

McCulloch, T. L. "The retarded child grows up: Psychological aspects of aging," *American Journal of Mental Deficiency* 62: 201–208, 1957.

Miles, C. C. "Influence of speed and age on intelligence scores of adults," *Journal of Genetic Psychology* 10: 208–210, 1934.

Miles, C. C., and W. R. Miles. "The correlation of intelligence scores and chronological age from early to late maturity," *American Journal of Psychology* 44: 44–78, 1932.

Nelson, E. N. P. "Persistence of attitudes of college students fourteen years later," *Psychological Monographs* 68: No. 2, 1954.

Owens, W. A. "Age and mental abilities: A longitudinal study," *Genetic Psychology Monographs* 48: 2–54, 1953.

Owens, W. A. "Age and mental abilities: A second phase of a longitudinal study," *Journal of Gerontology* 17: 472, 1962 (Abstract).

Terman, L. M., and C. C. Miles. *Sex and personality: Studies in masculinity and femininity.* New York: McGraw-Hill, 1936.

Tilton, J. W. "A measure of improvement in American education over a twenty-five-year period," *School and Society* 69: 25–26, 1949.

Toch, H. "Attitudes of the 'fifty plus' age group: Preliminary considerations toward a longitudinal study," *Public Opinion Quarterly* 17: 391–394, 1953.

Wechsler, D. *The measurement of adult intelligence.* Baltimore: Williams & Wilkins, 1939.

Wechsler, D. *Manual for the Wechsler Adult Intelligence Scale.* New York: Psychological Corporation, 1955.

Wechsler, D. *The measurement of and appraisal of adult intelligence.* (4th ed.) Baltimore: Williams & Wilkins, 1958.

Twenty-three

Creative Productivity between
the Ages of Twenty and Eighty Years

Wayne Dennis

Any developmental psychologist is likely to agree that ideally the Ss of a longitudinal study should be followed from birth to death, but he is probably not willing himself to undertake an investigation of such length. His reasoning is easy to understand.

Unless he can be given in advance an assurance of exceptional longevity, or of a posthumous corps of research assistants, he cannot be sure that his project will be completed.

Like most psychologists, he assumes that all data which he is to treat must be recorded by himself. In this respect he differs from scholars and the historians who expect to deal primarily with materials which have been recorded by others.

One of the purposes of this paper is to demonstrate that psychologists can make use of data which have been recorded by others. But the paper has primarily a substantive interest. It attempts to examine age changes in productivity. In pursuing this aim it finds documentary evidence most useful.

MATERIALS AND METHODS

THE PLAN OF THE STUDY

The paper presents data on the life curve of productivity of 738 persons, all of whom lived to age seventy-nine or beyond. It is concerned primarily with the following two questions: What are the differences among various fields of effort with respect to age curves of productivity? What are the probable causes of these differences?

The Ss of the study were persons engaged in scholarship, the sciences,

Dennis, W. "Creative productivity between the ages of twenty and eighty years," *Journal of Gerontology*, 1966, 21, 1–8. Reprinted with permission of the author and the publisher.

and the arts. In studying only persons interested in these areas, it is obvious that at the moment we are neglecting most of mankind. But no single study can encompass the world's population, and the investigator has the acknowledged right to undertake a limited task. He must, of course, be cautious in drawing general conclusions. Nevertheless, the possibility exists that the principles discovered in the study of particular groups may have a wider applicability.

SELECTION OF SUBJECTS

There were two general criteria for the selection of each person included in the study. First, both the year of his birth and the year of his death must be known and must be separated by eighty or more years. In most instances this requirement insures that the S lived to age eighty or beyond. However, since month and day of birth and death are not taken into consideration and often are not available, by subtracting year of birth from year of death to obtain ages, we have included among our Ss persons born in a certain month who died in earlier months of their eightieth year without reaching age eighty. However, in the interest of economy of expression, we shall frequently refer to all of our Ss as having reached age eighty, and undoubtedly most of them did. (Productivity before age twenty and after age eighty will not be included in our data.)

A second requirement for the inclusion of each S was that there must be available a relatively complete dated record of his works. This requirement, in most cases, limited us to persons born after 1600, since dated records of productivity are seldom available for earlier centuries. In some fields, which will be noted later, the Ss were even more limited with respect to the historical periods from which they were selected.

SOURCES OF DATA ON PRODUCTIVITY

Table 23.1 lists for each group of Ss its designation, the number of persons in it, and the total number of works produced by members of the group in each decade of life between ages twenty and eighty.

The composition of these groups will now be described.

For the most part, in forming groups of Ss and in obtaining data on output we have made use only of sourcebooks available in English. As a consequence, in several groups our subjects are entirely American or British.

The category of scholars includes English historians, English philosophers, and English scholars of Biblical, classical, and oriental literature. All scholars were British, since we used as a sourcebook the *Cambridge Bibliography of English Literature* Vol. II, 1660–1880 and Vol. III, 1800–

1900. For these scholars, the output for each person refers to the number of books produced, since other publications are seldom listed in the *Cambridge Bibliography*. Because the period covered by the sourcebooks ends at 1900, we could include in this group only scholars who reached age eighty before 1900.

TABLE 23.1   Percentage of Total Works between Ages 20 80 Which Were Done in Each Decade

				Age Decade			
	N	N	20s	30s	40s	50s	60s	70s
Scholarship	Men	Works						
Historians	46	615	3	19	19	22	*24*	20
Philosophers	42	225	3	17	20	18	*22*	20
Scholars	43	326	6	17	*21*	*21*	16	19
		Means	4	18	20	20	*21*	20
Sciences	Men	Works						
Biologists	32	3456	5	22	*24*	19	17	13
Botanists	49	1889	4	15	22	22	22	15
Chemists	24	2420	11	21	*24*	19	12	13
Geologists	40	2672	3	13	22	*28*	19	14
Inventors	44	646	2	10	17	18	*32*	21
Mathematicians	36	3104	8	20	20	18	19	15
		Means	6	17	*22*	21	20	15
Arts	Men	Works						
Architects	44	1148	7	24	*29*	25	10	4
Chamber Mus.	35	109	15	*21*	17	20	18	9
Dramatists	25	803	10	27	*29*	21	9	3
Librettists	38	164	8	21	*30*	22	15	4
Novelists	32	494	5	19	18	*28*	23	7
Opera Comp.	176	476	8	30	*31*	16	10	5
Poets	46	402	11	21	*25*	16	16	10
		Means	9	23	*26*	21	14	6

Note: Maximum values are shown in italics.

The scientists chosen for study were as follows: eminent biologists chosen from histories of biology (Locy, 1925; Nordenskiöld, 1935); eminent chemists, chosen from a history of chemistry (Hilditch, 1911); British and Irish botanists who were not among the eminent biologists but are listed in Britten and Boulger (1931), and geologists, mostly American, listed in Nickles (1923). In addition to these men we also placed in the science group inventors listed in the *Dictionary of American Biography* and eminent mathematicians whose names were obtained from histories of mathematics (Bell, 1937; Cajori, 1937; Smith, 1929).

The data on the productivity of biologists, botanists, chemists, and mathematicians were obtained from the *Royal Society of London's Catalog of Scientific Literature, 1800–1900*. Because the sourcebook was limited to a particular century, the Ss were necessarily limited to those whose lives from age twenty to age eighty fell within the century. Since the Royal Society's Catalog contains only scientific papers, other material such as popular articles, books, and monographs were necessarily excluded from the indices of scientific output. The records of output for inventors were obtained by searching the U.S. List of Patents.

As is shown by Table 23.1, persons listed as belonging to "the arts" included architects, composers of chamber music, dramatists, librettists, novelists, opera composers, and poets. The source of names and of data on productivity of architects was provided by Colvin (1954). Whenever possible, an architect's age with respect to one of his works was recorded as his age at the beginning of construction of this work. If this date was not known, the next subsequent known date, such as date of completion or of partial completion, was substituted.

For chamber music composers, our source of data was Cobbett (1929). This catalog is international with respect to its coverage.

The dramatists were English playwrights. The source was Nicoll (1952–1959). In determining a dramatist's age with regard to each of his works, we used either his age at the first printing of the work or at its first production, whichever was the earlier.

The New York Public Library has prepared a card index of librettists, which gives for each librettist dates of his birth and death and the operas for which he wrote librettos. His age at the first production of each opera was obtained by consulting Loewenberg (1955).

Data on opera composers were also obtained from Loewenberg (1955). This work, which is international in scope, covers all operas produced between 1600 and 1924. The composer's age with respect to each of his works refers to his age at the time of its first production.

Information concerning British novelists was obtained, from the *Cambridge Bibliography 1600–1900*. The author's age with respect to a novel was recorded as his age at the date of publication.

Our final group consisted of minor British poets, whose volumes of poetry are listed in the *Cambridge Bibliography*. Only books of original poetry were included in our data; that is, translations, selected works, and so on, were excluded.

Limitations in the Data

It will be noted that our different groups of Ss are not uniform with regard to their degree of eminence. Some sources, such as histories of science, are highly selective. On the other hand, Loewenberg attempts to include every person who ever composed an opera which was produced.

The units of productivity employed for the different groups also are unequal. Certainly the composition of an opera ordinarily requires greater effort than does the writing of a one-page scientific note, yet within its group each is counted as one unit of output. However, in no treatment of our data do we directly compare such unlike data. Despite differences with respect to selectivity of subjects and in size of unit of measurement, we believe it will be shown that certain trends in output were consistent.

RESULTS

AGE CHANGES IN PRODUCTIVITY

Our results are presented in two ways. Table 23.1 shows the percentage of the total output of each group between ages twenty and eighty which was produced in each of these six decades. This method of presentation makes it possible in the case of each group to compare decades with each other and also to make intergroup comparisons of the distribution of productivity regardless of differences in total output and in units of measurement.

Table 23.2 presents the same data in a different way. In this table, for

TABLE 23.2   Output of Each Decade Stated as a Percentage
of the Output of the Most Productive Decade

	N	N	20s	30s	Age Decade 40s	50s	60s	70s
Scholarship	Men	Works						
Historians	46	615	11	50	76	89	100	84
Philosophers	42	225	14	76	90	82	100	88
Scholars	43	326	26	78	100	100	73	90
Sciences								
Biologists	32	3456	22	90	100	81	69	55
Botanists	49	1889	17	67	100	98	100	70
Chemists	24	2420	45	87	100	80	49	53
Geologists	40	2672	12	47	81	100	70	53
Inventors	44	646	11	35	51	55	100	65
Mathematicians	36	3104	43	100	100	95	100	75
Arts								
Architects	44	1148	24	82	100	84	36	12
Chamber Mus.	35	109	70	100	78	96	87	43
Dramatists	35	803	31	93	100	73	29	9
Librettists	28	164	26	70	100	72	48	12
Novelists	32	494	20	69	64	100	80	24
Opera Comp.	76	476	26	97	100	50	30	16
Poets	46	402	46	87	100	65	65	40

each group the decade in which absolute output was the highest was given the value of 100 percent and the records of all other decades were stated proportionately. This method again makes it possible to observe readily the rise and fall of productivity regardless of the absolute number of contributions made in any field or in any decade.

Productivity in the Twenties   In each group the output of the twenties was considerably less than the output of the thirties. In most groups the productivity of the twenties was only half as great, or less than half as great as the output of the thirties. In each field these differences were statistically significant. (In this paper this term will be used to indicate a P of 0.01 or less.)

The number of contributions made in the twenties was a small part of the total output. In some groups it was as little as 2 percent of the total and in no group did it exceed 15 percent of the total. Among the scholars and scientists, age twenty to twenty-nine was the least productive period. For these persons this decade was exceeded in output even by the seventies. Only in some of the arts were the twenties more productive than the seventies.

Output before age twenty was omitted from our study because pilot studies showed that very little creative work was produced in the teens.

Productivity in Middle Age   For almost all groups, the period forty to forty-nine was either the most productive decade or else its record was only slightly below that of the peak decade. In only one field (chamber music) were the thirties higher in output than the forties. In all other cases in which the forties were not the most productive years, this honor went to a later period.

Productivity in the Seventies   For scholars as a whole, the seventies were as productive as the forties. The scientists, on the other hand, showed a significant decline in the seventies. The output of that decade was only 75 percent to 53 percent of the peak rate. Among our groups of scientists, the biologists, chemists, and geologists declined most with age, mathematicians and botanists least.

The decline in productivity in the arts was considerably greater than in science. Dramatists, architects, and librettists, respectively, produced in their last decade only 9 percent, 12 percent, and 12 percent as much as in their peak years. Persons in the arts who held up best were composers of chamber music and poets. They produced in their seventies 43 percent and 40 percent as much, respectively, as in their most productive periods. These percentages were below the lowest percentages recorded for any science. The contrast between the steady output of the scholars and the steep decline in arts is sharp.

To what extent did the decade of the seventies enable men to add to the output which they had achieved prior to age seventy? Between seventy to seventy-nine, scholars increased their preceding number of books by 25 percent. Scientists in their seventies increased their number of earlier articles by 18 percent. However, this decade enabled persons in the arts to add to their previous output by only 6 percent.

DISCUSSION

We have noted that artists in general "hit their stride" in output earlier than the other groups under consideration.

What are the causes of these differences in early productivity? An obvious possibility is that persons in arts need not undergo a long formal education before they begin to produce, whereas scholars and scientists are required to have a protracted preparation before productivity begins.

Another difference between the groups lies in the time required to produce a unit of work. The musician and the poet, by intense effort, may sometimes complete a contribution within a very short period of time, but a history of Rome or a record of the voyage of the *Beagle* cannot be dashed off in a week. Extensive gathering and evaluation of data usually must precede scholarly and scientific contributions. It may be significant that chemists and mathematicians were relatively more productive in their twenties than were other groups of scientists. Perhaps creative work in mathematics and in chemistry does not require the long period of study and of data gathering which characterizes other kinds of scientific work. Obviously further studies are needed to elucidate these points.

While our interpretation of the differences in the relative amount of work done in the early decades is not proposed as final, at present we are inclined to believe that the explanation lies primarily in the requirements for different kinds of contributions. Some careers require much more preparatory study and gathering of data than do others. This principle may prove to apply not only to differences between major fields, such as arts, scholarship, and sciences, but also to differences within fields, such as the apparent difference between chemists on the one hand and botanists and geologists on the other with respect to their early output.

We turn now to a discussion of the causes of the differences between the sharp decline in output in the later years which occurred in the arts, the moderate decline among scientists, and the almost steady maintenance of productivity among the scholars.

When one attempts to explain these differences, the possibility suggests itself that biological aging occurs earlier among artists than among scientists and earlier among scientists than among scholars. However, data on longevity do not support this hypothesis, nor do we know of any other data which would tend to confirm it.

A second possibility is that the groups may differ in the extent to which emotional problems in the later years may interfere with productivity. Pathological disorders increase sharply in the later decades. Raskin (1936) has shown that literary persons are more prone to psychological incapacitation than are scientists. This difference may be slight in the early decades, because mental disorders are rare in all groups at these ages, but they may loom large in later life: that is, the differences in productivity between persons in the arts and persons in scholarship and science in later years may increase with age as psychological difficulties become more prevalent among artists. At present few data indicate whether or not this interpretation is correct. In our opinion it deserves further study.

A third hypothesis is that group differences in age decrement in productivity may be related to the amount of effort required to create a unit of work. With increasing age the individual declines in the number of hours he can work, the energy he can expend, and the speed with which he can recover from fatigue. It can be argued that if in an area of creativity the traditional unit of output is characteristically small, if no prolonged expenditure of energy is required, and if no great fatigue is generated, there may be little reduction in output with age. Conversely, fields of activity which inherently require intense, prolonged, and fatiguing effort may be more susceptible to the effects of age.

This hypothesis does not seem to be supported by our results. It is assumed that the composition of an opera and the composition of an historical work each requires continued effort over a considerable period of time. Yet the output of composers declined markedly with age and the works of historians did not. Mere differences in energy requirements do not appear to explain this difference in decrement.

Finally, we come to those factors which, in our opinion, contribute most to the differences in age decrements. These are the contrasting ways in which productivity is achieved in the three fields of endeavor with which we are here concerned. The contrasts are various. For one thing, in some fields it is proper and customary for a person to have collaborators or assistants; in other fields, contributions must be primarily personal and individual. Second, in some kinds of work, a contribution requires the use of data or materials which must be collected, stored, and classified, in other areas, no contributions can be derived by such processes. Third, in some fields a person receives credit for modifying, or amplifying, or restating his previous views; in other areas, a contribution must consist of new work, not the revision of an earlier one.

With respect to each of these pairs of possibilities, it seems that the first alternative describes the situation which faces the scholar, while the second alternative describes what is expected in the arts. The scientist occupies an intermediate position.

Let us expand this point. In the first place, the utilization of assistants in collecting materials, in arranging them, and in assessing them is considered proper on the part of scholars and scientists, but is usually not considered appropriate in the arts.

To illustrate: Bancroft (1800–1891) (Nye, 1944), a famous American historian with considerable personal financial means, employed men to find and copy documents which he needed in connection with preparing his history of America. In addition to his paid assistants, diplomats and friends in foreign posts also supplied him with materials. For an historian this is entirely proper. In contrast, one can hardly envisage an opera composer or a poet paying for or requesting comparable assistance in the composition of an opera or a poem.

Pavlov (1849–1936) (Babkin, 1951), the famous Russian physiologist, exemplifies in the field of science the extent to which productivity may be enhanced and continued through aid from others. A review of Pavlov's work shows that the enormously time-consuming task of gathering data was done largely by others working under his supervision. Babkin, who worked with Pavlov for ten years and knew him well during the last thirty-five years of his life, gives a revealing picture of some aspects of Pavlov's productivity.

When Pavlov was director of the Institute of Experimental Medicine, and later when he was director of the School of Experimental Physiology, graduates of Russian medical schools came to his laboratory to complete the research thesis required for the M.D. degree.

> Each doctor, on joining the laboratory, was allowed a subject, which was usually the continuation of a problem worked on by his immediate predecessor. . . . Pavlov personally supervised the majority of the experiments. . . . With the help of these graduate students Pavlov, as it were, multiplied the results that could be produced by his own two hands (Babkin, pp. 70–71).
>
> After the Revolution of 1917, there were about 40 people working with Pavlov. When he was 74 he wrote, "My work progresses on a large scale. A great many workers have gathered, and I cannot accept all who want to come. . . ." Pavlov now resembled a chess player, playing several boards at once. The data obtained by one worker could be confirmed and carried a step further by the research of another worker (Babkin, pp. 113–114). Pavlov's research could not have been carried out on so large a scale without the assistance of so many co-workers (p. 123).

Such facilities rendered it possible for Pavlov to remain highly productive to the end of his life.

On the other hand, if the Russian government had offered to provide a dramatist or a novelist with a comparable staff of assistants in order that his output of plays or novels should not diminish, it is not likely that he could have made effective use of these helpers, or that the world would have given him credit for works produced in this fashion. The scholar and the scientist can utilize systematically stored information in ways which are not characteristic of persons in the "beaux arts."

The methods of Bancroft, the historian, again serve to illustrate this point. Bancroft rose at five or six in the morning to copy facts from his sources into his "daybooks," one book for every year to be covered. In another notebook, called a topic-book, he copied materials relative to topics such as The Army, Washington, Foreign Affairs, and so on. As Nye says (p. 289),

If he wished, for example, to write five pages on national finance in 1784, the proper page gave him all the facts in chronological order, cross-indexed and supported by reference to the original sources.

While Bancroft had an excellent memory, with such work-habits memory was scarcely necessary.

In six years of this routine, between ages seventy-six and eighty-two, Bancroft produced a completely new work, *The History of the Formation of the Federal Constitution.* At age eighty-nine he completed a life of Van Buren which was based upon notes made forty years earlier. It is no wonder his productivity in his eighties seemed to be as high as in any earlier decade.

If Wordsworth had attempted to write in his later years a poem on the French Revolution, using methods comparable to those of Bancroft, he would have referred to his revolutionary diary, written at ages nineteen to twenty-six, consulted all poems on the Revolution written by others, placed on his desk a rhyming dictionary, and started to work. Naturally Wordsworth did not work in this way, but if he had done so, his later poetry would not have been improved.

A successful instance of the methodical use of accumulated information is provided by Noah Webster (1758–1843) (Scudder, 1895), who first described the differences between the British and the American usages of the English language. He listened to Britons and Americans, he read newspapers, he read books, he took voluminous notes. Making use of these materials, between the ages of forty-eight and sixty-eight he produced his *American Dictionary of the English Language.* A revision was completed at age eighty-two, and he was at work on another revision at his death at age eighty-five. Neither a novelist nor a dramatist can proceed in quite the same way. They cannot create a novel or a play by compilation, nor do they usually find it feasible to revise a previously published work. In general, architects, music composers, and creative writers, if they are to be recorded as being productive in their later years, must produce genuinely new works on an individual basis, whereas the requirements for continued productivity on the part of the scholar and scientist are different.

COMPARISONS OF OUR PRODUCTIVITY DATA WITH THOSE OF LEHMAN

While Lehman in *Age and Achievement* (1953) was concerned primarily with the ages at which *outstanding* works were done, he has several tables which bear titles such as "Percentage of total contributions during each decade of life." If the reader compares these with our tables, he will find differences which are striking. For example, Lehman's Table 38 in Chapter 15 entitled "Percentage of total contributions to philosophy during each decade of the philosophers' lives" indicates that the contributions made in the seventies by philosophers constituted only 2 percent of their total, whereas the comparable value in our data is 22 percent. In Lehman's table the output of philosophers was highest in the thirties. A considerable decline occurred subsequently. Our data shows no such decline. Many other apparent discrepancies could be pointed out.

The explanation is simple. In his Chapter 15, Lehman was dealing with the number of contributions which were made in each decade of life without regard to the number of contributors who were surviving in each decade. Thus the fact that philosophers, according to Lehman, have produced more work in the thirties than in any other decade is primarily due to the fact that in the thirties more philosophers were still alive. In this respect Chapter 15 in *Age and Achievement* differs from most of Lehman's other chapters.

When we attempt to make comparisons between Lehman's findings and ours with regard to productivity, we encounter other difficulties. Lehman's data on output *per decade per survivor* ordinarily refer only to superior works, not to total output. But in a few tables he does refer to total output per decade per survivor. Thus Figure 29 shows the mean productivity of philosophers living in each decade of life. The data represented in this figure show a decline with age which is greater than we have found. Again the differences in findings are probably due to differences in methodology. In our tables, the men whose later records were examined are the same men whose early records were examined. In Lehman's data, however, only some of the persons represented in the thirties are represented in the seventies, since many were deceased by age seventy. So far as we can determine, Lehman has no tables which are made up in the same way as ours; that is, based on the unselected output of persons all of whom lived to be eighty. He has some data on men, all of whom reached eighty, but these data refer only to "notable" contributions, not to total productivity. In other words, no findings of Lehman's relative to productivity contradict our data, since his were not gathered in the same manner as ours. It is our view that no valid statements can be made concerning age and productivity except from longitudinal data involving no drop-outs due to death.

In the present paper we have not been concerned with the frequency of "outstanding" contributions at different ages, the subject which engages most of Lehman's attention. At this point we feel obliged to indicate whether or not our data may have some bearing on that topic.

On another occasion we have argued (Dennis, 1956) that the apparent superiority of certain decades with respect to the output of outstanding works may be in part the result of defects in methodology. Our views on this topic have not changed, but we have no desire to repeat our former arguments at this time.

It is relevant here to indicate that in each field in which there is a sharp decrease with age in *total* output there is also an age decline in *superior* works. For example, since, as of 1924, the date of Loewenberg's publication, only 24 operas had been written by men in their seventies, it is obvious that not many *great* operas can be found which have been composed in this decade of life. Our data tend to show that in the arts the age decline in productivity is sufficient to account for a considerable part of the decrement in "outstanding" contributions. However, lessened productivity cannot be a major factor in any decrement in superior works observed among scholars, since scholars continue to be productive. But a full consideration of the factors which may produce data showing a decline with age in works of high quality is beyond the scope of the present paper.

SUMMARY

Data have been presented from documentary sources showing trends with age in the productivity of 738 persons, each of whom lived to age seventy-nine or beyond. These men were engaged in 16 areas of work which were classified in three major groups as scholarship, the sciences, and the arts.

It was found that, in many groups, the decade of the twenties was the least productive period. However, persons in the arts produced a larger part of their life-time output in this decade than did scholars and scientists. Relatively speaking, persons in the fine arts and in literature were more productive in the twenties and the thirties than were the scholars and scientists. The highest rate of output, in the case of nearly all groups, was reached in the forties or soon thereafter. From age forty onwards the output of scholars suffered little decrement. After age sixty the productivity of scientists decreased appreciably and the output of persons in art, music, and literature dropped even more than did that of scientists. This brief summary, however, does not do justice to the differences which occur within each major category.

The interpretation proposed is that the output curve of the arts rises earlier and declines earlier and more severely because productivity in the arts depends primarily upon individual creativity. Scholars and scientists require a greater period of training and a greater accumulation of experience and of data than do artists. The use of accumulated data and the possibility of receiving assistance from others permit the scholar and scientist to make more contributions in their later years than do those in art, music, and literature.

REFERENCES

Babkin, B. P. *Pavlov: A biography*. London: Gollancz, 1951.

Bell, E. T. *Men of Mathematics*. New York: Simon & Schuster, 1937.

Britten, J., and G. S. Boulger. *A biographical index of deceased British and Irish botanists*. London: Taylor & France, 1931.

Cajori, F. *A history of mathematics*. New York: Macmillan, 1937.

Cobbett, W. W. *Cobbett's cyclopedia survey of chamber music*. London: Oxford University Press, 1929, 1930.

Colvin, H. M. *Biographical dictionary of British architects, 1660–1840*. London: Murray, 1954.

Dennis, W. "Age and achievement: A critique," *Journal of Gerontology* 11: 331–333, 1956.

Hilditch, T. P. *A concise history of chemistry*. New York: Van Nostrand, 1911.

Lehman, H. C. *Age and achievement*. Princeton, N.J.: Princeton University Press, 1953.

Locy, W. A. *The story of biology*. New York: Holt, Rinehart and Winston, 1925.

Loewenberg, A. *Annals of opera, 1597–1940*. Geneva: Societas Bibliographica, 1955.

Nickles, J. M. *The geological literature of North America, 1785–1918*. Washington, D. C.: U. S. Government Printing Office, 1923–1924.

Nicoll, A. *A history of English drama, 1660–1900*. Cambridge: Cambridge University Press, 1952–1959.

Nordenskiöld, E. *The history of biology*. New York: Knopf, 1935.

Nye, R. B. *George Bancroft: Brahmin rebel*. New York: Knopf, 1944.

Raskin, E. "A comparison of literary and scientific ability," *Journal of Abnormal and Social Psychology* 31: 20–40, 1936.

Scudder, H. E. *Noah Webster*. Boston: Houghton Mifflin, 1895.

Smith, D. E. *A source book in mathematics*. New York: McGraw-Hill, 1929.

Part Five

Social Behavior
and Adjustment

For a variety of historical reasons those complex aspects of human life subsumed under broad headings such as "personality," "social behavior," and "adjustment" have been devoted almost exclusively to infancy, childhood, and the college years. These historical reasons include the heavy influence of psychoanalytic theory, the needs of public schools to understand and aid their charges, and the convenient accessibility of young subjects. It is hoped that, eventually, major longitudinal studies like those of the University of California's Institute of Human Development will yield comprehensive descriptions and analyses of their subjects (who are now well into middle age). Meanwhile, we must content ourselves for the most part with descriptions and evaluations of segments of the lifespan, and hope to achieve at least a tentative view of the development and changes that occur.

A fundamental question, of concern to personality theorists and developmental psychologists alike, pertains to the stability of personality over time. Is there a basic, continuing "core" of personality that makes for individual consistency, or does the person primarily react to the situation in which he finds himself? The answer to this question is of fundamental importance, of course, and has implications for theory, research, and clinical practice. In the lead article of Part Five, Walter Mischel of Stanford University posits his view that it is not necessary to assume that all aspects of personality have deeply ingrained, complex, historical antecedents. Mischel clearly favors the view that the situation is of prime importance in determining behavior. In another paper by Alex Inkeles, Harvard sociologist, the socialization issue is examined. Inkeles' paper is unique, for it contends that socialization is a life-long process; it does not end, as most developmentalists imply, during the adolescent period. Inkeles puts heavy emphasis upon the demands and expectations that the societal structure places on the individual as he or she passes through life.

In our strongly environmentalistic psychology, we seem to assume explicitly or implicitly that the developing social organism, the child, is a rather passive creature to whom things are done. That is, the child is more or less a product of the sum total of his experiences. (See the two previous articles, for examples.) Curiously enough, little attention has been paid to the child—or adult for that matter—as a social stimulus, a particular organism that determines to some extent how other people are going to act toward him or interact with him. Richard Q. Bell, of the Child Research Bureau of the National Institute of Mental Health, reports research which reveals that the child is not only a recipient, but also a stimulator, of parent behavior. His report here is called "Stimulus Control of Parent or Caretaker Behavior by Offspring."

Although data are sparse, it appears that personality topics that are salient to researchers in childhood development have little relevance to adulthood (for example, cognitive and linguistic development, resolution of the Oedipal crisis). Dr. Bernice Neugarten examines this issue in her paper, "Continuities and Discontinuities of Psychological Issues into Adult Life." Dr. Neugarten, chairman of the Committee on Human Development, University of Chicago, emphasizes here the discontinuities of adult life, and calls for the formulation of a psychology of adulthood.

Dr. Neugarten continues to develop her thesis in the next article, "Adult Personality: A Developmental View." She discusses two major questions: What are the salient psychological phenomena that characterize adulthood? And which of these are developmental? The consideration of such questions is necessary to the evolution of an adequate understanding of adult personality.

Despite what appears to be an American preoccupation with sex, up until the recent past sex as an intrinsic aspect of life has been viewed in print as the exclusive property of the young. Recent research makes it clear that sex interest and activity remains important and influential into the later years. The late Isadore Rubin, who was editor of *Sexology* and member of the board of directors of the Sex Information and Education Council of the United States, calls for a change in public attitude for the good of all in his paper, "The 'Sexless Older Years'—A Socially Harmful Stereotype."

One of life's sex-related landmarks for females is the menopause. For some women it may signal relief from childbearing, or cessation of monthly discomfort, while for others it portends old age and the loss of sexual attractiveness. Despite the importance of menopause, it has stimulated little psychological research (as distinguished from medical or biological study). Bernice Neugarten and associates carried out a study bearing on this period of life, in which age differences in attitude were examined: "Women's Attitudes toward the Menopause." Their findings are surprising, in light of the assumptions made by many social scientists regarding the meaning of this event.

Anyone old enough to read this volume has lived long enough to recognize some changes in the meaning for himself of specific passages of time. A summer vacation day in the life of a ten-year-old boy is vastly different from a day spent during examination week in college. Then too, the concepts of past and future shift with experience. Robert Kastenbaum, Wayne State University professor and editor of the new journal, *Aging and Human Development*, contributes "On the Meaning of Time in Later Life." He surveys existing research, and his consideration of the phenomenon of the perception of time should stimulate further thinking on this important aspect of our lives.

We are all intrigued with newspaper reports of the views and recollections of someone born during the administration of, say U. S. Grant. Paul Costa, lecturer at Harvard, and Robert Kastenbaum have systematized and analyzed some thoughts and memories gathered by the Social Security Administration on persons who have lived beyond the century mark. Their paper is called "Some Aspects of Memories and Ambitions in Centenarians." Such a paper seems an appropriate closing piece for a book concerned with the psychology of the human lifespan.

Twenty-four

Continuity and Change in Personality

Walter Mischcl

The question of continuity and change in personality has enduring importance, and the position that one takes on this topic profoundly influences one's approach to most other issues in personality psychology. Almost no psychologist, myself included, would argue with the basic and widely shared assumption that continuity does exist in personality development (for example, Kagan, 1969). Indeed, few other phenomena seem to be so intuitively self-evident. The experience of subjective continuity in ourselves—of basic oneness and durability in the self—is perhaps the most compelling and fundamental feature of personality. This experience of continuity seems to be an intrinsic feature of the mind, and the loss of a sense of felt consistency may be a chief characteristic of personality disorganization.

Clinically, it seems remarkable how each of us generally manages to reconcile his seemingly diverse behaviors into one self-consistent whole. A man may steal on one occasion, lie on another, donate generously to charity on a third, cheat on a fourth, and still construe himself readily as "basically honest and moral." Just like the personality theorist who studies them, our subjects also are skilled at transforming their seemingly discrepant behavior into a constructed continuity, making unified wholes out of almost anything.

It might be interesting to fantasize a situation in which the personality theorist and his subjects sat down together to examine each subject's data on behavioral consistency cross-situationally or over time. Actually it might not even be a bad idea for psychologists to enact such a

Walter Mischel, "Continuity and change in personality," *American Psychologist* 24, 1969, pp. 1012–1018. Copyright 1969 by the American Psychological Association, and reproduced by permission of the publisher and author.

This article is based on a paper presented at the symposium "Behavioral Continuity and Change with Development," held at the meeting of the Society for Research in Child Development, Santa Monica, California, March 27, 1969. Preparation of this paper was facilitated by Grant M-6830, from the National Institutes of Health, United States Public Health Service.

fantasy. In inspecting these data the theorist would look for genotypic unities that he is sure must be there; his subject would look for genotypic unities and be even more convinced that they exist and would proceed to find his own, often emerging with unities unknown to the theorist. But the consistency data on the IBM sheets, even if they reached statistical significance, probably would account for only a trivial portion of the variance, as Hunt (1965) has pointed out. A correlation of .30 leaves us understanding less than 10 percent of the relevant variance. And even correlations of that magnitude are not very common and have come to be considered good in research on the consistency of any noncognitive dimension of personality.

How does one reconcile our shared perception of continuity with the equally impressive evidence that on virtually all of our dispositional measures of personality substantial changes occur in the characteristics of the individual longitudinally over time and, even more dramatically, across seemingly similar settings cross-sectionally? I had the occasion to broadly review the voluminous evidence available on this topic of consistency and specificity (Mischel, 1968). In my appraisal, the overall evidence from many sources (clinical, experimental, developmental, correlational) shows the human mind to function like an extraordinarily effective reducing valve that creates and maintains the perception of continuity even in the face of perpetual observed changes in actual behavior. Often this cognitive construction of continuity, while not arbitrary, is only very tenuously related to the phenomena that are construed.

To understand continuity properly it is necessary to be more specific and to talk about types of variations and the conditions that regulate them. In this regard it may be useful to distinguish between consistency in various types of human activity.

There is a great deal of evidence that our cognitive constructions about ourselves and the world—our personal theories about ourselves and those around us (both in our roles as persons and as psychologists)—often are extremely stable and highly resistent to change. Data from many sources converge to document this point. Studies of the self-concept, of impression formation in person perception and in clinical judgment, of cognitive sets guiding selective attention—all these phenomena and many more document the consistency and tenacious continuity of many human construction systems (Mischel, 1968). Often these construction systems are built quickly and on the basis of little information (for example Bruner, Olver, and Greenfield, 1966). But, once established, these theories, whether generated by our subjects or ourselves, become exceedingly difficult to disconfirm.

An impressive degree of continuity also has been shown for another aspect of cognition: These are the features of problem solving called cognitive styles. Significant continuity often has been demonstrated on many cognitive style dimensions (for example, Kagan, 1969; Witkin, Goodenough, and Karp, 1967). The current prolific cognitive style explorations on this topic provide excellent evidence of developmental continuity. In this case the research also reveals a welcome continuity in our professional developmental history. Research into consistent individual differences in cognition has had deep roots and a long and distinguished history in experimental psychology. Simple cognitive measures like reaction time and response speed and duration have intrigued psychologists since the earliest laboratory work on mental measurement began more than seventy years ago. Individual differences on specific measures of problem solving, such as speed of reaction time and weight judgments, began to be explored in 1890 by James McKeen Cattell and others. Their studies of responses on specific cognitive and ability measures in the early laboratories were neglected when the development of practical intelligence testing started in this century. At that time, Binet and Henri shifted attention to the measurement of generalized intelligence by studying individual differences in more complex global tasks. Now it is refreshing to witness the reawakened interest in such enduringly important topics as reaction time and "conceptual tempo" and it is good to see sophisticated consistency evidence for it (Kagan, 1969). The generality and stability of behaviors assessed by these cognitive measures often have been found to be among the best available in personality research.

Some puzzling problems may arise, however, from the correlations found between some of the most promising new cognitive style measures and the traditional measures of generalized intelligence such as the performance I.Q. on the WISC. That is, correlations between measures of generalized intelligence and cognitive style such as Witkin's field dependence raise the question of the degree to which the consistency of cognitive styles may be due to their associations with intellectual abilities. The obtained generality and stability, as well as the external personality correlates, of at least some cognitive style measures thus may rest in part on their sizable correlations with indexes of more generalized intelligence and achievement behavior, as has been found in other studies (for example, Crandall and Sinkeldam, 1964; Elliott, 1961). To illustrate, the Witkin measures of cognitive style are strongly related to performance I.Q. ability indexes. Indeed the relationship between the Witkin Embedded Figures Test and the Wechsler Intelligence Block Design subtest is so strong that Witkin (1965) has indicated he is willing to use Block Design scores when available as a substitute for other field-dependence measures. When such cognitive styles as field independence and such coping patterns as "intellectualization" are substantially correlated with I.Q. then the stability reported for them and their correlates (for example, by Schimek, 1968) may partly reflect the stability of the I.Q.

This issue might also constitute a problem in interpreting such cognitive styles as Kagan's conceptual tempo. To the extent that conceptual tempo involves reaction time, and fast reaction time is a determinant of generalized performance I.Q. one would have to be alert to their interrelations, as has been pointed out by Campbell and Fiske (1959). It will be interesting to continue to explore exactly how conceptual tempo and other cognitive styles based on performance indexes such as response speed and accuracy take us beyond generalized ability measurement and into the domain of personality traits. Ultimately research on cognitive styles surely will provide a clearer analysis of intellective behavior. The implications of cognitive styles for the concept of general intelligence (as well as the reverse relation) should then become more explicit than they are now. In the course of these explorations the meaning of intercorrelations among diverse cognitive style measures—such as conceptual tempo, field dependence-independence, leveling-sharpening, and so on—will become clearer. At the same time our understanding of the interactions among cognitive and noncognitive personality dimensions hopefully will improve.

When we turn away from cognitive and intellective dimensions to the domain of personality and interpersonal behavior, consistency evidence is generally much harder to establish, at least whenever we use conventional tactics and the correlation coefficient (for example, Maccoby, 1969). On the basis of past literature on this topic, one should no longer be surprised when consistency correlations for social behavior patterns turn out to be quite low. Theoretically, in my view, one should not expect social behavior to be consistent unless the relevant social learning and cognitive conditions are arranged to maintain the behavior cross-situationally. On theoretical as well as on empirical grounds, much of the time there is no reason to expect great consistency in the social behaviors comprising most of our personality dimensions.

It is not possible to even begin to cite here the extensive evidence that I believe supports this point, namely, that noncognitive global personality dispositions are much less global than traditional psychodynamic and trait positions have assumed them to be (Mischel, 1968). A great deal of behavioral specificity has been found regularly on character traits such as rigidity, social conformity, aggression, on attitudes to authority, and on virtually any other nonintellective personality dimension (Mischel, 1968; Peterson, 1968; Vernon, 1964). Some of the data on delay of gratification with young children, emerging from our current studies at Stanford, are illustrative. In an ongoing longitudinal study on this problem we have obtained evidence that delay of gratification has some developmental consistency and increases with age, up to a point.[1] Much more impressive in my view, however, is our finding that within any child there exists tremendous variability on this dimension. Now we are studying how long preschool children will actually sit still alone in a chair waiting for a preferred but delayed outcome before they signal to terminate the waiting period and settle for a less preferred but immediately available gratification. We are finding that the same three-and-one-half-year-old child who on one occasion may terminate his waiting in less than half a minute may be capable of waiting by himself up to an hour on another occasion a few weeks earlier or later, *if* cognitive and attentional conditions are appropriately arranged. Our conclusion is that some significant predictions of length of voluntary delay of gratification certainly can be made from individual differences data; but the most powerful predictions by far come from knowledge of the cognitive and incentive conditions that prevail in the particular situation of interest.

These results are not at all atypical. A tribute to the interaction of person and environment is usually offered at the front of every elementary textbook in the form of Kurt Lewin's famous equation: Behavior is a function of person and environment. In spite of such lip service to the stimulus, most of our personality theories and methods still take no serious account of conditions in the regulation of behavior. Literally thousands of tests exist to measure dispositions, and virtually none is available to measure the psychological environment in which development and change occurs.

Evidence on observed instability and inconsistency in behavior often has been interpreted to reflect the imperfections of our tests and tools and the resulting unreliability and errors of our measurements, as due to the fallibility of the human clinical judge and his ratings, and as due to many other methodological problems. Undoubtedly all these sources contribute real problems. Some of these have been excellently conceptualized by Emmerich (1969). His emphasis on the need for considering rate and mean changes over age if one is to achieve a proper understanding of continuity, growth, and psychological differentiation is especially important. Likewise, his call for longitudinal, multimeasure, and multivariate studies needs to be heeded most seriously.

I am more and more convinced, however, hopefully by data as well as on theoretical grounds, that the observed inconsistency so regularly found in studies of noncognitive personality dimensions often reflects the state of nature and not merely the noise of measurement. Of course, that does not imply a capriciously haphazard world—only one in which personality consistencies seem greater than they are and in which behavioral complexities seem simpler than they are. This would, if true, be extremely functional. After all, if people tried to be radical behaviorists and to describe each other in operational terms they would soon run out of breath and expire. It is essential for the mind to be a reducing valve—if it were not it might literally blow itself!

Perhaps the most widely accepted argument for consistency in the face of seeming diversity is the one mentioned so often, the distinction between the phenotypic and the genotypic. Thus most theorizing on continuity seems to have been guided by a model that assumes a set of genotypic personality dispositions that endure, although their overt response forms may change. This model, of course, is the one shared by traditional trait and dynamic dispositional theories of personality. The model was well summarized in the example of how a child at age twelve may substitute excessive obedience to a parent for this earlier phobic reaction as a way of reducing anxiety over parental rejection (Kagan, 1969). At the level of physical analogy Kagan spoke of how the litre of water in the closed system is converted to steam and recondensed to liquid.

This type of hydraulic Freudian-derived personality model, while widely shared by personality theorists, is of course not the only one available and not the only one necessary to deal with phenomena of continuity and change. Indeed, in the opinion of many clinical psychologists the hydraulic phenotypic-genotypic model applied to personality dynamics, psychotherapy, and symptom substitution has turned out to be a conceptual trap leading to some tragic pragmatic mistakes in clinical treatment and diagnosis for the last fifty years (for example, Mischel, 1968; Peterson, 1968). I am referring, of course, to the unjustified belief that seemingly diverse personality problems must constitute symptoms of an underlying generalized core disorder rather than being relatively discrete problems often under the control of relatively independent causes and maintaining conditions.

The analysis of diverse behaviors as if they were symptomatic surface manifestations of more unitary underlying dispositional forces also is prevalent in our theories of personality development (for example, Kagan, 1969; Maddi, 1968). But while diverse behaviors often may be in the service of the same motive or disposition, often they are not. In accord with the genotype-phenotype distinction, if a child shows attachment and dependency in some contexts but not in others one would begin a search to separate phenotypes from genotypes. But it is also possible that seeming inconsistencies, rather than serving one underlying motive, actually may be under the control of relatively separate causal variables. The two behavior patterns may not reflect a phenotype in the service of a genotype but rather may reflect discrimination learning in the service of the total organism. Likewise, while a child's fears sometimes may be in the service of an underlying motive, most research on the topic would lead me to predict it is more likely that the fear would involve an organized response system with its own behavioral life, being evoked and maintained by its own set of regulating conditions (for example, Bandura, 1969; Paul, 1967).

When we observe a woman who seems hostile and fiercely independent some of the time but passive, dependent, and feminine on other occasions, our reducing valve usually makes us choose between the two syndromes. We decide that one pattern is in the service of the other, or that both are in the service of a third motive. She must be a really castrating lady with a facade of passivity—or perhaps she is a warm, passive-dependent woman with a surface defense of aggressiveness. But perhaps nature is bigger than our concepts and it is possible for the lady to be a hostile, fiercely independent, passive, dependent, feminine, aggressive, warm, castrating person all-in-one. Of course which of these she is at any particular moment would not be random and capricious—it would depend on who she is with, when, how, and much, much more. But each of these aspects of her self may be a quite genuine and real aspect of her total being. (Perhaps we need more adjectives and hyphens in our personality descriptions. That is what is meant, I think, by "moderator variables.")

I am skeptical about the utility of the genotype-phenotype distinction at the present level of behavioral analysis in personality psychology because I fear it grossly oversimplifies the complexity of organized behavior and its often nonlinear causes. The genotype-phenotype oversimplification may mask the complex relations between the behavior and the organism that generates it, the other behaviors available to the organism, the history of the behavior, and the current evoking and maintaining conditions that regulate its occurrence and its generalization.

The question of the nature of the similarity or dissimilarity among the diverse responses emitted by a person is one of the thorniest in psychology. Even when one response pattern is not in the service of another the two of course may still interact. No matter how seemingly separated the various branches of behavior may be, one can always construe some common origins for them and some current interactions. At the very least, all behavior from an organism, no matter how diverse, still has unity because it is all generated from the same source—from the same one person. At the other extreme, incidentally, few response patterns are ever phenotypically or physically identical: Their similarity always has to be grouped on some higher-order dimension of meaning. To make sense of bits of raw behavior one always has to group them into larger common categories. The interesting theoretical issue is just what the bases of these groupings should be. Dispositional theories try to categorize behaviors in terms of the hypothesized historical psychic forces that diverse behaviors supposedly serve; but it is also possible to categorize the behaviors in terms of the unifying evoking and maintaining conditions that they jointly share.

Moreover, few potent response patterns can occur without exerting radical consequences for the other alternatives available to the person. Thus an extremely "fast-tempo" child may be so active that, in addition to fatiguing his parents, he may as Kagan (1969) found, smile less. Perhaps that happens because he is too busy to smile. My comment about how fast-tempo children may be too busy to smile is not really facetious. One of the intriguing features of any strong response syndrome is that it soon prevents all kinds of other intrinsically incompatible behaviors. If a child darts about a lot and is fast there are all sorts of other things he automatically cannot do. His speed in living, his pace, not only automatically influences his other possible behavior, it also soon starts to shape his environment. I now expect my fast-tempo children to be fast tempo, and currently it takes almost no cues from them to convince me I am right about them.

It would have been relatively simple to assess and predict personality if it had turned out to consist mainly of stable highly generalized response patterns that occur regularly in relation to many diverse stimulus constellations. The degree and subtlety of discrimination shown in human behavior, however, is at least as impressive as is the variety and extensiveness of stimulus generalization. What people do in any situation may be altered radically even by seemingly minor variations in prior experiences or slight modifications in stimulus attributes or in the specific characteristics of the evoking situation. From my theoretical perspective this state of affairs—namely, the enormously subtle discriminations that people continuously make, and consequently the flexibility of behavior—is not a cause of gloom. Instead, the relative specificity of behavior, and its dependence on environmental supports, is the expected result of complex discrimination learning and subtle cognitive differentiation. When the eliciting and evoking conditions that maintain behavior change—as they generally do across settings—then behavior surely will change also. While the continuous interplay of person and condition may have been a surprise for faculty and trait psychology it should come as no upset for us now. If one pays more than verbal tribute to the dependency of behavior on conditions, and to the modification of behavior when situations change, then the so-called negative results of dispositional research on behavioral continuity appear attributable largely to the limitations of the assumptions that have guided the research. From the viewpoint of social behavior theory the findings of behavioral specificity, rather than primarily reflecting measurement errors, are actually congruent with results from experimental research on the determinants and modification of social behavior (Mischel, 1968). When response consequences and valences change so do actions; but when maintaining conditions remain stable so does behavior.

The last decade has seen an exciting growth of research on cognitive styles and many researchers have begun to study the person as an information-processing and problem-solving organism. Generally, however, these processes have been viewed in dimensional and dispositional terms and quickly translated back to fit the consistency assumptions of traditional global trait and psychodynamic theory. Individual differences on dimensions such as conceptual tempo, field dependence, leveling-sharpening, and so on, have been isolated with some promising results. Less progress has been made in applying the concepts and language of information processing and cognitive styles to forming a better theoretical conception of personality structure itself. It has become fashionable to speak of the organism as creating plans, generating rules, and, depending on his needs and situations, devising strategies. These tactics yield payoffs and consequences, and in light of these the person modifies his plans accordingly. But when contingencies change stably, what happens? For example, what happens when the mother-dependent child finds that his preschool peers now consistently have little patience for his whining, attention-getting bids, and instead respect independence and self-confidence? Generally the child's behavior changes in accord with the new contingencies, and if the contingencies shift so does the behavior—if the contingencies remain stable so does the new syndrome that the child now displays. Then what has happened to the child's dependency trait?

One might argue that the basic genotype remained but its manifestation phenotypically has altered. But is this just a "symptom" change leaving unaffected the psyche that generated it and the lifespace in which it unfolds? A vigorous "No!" to this question comes from much research on behavior change in the last few years (for example, Bijou, 1965; Fairweather, 1967; Mischel, 1966; Patterson, Ray, and Shaw, 1969).

What would happen conceptually if we treated the organism as truly active and dynamic rather than as the carrier of a stable dispositional reservoir of motives and traits? Might one then more easily think of changes in the developing organism not as phenotypic overlays that mask genotypic unities but as genuinely new strategies in which many of the person's old plans are discarded and replaced by more appropriate ones in the course of development? (Perhaps Gordon Allport's idea of functional autonomy needs to be rethought.) Can the person even become involved in plans to change what he *is* as well as what he *does*? George Kelly and the existentialists in their search for human nature noted that existence precedes essence. According to that position, to find out what I *am* I need to know what I *do*. And if my actions change do they leave me (the "real me") behind? Or perhaps they just leave some of my discarded psychological genotypes behind?

A search for personality psychology that has conceptual room for major variability and changes within the individual's dispositions can easily be misinterpreted as undermining the concept of personality itself.

That would be an unfortunate misconstruction. Instead, we do need to recognize that discontinuities—real ones and not merely superficial or trivial veneer changes—are part of the genuine phenomena of personality. If one accepts that proposition, an adequate conceptualization of personality will have to go beyond the conventional definition of stable and broad enduring individual differences in behavioral dispositions. We may have to tolerate more dissonance than we like in our personality theory. To be more than nominally dynamic our personality theories will have to have as much room for human discrimination as for generalization, as much place for personality change as for stability, and as much concern for man's self-regulation as for his victimization by either enduring intrapsychic forces or by momentary environmental constraints.

NOTE

[1] W. Mischel, E. B. Ebbesen, and A. Raskoff. In progress research report, Stanford University, entitled "Determinants of Delay of Gratification and Waiting Behavior in Pre-school Children."

REFERENCES

Bandura, A. *Principles of behavior modification.* New York: Holt, Rinehart and Winston, 1969.
Bijou, S. W. "Experimental studies of child behavior, normal and deviant." In: Krasner, L., and L. P. Ullmann (eds.), *Research in behavior modification.* New York: Holt, Rinehart and Winston, 1965.
Bruner, J. S., R. R. Olver, and P. M. Greenfield. *Studies in cognitive growth.* New York: Wiley, 1966.
Campbell, D., and D. Fiske. "Convergent and discriminant validation by the multitrait-multimethod matrix," *Psychological Bulletin* 56: 81–105, 1959.
Crandall, V. J., and C. Sinkeldam. "Children's dependent and achievement behaviors in social situations and their perceptual field dependence," *Journal of Personality* 32: 1–22, 1964.
Elliott, R. "Interrelationships among measures of field dependence, ability, and personality traits," *Journal of Abnormal and Social Psychology* 63: 27–36, 1961.
Emmerich, W. *Models of continuity and change.* Paper presented at the meeting of the Society for Research in Child Development, March 27, 1969, Santa Monica, California.
Fairweather, G. W. *Methods in experimental social innovation.* New York: Wiley, 1967.
Hunt, J. McV. "Traditional personality theory in the light of recent evidence," *American Scientist* 53: 80–96, 1965.
Kagan, J. *Continuity in development.* Paper presented at the meeting of

the Society for Research in Child Development, March 27, 1969, Santa Monica, California.

Maccoby, E. E. *Tracing individuality within age-related change.* Paper presented at the meeting of the Society for Research in Child Development, March 27, 1969, Santa Monica, California.

Maddi, S. R. *Personality theories: A comparative analysis.* Homewood, Ill.: Dorsey Press, 1968.

Mischel, W. "A social learning view of sex differences in behavior." In: Maccoby, E. E. (ed.), *The development of sex differences.* Stanford: Stanford University Press, 1966.

Mischel, W. *Personality and assessment.* New York: Wiley, 1968.

Patterson, G. R., R. S. Ray, and D. A. Shaw. "Direct intervention in families of deviant children," *Oregon Research Institute Bulletin, 8(9):* 1–62, 1969.

Paul, G. L. "Insight versus desensitization in psychotherapy two years after termination," *Journal of Consulting Psychology 31:* 333–348, 1967.

Peterson, D. *The clinical study of social behavior.* New York: Appleton-Century-Crofts, 1968.

Schimek, J. G. "Cognitive style and defenses: A longitudinal study of intellectualization and field independence," *Journal of Abnormal Psychology 73:* 575–580, 1968.

Vernon, P. S. *Personality assessment: A critical survey.* New York: Wiley, 1964.

Witkin, H. "Psychological differentiation and forms of pathology," *Journal of Abnormal Psychology 70:* 317–336, 1965.

Witkin, H., D. R. Goodenough, and S. A. Karp. "Stability of cognitive style from chidlhood to young adulthood," *Journal of Personality and Social Psychology 7:* 291–300, 1967.

Twenty-five

Social Structure and Socialization

Alex Inkeles

Social structure impinges on, and in many ways determines socialization. In its turn, socialization may have substantial effects on social structure. This relationship is not necessarily one of discrete interactions, but may take the form of cycles or other sequences prolonged over substantial periods of historical time. We do not deal here with exchange between two self-contained and more or less independent systems of action, but rather with a part-whole relationship, a substructure of socialization being one of the functional requisites of any social system (Levy, 1952). Understanding this relationship is further complicated by the fact that the part is not entirely subsumed under the whole, but rather includes elements dependent on still other systems.

Like a distressingly large number of other social science terms, "socialization" applies to an exceedingly large range of phenomena. It simultaneously describes a process, or input, external to the person, the individual's experience of the process, and the end product or output. In its broadest conception, socialization refers to the sum total of past experiences an individual has had which, in turn, may be expected to play some role in shaping his future social behavior. From the perspective of different disciplines one would, of course, take a stronger interest in some of these early experiences, and be concerned more with certain later behaviors, than others. Encounters unique to a given person, yielding decidedly idiosyncratic behavior, may be of great significance to the clinical psychologist, but they have little interest for the anthropologist, whose concern is more with shared and culturally patterned experiences. Some practices which are widely diffused and personally important to many people, let us say the trimming of the Christmas tree, may be prime data for the anthropologist, but will meet relative indifference on the part of the sociologist because these experiences do not in major degree shape one's selection of or performance in importantly differentiated social roles.

Alex Inkeles, "Social Structure and Socialization," in David T. Goslin (ed.), *Handbook of Socialization Theory and Research*, © 1969 by Rand McNally and Company, Chicago, pp. 615–632.

From the sociological point of view, socialization refers to the process whereby individuals acquire the personal system properties—the knowledge, skills, attitudes, values, needs and motivations, cognitive, affective and conative patterns—which shape their adaptation to the physical and sociocultural setting in which they live (see Inkeles, 1966b). The critical test of the success of the socialization process lies in the ability of the individual to perform well in the statuses—that is, to play the roles in which he may later find himself. The subtlety and complexity of the problem of socialization stems, in good part, from the diversity of these statuses and from the uncertainty as to the roles which may be associated with them.

Some of the statuses a person will acquire are definitely known and more or less invariant. His sex and age group, usually his religion and ethnicity, fall in this category. Socialization to the roles appropriate to such statuses may, therefore, be relatively less problematic. Other positions the person will later occupy may be much less definite. Thus, it may at the outset be quite uncertain whether a man will be chief or follower, doctor or boot black, married or single. Some of the statuses a man can find himself in as an adult may not even have existed when he was a boy, as for example the airplane or space pilot. Some roles a man plays he may largely have invented or created himself as in the case of Young Man Luther (Erikson, 1958). The problems of socialization will therefore be markedly different in societies with more complex status differentiation than in those of more simple structure, and will be much more perplexing in those societies which assign individuals to their statuses predominantly on an ascriptive basis as against those which leave the attainment of most status positions more open to achievement or to accident.

Whatever the degree of rigidity or flexibility in their usual pattern of allocating individuals to status positions, societies will also vary in the degree to which they are experiencing stability or undergoing rapid change. When rapid change is experienced by a society, the established system for assigning individuals to recognized statuses may break down. Wholly new statuses may come into existence, sometimes in large numbers and requiring many status incumbents. An outstanding example would be the shift in the modern Euro-American nations from a predominantly rural resident, agriculturally employed, labor force to an overwhelmingly urban resident, industrially employed, population. Quite substantial strain may thus be put on the existing system of socialization, and many individuals may find themselves inadequately socialized to the demands of the roles they are now called on to play. Late or adult socialization, and other forms of re-socialization, may therefore be required. Even if such measures are adopted, many individuals may feel themselves dislocated, unfit, or inadequately prepared for the tasks facing them in the statuses they have acquired or to which they have been assigned. In turn, the society or major institutional units within it, may find the discharge of their social responsibility for production, security, governance or whatever, impaired by the inadequate role performance of the individuals they have recruited or have had assigned to them.

To fashion a very simple analogy, then, we may say that the problem of socialization is rather like that of putting on a play. Under the simplest of circumstances the script is fixed and has been handed down unchanged through generations. There are very few parts to play, and it is well known exactly who will play which part, so that firm preparation may be made long in advance. Everyone knows all the parts very well, so that interaction among the actors is smooth and satisfying. Since the play has been played many times before, all the props and necessary accoutrements are in ample supply and well-tried. The performance runs smoothly and continuously. This presumably is the condition which prevailed in relatively stable and isolated pre-literate or "primitive" societies before their encounter with modern "civilization."

At the other pole we have the style of play which is more like a modern "happening." In this case there is no specification as to exactly how many "actors" there are, and their "parts" are only vaguely defined. Indeed, there may be no fixed script at all. The play may never have been put on in quite this form before, so that the interplay of the actors is uncertain and difficult, and neither the other players nor the audience can do much to influence the course of the action in the "right" direction. New parts are being added all the time, and old ones cut out. No one is too firmly assigned to any given part, and the actors often improvise their roles as they go along. This more extreme, and perhaps limiting, case highlights the point that socialization, or training for role performance, would be quite a different task for the directors, the actors, and the audience in this more unstructured social drama than would be the case in the more traditional pattern. Of course, no society and very few organizations could for very long endure in a state so fluid as to approximate the model of a happening. Nevertheless, due to the rapidity and extensiveness of social change, many parts of the world in the mid-twentieth century more nearly approximate the happening than they do the fixed and frozen drama of the traditional model of socialization.

Every individual, as we encounter him, is the outcome, the "product," in a sense, of a given socialization process. This outcome, "successful" or "unsuccessful" from the perspective of a particular social system in a given time and place, will, in fact, depend on a series of inputs. Most apparent, it will depend on the original genetic potential of the individual and the degree to which climate, alimentation, nurturance, and other factors have permitted the realization of that potential. The outcome will also have been influenced, in part, by the personality of the individuals with whom the socializee has been in significant contact, such as relatives, teachers, friends, co-workers, and bosses. Finally, the result of socialization will be shaped by certain more or less fixed or regular aspects of the network of social relations in which the individual lives. It is these regular, recurrent, or socially structured aspects of the individual's experience which are of particular interest to the sociologist as "inputs" in the socialization process. Similarly, on the "output" side the sociologist is concerned mainly with those actions of the individual which play some part in the maintenance and change of those patterned and recurrent modes of social interaction which provided the systematic property of the concept of social structure.

The complexity of both social structure and socialization is such that we cannot hope to deal with the impact of the former on the latter without considerable differentiation of the elements of both. For socialization we shall adopt the conventional division of the life cycle into periods of infancy and childhood, youth and adolescence, the middle years, and old age. Each period may be conceived as focused on some issue central to that stage of personal development, thereby rendering certain aspects of social structure more relevant as influences on the socialization process.

For each period we will identify and discuss four main elements in the socialization matrix: the main socialization *issue*, that is, the typical life condition or social demand which dominates the attention of socializee and socializers and becomes the characteristic or defining aspect of any given stage of individual bio-social development; the *agents* of socialization, those individuals and social units or organizations which typically play the greatest role in the socialization process in the several stages of development; the *objectives* which these agents set as the goals for successful socialization in each period, that is, the qualities they wish to inculcate and the conditions under which they prefer to train the socializee; and the main *task* facing the socializee, that is, the problem to be solved or the skill learned as it confronts the socializee from his internal personal perspective.

Several broad aspects of social structure may be considered of likely relevance to the socialization process at any stage of the life cycle, and may therefore serve us as a checklist to insure systematic coverage. The dimensions we will consider are:

1. *Ecological:* here we note the size, density, physical distribution, and social composition of the population; its relationship to its resource base, and to surrounding populations.

2. *Economic:* this has special reference to the social forms for defining, producing, and distributing goods and services. The type and amount of material resources available in a given society are also

important. Economic and ecological elements may be intimately related.

3. *Political:* the political subsystem encompasses the structure of power: its distribution, forms, and application, along with the institutional arrangements for generating, legitimating, and exercising it.

4. *System of values:* economic and political institutions of course embody values, but many important values which guide socialization efforts are not most visible or effective in their more institutionalized forms. In addition, a large part of socialization consists in the simple effort to inculcate values.

These four dimensions are not all of the same type. The ecological concerns primarily the population; the system of values mainly expresses what is sometimes called culture; and the economic and political dimensions have reference mainly to institutions. These latter dimensions of social structure may, however, also be understood as referring not so much to concrete institutions as to functions which must be fulfilled. Our four aspects of social structure are, therefore, not exhaustive. Others might well be selected, either to enlarge this list or to reflect a different perspective. We might have included or substituted other dimensions of social structure such as the system of stratification, or the modes of social control and the patterns of deviance; or other institutional complexes such as the kinship organization, the family, the school, or the work unit. Our objective here, however, is not to provide a catalogue of social institutions. Rather it is to present a mode of analysis, which can serve to *illustrate* the relations between selected elements of social structure and the socialization process. We mean to provide a model which can easily be carried over for use in the analysis of socialization issues as they arise in other institutions and social structures.

The elements of social structure are not mutually exclusive, but interpenetrate in a complex web of relations constituting the total sociocultural system. The degree of consistency or coherence among the socialization pressures generated in each of the realms of social structure, presents special problems for analysis. For example, socialization to the demands of the economic system may require the traits of aggressiveness, including the objective of eliminating your competitors; secrecy, as in the preservation of trade secrets; and the minimization of charitable impulses in favor of the maximization of profits. By contrast, the effective functioning of the political system may require individuals to display tolerance for opposition and especially for defeated opponents; to disclose fully and publicly all relevant facts at one's disposal; and to assign priority to humane and welfare goals over other considerations. Where there are such discordances among the role requirements of different elements of the same society, the agents of socialization face difficult choices of emphasis, just as, in time, the socializee will face difficult choices both in selecting his own course of action and in reconciling the action in one role subsystem with that in another.

There are, in addition, major complications which arise from the problem of levels of complexity in social structure. In the economic subsystem of modern society, for example, the elements of structure will range from a unit relatively as simple and concrete as the role of industrial worker, through institutions such as the factory, and networks of institutions such as "an industry," to something so large and complex as "the economic system of the United States." There may be a tendency for certain qualities to be widely diffused across all the levels of any given subsystem of society, thus rendering the same personal attributes adaptive throughout. It is often the case, however, that the performances typically required at different levels of the *same* subsystem are less alike than those for comparable levels of *different* subsystems. Both the assembly line worker and the factory manager or entrepreneur may require a strict sense of time to fulfill their obligations. But for effective performance of his role the entrepreneur will also need high achievement aspirations, initiative, and autonomy in far greater degree than is required of the ordinary worker. For his part, the production line worker job may require manual dexterity, or interpersonal skills in handling his relations with supervising personnel and fellow workers, which in nature or degree are quite different from the skills appropriate to management roles. Whether these appropriate special qualities are developed during early life or through later training "on the job," different socialization practices clearly will be required to equip men to perform these different roles which are nominally part of the "same" societal subsystem.

We cannot here systematically review the problem of socialization as it presents itself at each of the levels in each realm or subsystem of society. Rather, we must restrict ourselves to illustrations which can serve as models of a mode of analysis appropriate to the complexity of the social structure of large scale societies.

INFANCY AND EARLY CHILDHOOD

These two periods of the life cycle are distinctive enough to warrant separate treatment, but for reasons of economy we treat them together. The central *issue* of this period is the infant's helplessness, his utter dependence on adults for sustenance in life. The main *objective* of the socialization agents, apart from keeping the child alive, is to move him on the next stage of his development. The main *task* of the child is to master control of his body, to adjust it to rhythms and other modes of action acceptable to the adults around him, and to secure himself against the most obvious physical dangers of his environment. Problems of physical mastery and control occupy a great deal of everyone's attention, socializer and socializee alike, not only in toilet training, but in bringing the child to the point where he can walk, talk, feed and dress himself, and learn to stay out of the most obvious and immediate dangers such as fire and water.

Probably the single most important acquisition of this period, however, is language, the control of which is only partly a matter of physical development. Language is the critical symbol system which will mediate the greater part of all later experience. Indeed, the other processes of physical training also have significance far beyond their immediate and obvious purpose, since they inevitably affect the psychic qualities of the child as well. The simplest and most efficient way to keep a child out of danger might be physical restraint, and beyond that would be to issue constant and fearful warning cries as he moved about. While this would keep him out of danger, it would also greatly restrict his exploratory behavior and encourage passivity and timidity. A society which requires its males to be assertive and fearless, as for example in hunting big game, could hardly afford paying that price in order to keep its children out of trouble around the campgrounds.

Taking into account the roles girls will later play, parents almost everywhere seem to train them more vigorously in obedience, responsibility, and nurturance, whereas boys are more socialized to be achieving, self-reliant, and independent. Where the conditions of life seem to require the utmost of males in this respect—as in cultures herding large animals— the emphasis on these "manly" qualities is generally heightened. It is interesting to note, however, that in societies in which self-reliance is maximally adaptive, the mode is often not strictness of early training, but rather the giving of great freedom to young boys. Apparently spending a good deal of time away from home in the company of other boys in autonomous gangs is recognized in many societies as likely to be the most effective method for early inculcation of independence, self-reliance, and the drive for achievement (see Aberle, 1961).

ECOLOGICAL

The direct impact of ecological patterns on infant socialization practices has not been much studied. The world is experiencing increasing densities of population both because of absolute population growth and the ubiquitous spread of urbanism. Freedom to explore one's environment over a wider range, and to do so less under the immediate surveillance of adults, would seem to be greater in rural settings, although this might be less true of an Indian peasant village, with its extremely high density, than of an

American urban slum. The shift to smaller families, and to physical separation of the nuclear conjugal family from extended kinship groups, should bring greater concentration and perhaps intensity of interaction between socializing agents and the child. Instead of passing from hand to hand and encountering one face after another, the child is limited, the greater part of the time, to interaction with the mother, and perhaps to one or two siblings. If the mother is inadequate to the task of child-rearing, the consequences will obviously be more serious than in environments in which there are many parent surrogates who can step in to supplement the mother's efforts. Barring this contingency of an inadequate mother, however, the intense pattern of interaction of mother and child is likely to lead to precosity and early development of speech and perhaps motor skills.

Little truly systematic research has been done comparing socialization in rural as against urban families. Studies in France indicate there are indeed important differences which rest, in part, on the dissimilarity in occupation and in part on the ecology of rural versus urban life. The greater isolation and family centeredness of rural life means a slow social awakening for the rural children, greater fear of strangers, and slower development of imagination and language skills. By contrast, both the responsibilities early assigned to them, and their contact with animals and nature, seem to yield the rural children early and highly developed sensory motor functioning (Lanneau and Mabrieu, 1957).

The movement from countryside to city typical of the first part of the twentieth century has in mid-century been followed by a new movement from the city to the suburbs. From an ecological point of view, however, we should note that in suburban living the physical separation of the father's place of work, and hence his absence from the home, is even more marked than in the city. The physical separation of each family homestead from every other is also notable. In addition the suburbs, at least in the United States, are characterized by higher degrees of homogeneity in their ethnic, religious, and socioeconomic composition than are most cities. As yet, however, we do not know what effect, if any, these and other ecological consequences of suburbanization will have on infant and child socialization.

Economic

In systems in which scarcity of food or shelter is acute, the helplessness of the child may lead to the extinction of all but the most hardy. In certain districts of East Pakistan, for example, when there are food shortages it is expected that the distribution will be arranged to insure the survival and effectiveness of the male head of the household. Even barring such extreme circumstances, the resources generally available in any given society will obviously influence the conditions of socialization of the child. These effects will, in part, be mediated by the system of stratification. McClelland (1961) argues that the increased wealth of the more ambitious and rising classes of society leads them to acquire servants to whom they turn over the rearing of their children, with a consequent reduction in the children's need for achievement (*n ach*) and subsequent decline in their entrepreneurial effectiveness.

The most direct effects of the economic order on the conditions and patterns of infant socialization are related to the work and production patterns characteristic for a given society. If young women are expected to work in the fields during child-bearing years, as among the Tanala as described by Kardiner (1939), the prospect is increased for relative neglect of children who must either be left at home unsupervised or brought to the fields. The nature of the father's employment will greatly affect the degree to which children grow up in a father-absent home. Certain occupations such as fishing or itinerant selling, or the lack of local employment opportunities, may require long periods of absence by the father. Indeed, the fathers may be away almost all the time, and the children grow up in what is, in effect, a female-headed household more or less exclusively subject to the socialization efforts of women. This condition is extremely common among American Negro families, and in the famous Moynihan Report it was identified as perhaps the most critical problem of the American Negro family (see Rainwater and Yancey, 1967). One popular psychoanalytically oriented theory holds that in the father-absent home the young boy develops too close an identification with the mother. Fear of this feminine identity later leads to a reaction formation, characterized by overassertion of masculine traits, which in slum conditions easily finds expression in male delinquency. Evidence in support of this hypothesis drawn from American cities is not substantial, although research on nonliterate societies seems to support the theory (see Whiting, 1965; Bacon, Child and Barry, 1963).

The nature of a man's experience on the job may also produce effects carried over into the home and expressed in his treatment of his children. Miller and Swanson (1960) made an important contribution in their investigation of the consequences for child training—in systems of expressive behavior, styles of punishment, and cognitive modes—as these may depend on whether the father is employed in a bureaucratic (mass) or entrepreneurial (small scale) setting. They reasoned, for example, that middle class "bureaucrats"—meaning mainly people working for large scale organizations—are currently faced by diminishing needs to internalize traditional standards, and hence would less often emphasize defenses such as turning on the self and reversal. By contrast, the working-class men in these bureaucratic atmospheres, faced by work ever more simple and routine, were expected to resort increasingly to the defense of denial and to the cultivation of leisure activities.

Miller and Swanson were far from successful in accurately rating the quality of the occupational setting of their subjects as "entrepreneurial" or "bureaucratic," although the bulk of their evidence supports the theoretical importance of the distinction they introduced. Further support for the idea that the quality of the father's work setting will significantly influence his child-rearing efforts was provided by McKinley (1964), who obtained fairly precise measures, separately for the middle and working class, of the degree to which the father was closely supervised on the job. He found, in both social strata, that fathers who were more closely supervised—as against those with more autonomy on the job—tended to be more punitive in the treatment of their sons. These relationships apparently are quite general. Thus, Pearlin and Kohn (1966) demonstrated that in Italy as well as the United States the closeness of supervision, the amount of self-reliance required on a job, and the extent to which it involves working with people or things all seem to influence the values which fathers emphasize in their child-rearing.

An ever-increasing proportion of the population is employed by large, often vast, corporate enterprises, and the labor force within them is believed to be experiencing increasingly common conditions of work. Along with ever rising personal income, the equalizing effects of social welfare legislation, and of benefits providing more widespread leisure, these changes in the occupational structure may in time bring about a high degree of homogenization in the work and work-related life experience of all members of the labor force. If it is correct that differentiated work conditions for the breadwinner yield differentials in their respective child-rearing practices, it should follow that homogeneity in work and work-related conditions will in time increase the homogeneity of socialization practices.

POLITICAL

Just as in the control of economic resources, so in the command of power the dominance of the adult over the infant is virtually absolute. With increasing motility the child can escape surveillance, restraint, and punishment, but only to very limited degrees. The forms in which parental power is exercised are believed to play a major role in determining the patterns of political action later to characterize the adult. Fromm (1941) early explored this relationship in the development of authoritarianism in the German family, and his work influenced the classic study of *The Authoritarian Personality* (Adorno, 1950). Erikson (1950) has interpreted the structure of power relationships in the American family as a training ground in compromise particularly suited to the pattern of political life in the United States.

Just how far the political system as a whole, and the parent's position in it, influence the forms and content of the socialization of his children is not well documented. The research of McKinley, and Miller and Swanson cited above, would suggest a model of direct carry-over—the more authoritarian, or democratic, the larger political milieu in which the parent participates, the more authoritarian, or democratic, his treatment of his own children. There is room, however, for a predictive model based on the anticipation of a reaction in the home against the predominant pressures of the political system outside it.

THE VALUE SYSTEM

Limits on the child's grasp of language and relevant experience make the direct and explicit inculcation of values of minor importance in infancy and early childhood. The indirect influence of cultural values in shaping socialization practices is, however, enormous, and the child's acquaintance with the values implicit in various socialization practices may be more important than the manifest content he learns from these activities.

Feeding schedules, for example, should be understood not exclusively in terms of nurturance or in relation to the cycle of hunger pains in the infant, but also as expressing, and presumably communicating to the child, some of the emphasis his culture places on the ordering of events in strict accord with a clock-paced schedule. Such an orientation to time, if it exists, will, of course, be communicated to the child not through one infant-care discipline alone, but through several. The orientation to time will thus be expressed not only in feeding schedules, but in toilet training, play and sleeping arrangements, procedures for dress, and in numerous other ways. Similarly, the handling of the child and the procedures adopted for or imposed on his care will express his society's values concerning orderliness and abstemiousness, aggressiveness, expressive behavior, striving, intellect and so on through the register of culturally important value themes.

Of course, the values being thus expressed and inculcated in the child will not exist in a social vacuum. The value system of any culture has a certain integrity and autonomy which permits it to persist relatively intact over time, but it must inevitably reflect the influence of conditions in the social structure in which the values are operative. Thus, in an emerging industrial society we may well expect the value of timeliness to be fostered and reinforced as an element in child rearing. In a society in which the scientific and technical mode of action is predominant, the values expressed in child rearing may come to be greatly influenced by the diffusion of the opinion of "experts." This will be especially marked in the segments of the population more highly educated and more inclined to follow the latest scientific advice. Thus, Urie Bronfenbrenner (1958) notes that in the

United States over a quarter-century, from 1930 to 1955, American mothers became consistently more flexible with regard to infant feeding and weaning. Most notable, however, was the extent of the shift of the middle-class mother from great strictness to much greater permissiveness with regard to both feeding and toilet training. These changes, according to Bronfenbrenner, "show a striking correspondence to the changes in practices advocated in successive editions of the U.S. Children's Bureau bulletins and similar sources of expert opinion." Furthermore, the changes seemed most marked "in those segments of society which have the most ready access to the agents or agencies of change."

LATE CHILDHOOD AND ADOLESCENCE

In this period changes in physical and mental capacity—as for example, sex and aggression—stemming from the maturational process, interact with changes in society's response to and expectations about the individual as a potential member of society. The capacity of the individual to adjust to these changes, and the flexibility of the society in adapting to the impact of these adjustments, presents the central *issue* of this stage of the life cycle. The individual increasingly takes on roles which foreshadow, are directly supplemental to, or already fall within the realm of adult roles. The acquisition of such roles is, however, uneven. In our society, for example, adolescents are able to assume adult economic roles long before they are permitted to assume comparable roles in the political realm. The *task* of the adolescent is to manage the changes in himself, and the changing expectation of society towards him, without too vigorously disrupting the patterns of adult control and dominance. The *objective* of the socializers is to move the adolescent as effectively as possible toward the eventual assumption of adult roles by getting him to give up earlier gratifications and to train for new obligations. At the same time, society is often not yet ready to grant the adolescent full responsibility or authority. Substantial strain may result from the discrepancy between the adolescent's sense of his capacity, and his awareness of his increased responsibility, on the one hand, and his still sharply circumscribed autonomy and authority, on the other.

From a structural point of view the central feature of this period is the gradual replacement of the family and adult kin by other *agents* and *agencies* of socialization: schools, teachers, peer groups, tribal or political authorities, local heroes, religious specialists, actors and other public figures, and so on. So far as the content of socialization is concerned, problems of physical management, while far from irrelevant, become decidedly secondary to others such as: the acquisition of values; learning specific adult relevant skills, especially those connected with earning a living; managing mature heterosexual relations; and manifesting increased readiness to accept responsibilities relevant to adult status such as marriage and parenthood.

ECOLOGY

The age structure of a population can never make its effect felt entirely independently of the society's cultural and institutional framework, but within those limits it may greatly influence the extent to which other youths rather than adults socialize the adolescent. The postwar baby boom in the United States, for example, may not be entirely unrelated to the problems of controlling the human wave tactics adopted by adolescent crowds in various American cities and resorts. The role of other youths, as against adults, as agents of socialization will also be affected by the degree of age grading in a society, especially with regard to residence. When the youth population is diffused in small numbers in families, or comparable settings, which are heterogeneous in age composition, the role of adults in the socialization process can be very much greater. Even within the home, the proportion of youths versus adults may influence the opportunities for adult control, the bases of solidarity in opposition to parental desires, and the prospects for introducing new role models.

ECONOMICS

Within the constraints imposed by the system of power and the structure of values, the system of resource allocation can greatly influence the socialization process in adolescence. Where adults, and especially the father, more or less totally control access to a livelihood through control of scarce resources such as land, hunting rights, and special artisan skills, then the influence of the father and other adults as socialization agents and as models to be emulated will be proportionately great. At the other pole, an open labor market permitting youth to earn independently, the availability of an abundance of material goods, and certain values concerning youth and its rights, have combined in many of the advanced industrial countries to free adolescents from adult economic control and increased the importance as socializing agents of peers, mass media, youth leaders, and others outside the family and extended kin network (see Coleman, 1961).

In modern technologically based societies the socialization experience of adolescents has also been deeply affected by the patterns of occupational recruitment and practice characteristic of an industrial order. When farming and crafts were the main occupations, a young man was socialized to his occupational role by working with his father or by being apprenticed to an artisan. Even the professions, such as law and medicine, socialized their members through an apprenticeship system. Except in rare cases, this system has been replaced by training in specialized schools for the professions and for most nonprofessional but technical occupations. Consequently, socialization to occupations is now much more the responsibility of formal, more impersonal, specialized agencies. In certain ways this has widened access to these desirable occupational positions, since acceptance for training is decided by objective criteria in more or less open competition. In other ways, however, it means more restricted access, because young people must fairly early decide on the career they will follow, and get on the right "track." It is no longer possible for a young Abe Lincoln to come out of the backcountry with only a few years of schooling, and that mostly self-instruction, and apprentice himself to a local lawyer as the start of a career in law.

POLITICS

As in the case of goods and services, the pattern of socialization depends on the distribution of resources—in this case, power. Where young people, in their own right or through youth groups, are able effectively to represent their "interests" in the political arena, the forms and content of socialization will be different from the case where youths are politically dependent and subordinate. The issue has recently been dramatized throughout the world by the role of the Red Guards youth groups in Communist China. Under conditions of very rapid change, or under the stimulus of revolutionary leadership, the values and action patterns of the older generation may become unacceptable, and the youth may act independently of established traditions or use established political arrangements to introduce new social policies. At this stage anomalies in social structure may be glaring, as exemplified in the requirement that eighteen-year-olds in the United States serve in the army even though they are not entitled to vote.

VALUES

Late childhood and early adolescence are the prime periods for the formal inculcation of social values. In good part the indoctrination is specific, explicit, and didactic. Teachers, wise men, sages, oracles, and every other adult, if prepared to, practice telling the young the principles of right living. Folk sayings, proverbs, fairy tales, and a variety of exhortatory literature will spread the doctrine of the particular culture: Love thy neighbor, or, the only good *whoever* is a dead *whoever*; turn the other cheek, or, an eye for an eye; a man should try to get all he can, or, moderation in all things. The values taught will usually cover all things—they will define the goals of life, specify the legitimate paths to their attainment, elaborate the rewards, judge the deviant, abstractly resolve conflicts, and even suggest the appropriate compensations for failure.

The adolescent cannot, however, be expected automatically and passively to accept the values disseminated by society and by the adults with whom he is most intimately in contact. Indeed, if those values are not consistent, and often they are not, the adolescent is forced to choose between them, or he may be forced into passivity and inaction in order to avoid violating the norms of one or another of his mentors. In a more critical culture, and in a more self-conscious age, adolescents may also become acutely aware of the contrast between the values expressed by adults and those implicit in their actual behavior. The sensitive adolescent may, then, reject the formally espoused values, or act mainly in accord with what he judges to be the covert values. In either case, conflict with adults and representatives of adult society is likely to result.

In times of rapid social change, furthermore, adults may become uncertain as to the validity and appropriateness of the existing values as standards to guide the rearing of youth. A study of child-rearing values in three generations of Soviet Russian families showed marked shifts in the values each generation of parents deemed appropriate for raising their children. In guiding their children's preferences in the occupational realm, for example, those who raised them in the Tsarist period give greatest emphasis to material rewards (41 percent) and to tradition, that is, following in father's footsteps (35 percent). These value emphases declined markedly among the parents who raised their children in the Revolutionary era. Among them there was an understandable sensitivity to the political implications of having a child hold any given job. By the time the generation of parents which would raise its children after World War II came forward, there was a total transformation in the relative standing of the value emphases in socializing the child's occupational aspirations. In this new generation of parents, encouraging the child to seek a job assuring "self-expression" emerged as the overwhelmingly dominant emphasis (62 percent), whereas "rewards" and "tradition" as themes now were emphasized by a mere 24 percent *combined* (Inkeles, 1955).

ADULT YEARS AND OLD AGE

Trusting that the fuller account of the interaction of socialization and the social structure in infancy, childhood, and adolescence will suffice to exemplify the recommended mode of analysis, I will restrict myself to a much briefer statement concerning the adult years and old age. This account is necessarily a simplified analysis, which neglects important variation in different cultures, strata, and historical epochs. Instead, the picture given is that appropriate mainly to large scale industrial societies.

In adulthood the main *issue* is the degree to which the individual accepts, and the quality of his performance in, the whole panoply of roles which accompany the statuses of adulthood such as husband, father, earner of a living, member of a religious community, warrior, citizen of a polity, and so on.

The *task* of the socializee is to "fit in," "to take hold," in the large set of new statuses into which he is now thrust. What in adolescence may be a forgivable relapse, indicating that one is not quite yet ready but must try again, becomes in adulthood a serious failure which is not easily forgiven and may not admit of repeated trials. The adolescent must "put off childish things," renounce adolescent indulgences, and accept serious adult responsibilities. In contrast to earlier periods, the content to be learned shifts from general dispositions, such as reliability which is applicable to a wide variety of roles, to highly specific skills and responsibilities especially relevant to more specialized roles. The agents of socialization shift markedly. Kin and others whose role is personally or professionally protective, such as teachers, are replaced by formal, impersonal, organizational agencies which less "treat" and more "handle" and "process" the individual. Peers continue to be important, but they can less be selected by preference. Instead, they come more to be "built into" one's situation, having been previously recruited by others, as in the work gang. These more impersonal and professional agents of socialization tend to have highly specific and delimited objectives in their efforts to socialize the adult socializee. They teach one to drive, to do a certain job, to follow a certain path, without necessarily having any concern with our performance in related tasks, without the obligation to show strong interest in our future roles, and generally without profound consideration of the relation of these new roles to our general social and emotional adjustment.

In old age there is an analogue to the main *issue* of infancy and early childhood, in that adjustment to physical changes comes again to oblige us to acquire new skills and change established habits. The older person must, of course, accommodate not only to actual transformations in his physical capacity, but to changes in the expectation of others with regard to a person of his age. A main part of the older individual's *task* is to learn to renounce or abandon previously held positions, and especially the power, prestige, and economic rewards associated with them. The learning of entirely, or largely, new roles may be also required, such as dealing with total leisure. Except insofar as old people are confined in specialized institutions, the more formal and bureaucratic agencies of socialization become relatively unimportant as progressive withdrawal from the standard roles of the middle years occurs. Again as in adolescence, one's peers become very important socializing agents, and in a peculiar reversal, so do one's own children and other juniors and subordinates whom one earlier had socialized. The *objective* of the socializers tends to be short term, in the nature of the case, not so much to teach specific skills as to encourage acceptance of a new status. New skills, however, are not unimportant as an area of concern, as older men may learn to garden, work in pottery, paint and otherwise acquire skills which are appropriate to full-time leisure.

This sketch of individual socialization through the life cycle, brief, unsystematic and necessarily very incomplete, may yet serve to highlight a number of general issues which are critical in defining the influence of social structure on socialization. We may note the following seven main points:

1. Effective socialization is a pre-condition of organized social life. Every social organization must be prepared to do some socialization of its constituent members, partly to teach ways of acting distinctive to its needs and new to the socializee, and partly to reinforce established patterns, thus insuring minimal drift away from expectations and norms. Every socal organization, even a dyad, is, therefore, to some degree an agent, or "producer" of socialization. Most organizations have other, more specific, functions to perform, however, such as production, consumption, policing, and the like. Therefore, they must inevitably rely on other social units to provide the basic socialization of their ultimate constituents or members. Each social unit is, therefore, also a "consumer" of the products of prior socialization by other social agencies. The family is clearly the prime and most ubiquitous agency serving this function of general socialization. Where formal schooling exists, it generally is second in importance in fulfilling the function of general socialization.

2. The concentration of attention on the family as a socialization setting may be justified on the grounds that its impact is prior to that of other efforts, and presumably comes at a time when the individual is most malleable. From a sociological perspective, however, much of the research on infant and child socialization is too exclusively concerned with problems in the mastery of basic disciplines such as toilet training, control of aggression, and understanding and speaking the native language. Child-rearing research has relatively neglected to study the acquisition of other skills relevant to social functioning such as competition, cooperation and sharing, mutual support, observance of group rules and the like. Even within its accepted frame of reference, child socialization research has considered as major independent variables mainly the in-built maturational potentials of the child, and the personality and child-rearing techniques particular to a given parent. Social class and ethnic differences in child rearing have been carefully noted, but not systematically and theoretically integrated as an element in research design. The significance of gross occupational differences has been somewhat studied, but in very modest degree relative to research which does not systematically take account of this basic attribute.

Other aspects of social structure which may impinge on the parent-child relation have been studied hardly at all. Although the observation of everyday life makes it apparent that much of what each parent does to the child is guided by reference to some image of "what he must be like to get on in life and the 'world' later," we know very little of what these images are, where parents get them, and how successfully they translate them into action in their socialization practices. Ethnic and regional differences, the styles of child rearing in homes of parents of varying political persuasion, and the effects which social conditions such as depression and war have on the modes and content of infant and early child socialization have been little studied.

3. Despite the massive importance of the earliest years in the development of the individual, we must recognize that socialization is a process that goes on continuously throughout the life cycle. By the age of four the brain may have attained 90 percent of its potential weight, but at that age the individual, as a member of society, has probably acquired not much more than 10 percent of the repertoire of social roles he will later play in life. We can be easily misled by the assertion that in the development of our general intelligence, 50 percent takes place between conception and age four, and 30 percent more between ages four and eight. Social development, admittedly harder to measure, must be recognized as mainly occurring at later ages. A child may at age eleven know 50 percent of all the words he will know, but so far as other forms of socially relevant knowledge are concerned, virtually all his learning and development are still ahead of him. Realistic vocational interests, for example, seem to develop mainly in the period between fourteen and twenty. Certainly at age four, whatever may be the weight of his brain, almost everything an individual is to learn of team work, of sports, of politics, of large scale organization and bureaucracy, of earning a living, of war, and of heterosexual love is yet to come.

Not only do new socialization problems and issues come to the fore as the individual moves through the life cycle, but the processes of social change may transform his situation within a short space of time and require of him profound new learning within the space of a single phase of the life history. Major economic changes such as depression, radical political changes such as war and revolution, and technological innovation are among the more dramatic instances, yet in dynamic industrial societies less dramatic but cumulatively substantial processes of culture change requiring individual adaptation have become endemic.

4. Recognizing the life-long continuity of socialization requires us to acknowledge the importance of social units other than the nuclear family as socializing agencies. The school is generally so recognized, but largely with regard to the content of the formal instruction which it dispenses. Inadequate attention has been paid to its role as a training ground in dealing with peers, as the child's first introduction to formal authority and bureaucratic organization, and as inculcator of cultural values, not necessarily explicitly taught but latent in the forms and content of teaching (see Dreeben, 1967). The factory and office, the government bureaucracy, the church, the hospital, the political party, the army, are examples of but a few of the agencies which

both set expectations as to what is adequate socialization and themselves engage in various forms of socialization (see Inkeles, 1966b). These formal agencies, furthermore, may include much less formally organized subunits, each of which also sets standards and socializes both new and old members. These subunits include the office or shop work group, the army platoon or crew, and the neighborhood community. Still other relatively enduring social units quite outside any formal organization also play important roles in socialization, such as the neighborhood gang, and the circle of friends.

5. The complexity of social structure, even within nominally "simple" societies, predetermines a necessary complexity of the socialization experience of the individual. The fact that any individual learns to live in society and meet its myriad demands is no less awesome than that any child learns to understand and use a language. Some results of socialization, of course, are appropriate to participation in all or a great variety of the statuses an individual will occupy. The most obvious of these is knowledge of a language. The predisposition to conform to social rules has been cited as perhaps the single most important quality a person must learn in any society. Certain qualities which suffuse the culture and permeate the social structure—for example, tendencies toward dominance or submission, or emotional expressiveness—may be adaptive in a wide variety of situations. Ultimately, however, specific roles come down to fairly specific behaviors, and the number of those which must be learned by the individual in the course of his socialization is very great. Just how great this repertory of potential behavior is in the average individual has not been systematically studied.

6. The integration of the individual as a psychic or personality system and the integration of society as a social system set limits on the variability of socialization within any given sociocultural system. If the socialization demands of different parts of the social system are too disparate, individuals may be subject to unendurable pressure or conflict. This is certainly one element contributing to what the anthropologists have noted as the "strain toward consistency" in cultures. If the individuals whom the system must integrate in coordinated social action have been too diversely socialized, then, as in Yeats' "Second Coming," the falcon cannot see the falconer, the center cannot hold, and things fall apart.

7. Although socialization is one of the most important mechanisms giving society stability and continuity, it may also serve as a major vehicle for change. Some individuals, indeed whole segments of a population, may be socialized to play the role of creative, innovative, change-inducing catalysts. The entrepreneur in economic development represents this tendency, and McClelland (1961) argues it requires specific socialization of a strong need for achievement. When structural change has taken place due to major innovations in economic and political forms, in the extreme case through revolution, sustaining the revolution and adapting people to it both in the adult and the upcoming generation depend on socialization to new role demands. Unless it can successfully effect this re-socialization, no major process of social change can hope to endure.

REFERENCES

Aberle, D. F. "Culture and socialization." In: Hsu, F. L. K., *Psychological anthropology.* Homewood, Ill.: Dorsey Press, 1961.

Adorno, T. W., et al., *The authoritarian personality.* New York: Harper, 1950.

Bacon, Margaret K., I. L. Child, and H. Barry, III. "A cross-cultural study of correlates of crime," *Journal of Abnormal and Social Psychology* 66 (3), 1963.

Bronfenbrenner, U. "Socialization and social class through time and space." In: Maccoby, Eleanor E., T. M. Newcomb and E. L. Hartley (eds.), *Readings in social psychology.* New York: Holt, Rinehart and Winston, 1958.

Coleman, J. S., et al. *The adolescent society: The social life of the teenager and its impact on education.* Glencoe, Ill.: Free Press, 1961.

Dreeben, R. "The contribution of schooling to the learning of norms," *Harvard Educational Review* 37 (2), 1967.

Erikson, E. H. *Childhood and society.* New York: Norton, 1950.

Erikson, E. H. *Young man Luther: A study in psychoanalysis and history.* New York: Norton, 1958.

Fromm, E. *Escape from freedom.* New York: Holt, Rinehart and Winston, 1941.

Inkeles, A. "Social change and social character: The role of parental mediation," *Journal of Social Issues* 11: 12–23, 1955.

Inkeles, A. "The modernization of man." In: Weiner, M. (ed.), *Modernization.* New York: Basic Books, 1966. (a)

Inkeles, A. "Social structure and the socialization of competence," *Harvard Educational Review* 36 (3), 1966. (b)

Kardiner, A. *The individual and his society: The psychodynamics of primitive social organization.* New York: Columbia University Press, 1939.

Lanneau, G., and P. Mabrieu. "Enquête sur l'éducation en milieu rural et en milieu urbain," *Enfance* 4: 465–485 (Sept.–Oct.), 1957.

Levy, M. J. *The structure of society.* Princeton: Princeton University Press, 1952.

McClelland, D. C. *The achieving society.* Princeton, N. J.: D. Van Nostrand, 1961.

McKinley, D. G. *Social class and family life.* New York: Free Press, 1964.

Miller, D. R., and G. E. Swanson. *Inner conflict and defense.* New York: Holt, Rinehart and Winston, 1960.

Pearlin, L. I., and M. L. Kohn. "Social class, occupation, and parental values: A cross-national study," *American Sociological Review* 31, 1966.

Rainwater, L., and W. L. Yancey. *The Moynihan report and the politics of controversy: A trans-action social science and public policy report.* Cambridge: M.I.T. Press, 1967.

Whiting, Beatrice B. "Sex identity conflict and physical violence. A comparative study," *American Anthropologist* 67 (6), Part 2, 1965.

Twenty-six

Stimulus Control of Parent
or Caretaker Behavior by Offspring

Richard Q. Bell

The child's contribution to parent-child interaction has been equated in the past with congenital and genetic factors, and thus neglected by socialization theorists who reacted with extreme environmentalism against instinct theory and other biological extensions of the theory of evolution. As a result, most investigators have only considered the child an object on which parental actions are registered, rather than a participant in a social system, stimulating as well as being stimulated by the other. As a corrective, a way of thinking about the child's stimulus effects is advanced and applied to parent-child interactions observed in home settings.

Since 1961 eight different investigators have commented on the oddity that the child's contribution to parent-child interaction is overlooked (Bell, 1968; Gewirtz, 1961; Kessen, 1963; Korner, 1965; Rheingold, 1969; Stott, 1966; Wenar and Wenar, 1963). Child behavior is seldom an independent variable, parent behavior a dependent variable, even if the child is acknowledged by a formal place in theories. Until very recently we have had no hypotheses concerning the child's stimulating effects on the parent. Accordingly data are not gathered so that the effects of the child can be identified, and most of the relevant findings are accidents generated by research directed to other purposes. Parent and child are clearly a social system and in such systems we expect each participant's responses to be stimuli for the other. Why, then, is the child's own contribution to an interaction overlooked by social scientists, and what can be done about the problem?

R. Q. Bell., "Stimulus control of parent or caretaker behavior by offspring," *Developmental Psychology* 4, 1971, pp. 63–72. Copyright 1971 by the American Psychological Association, and reproduced by permission of the publisher and author.

This article is a condensed version of an invited address, Division of Developmental Psychology, presented at the meeting of the American Psychological Association, San Francisco, September 1968. The author wishes to express his appreciation to colleagues Howard Moss, Lawrence Harper, and Charles Halverson, whose interest stimulated the discussions out of which the article developed.

HISTORICAL DETERMINANTS

One possible historical determinant is our American political and social philosophy which emphasized egalitarianism and opportunity, as a reaction to hereditary determination of position in European societies. If one is committed too zealously to our philosophy it is difficult to admit the diversity of human existence contributed by the individual's own nature. Thus the general climate of our American political and social philosophy seems, at first glance, to provide an answer to the question of historical determinants. Closer examination of this possibility, however, reveals that considerable fluctuations in the views of social scientists have occurred in periods when major changes in the general climate of opinion have not been identified.

Clausen (1967) has pointed out changes in the views of sociologists. An overeager and uncritical acceptance of the theory of evolution in the formative years of sociology, around the end of the last century, led to an equally strong counterreaction. When instinct theory and the search for innate differences between social classes and national groups was abandoned, it became unfashionable to hold or express views on biological contributors to development. To hold such views was to be a conservative. The liberal sociologist believed in educability and thus turned his attention to society's values, institutions, and child-rearing techniques. This shift in perspective led to overlooking the child's contribution to the interaction in socialization theory; then, as today, the contribution of the child was equated with the operation of genetic or congenital factors.

Some of the same factors which affected sociology also affected psychology, but there were additional elements. The backlash of behaviorist reaction to introspection negated consideration of the child as a source of stimuli. Here again, as in sociology, the contribution of the child to development was equated with genetic and congenital factors and dismissed with them.

From the historical review it is evident that, for both sociology and psychology, the importance accorded the child's contribution to behavior development was linked with the value placed on biological factors during any given era. The equation of the child's contribution with biological factors led to a fundamental error in socialization theory. The possibility has been overlooked that the child could process and integrate experience during one period, and subsequently manifest to the parent new products of this integration at a later period, even if there were no maturational contributions to changes in behavior and no individual differences on congenital or genetic grounds. It seems evident that the child, after exposure to parent behavior, can present the parent with emergent behaviors which, in turn, modify subsequent parent behaviors. Thus, it is necessary to assert the view that stimulus effects of children on parents deserve treatment regardless of the question of genetic, congenital, or experiential contributors and their differential weights. We may study child behavior directly without necessarily being concerned with its origins; we may observe it, manipulate it, and thus move toward specifying its effects on the parent in various developmental periods.

PRESENT STATUS OF SOCIALIZATION RESEARCH

Since it is difficult to identify the operation of biological factors, let alone experimentally manipulate their contribution, it is quite understandable that even investigators who recognized the importance of the child in socialization (Nowlis, 1952; Sears, Maccoby, and Levin, 1957) simply proceeded with their studies, hoping the omission of the child's effects on the parent would not be overly damaging. If a substantial body of dependable findings had emerged from the study of socialization over the last twenty-eight years, the omission of the child's stimulating effects could be overlooked, along with the shortcomings of the interview and questionnaire approaches used. However, several reviewers have noted that the approach has been barren of results (Becker and Krug, 1965; Caldwell, 1964; Orlansky, 1949; Sewell, 1963; Yarrow, Campbell, and Burton, 1968). Authors of the last mentioned review concluded that the case for positive findings can only be maintained by relying on studies in which both parent and child behavior was reported by the same informant, by interpreting consistency in nonsignificant correlations, or by ignoring contradictory data.

Even if we accept some of the disputed findings, it is possible to provide quite plausible interpretations in terms of effects of children on parents. This is because in almost all cases the findings are based on a simple correlation between parent and child characteristics, and a correlation does not indicate direction of effects. Too often it has been assumed that it is most parsimonious to interpret correlations between parent and child characteristics in terms of a unidirectional effect from parent to child in spite of Sears et al.'s (1957) original caveats on this issue. To counter this tendency, explanations in terms of child effects on parents have been offered for major findings of socialization studies (Bell, 1968). Furthermore, if the criterion for explanation is to be parsimony, it should be noted that the explanations based on child effects on parents were developed from a single set of propositions which is not overly complicated or difficult to defend. This alternative system of explaining correlations between parent and child characteristics is detailed later in this article and expanded into a system for dealing with ongoing interactions.

DISENTANGLING THE DIRECTION OF EFFECTS

It is difficult to discriminate child and parent effects when each participant is reacting in turn to the other in an ongoing process, and the process is not subjected to an experimental intervention. It does not follow, however, that we are limited to a purely descriptive approach such as simply recording the event sequences. For example, predictions concerning interaction contingencies differ, depending on whether we expect it to be the child or the parent who initiates interactions. Data bearing on these different predictions are reported later. In addition, logical systems applicable to a wide variety of research problems, have been developed for making inferences about causal relations in naturally occurring sequences. The underlying concept is that a change by one participant, in the direction of the other's base line, supports the influence of the other. The bases for conclusions about causation are not as firm as would be the case with direct manipulation of the behavior of participants, but there is an increase in the confidence which can be attached to conclusions. A review of statistical techniques appropriate to this approach is available (Yee and Gage, 1968). A related approach has been used by Hinde and Spencer-Booth (1968) in analyzing changes in rhesus mother-infant pairs over a period of several weeks.

Better control over possible unrecognized contributors can be achieved by manipulating behavior of one of the participants and noting the effect on the other. A previous article has listed studies which have isolated effects of experimentally manipulated adult behavior on children and vice versa (Bell, 1968). In some the experimenter altered the mother's attitude toward her child's performance, then noted the shift. The most recent example is a study by Hilton (1967), who experimentally altered the mother's attitude and produced dramatic shifts in the interaction characteristics of mothers of first- and later-born children. In other studies, children with different behavioral capabilities were brought into interaction with adults with whom they had had no previous experience: The effects on the adults of the behavior of these children was measured.

White (1969) and McCall and Kagan (1967) have carried out sustained manipulations of infant behavior by changing (or having mothers change) their environment for several weeks. If the subsequent impact on the caretakers of the infant's changed behavior had been measured, we would have had the necessary elements for identifying stimulus effects of the infant on the parent.

A more complete form of the research approach advocated occurred serendipitously in a learning study carried out by Etzel and Gewirtz (1967). Levels of crying and smiling were manipulated in the case of two

6- to 20-week-old infants by removing them each day from a boarding nursery to an experimental room for a short training session, then returning them to the nursery. Informal observations indicated that the increased level of smiling shown in the nursery by one infant affected the behavior of several caretakers. This change in the infant resulted from reinforcement produced in the experimental room. The caretakers who previously spent little time with the infant now spent considerable time responding to his smiles.

To recapitulate, several studies have been described to document the point that the inclusion of the offspring in our theories of socialization need no longer be held up for lack of available research approaches. Stimulus effects of both the young and their caretakers can be differentiated.

A BIDIRECTIONAL MODEL APPLIED
TO ONGOING INTERACTIONS

Up to the present point in this article very general points have been made concerning history, previous theory, and methods. If these points have served to provide perspective, a state of discontent with our hypotheses, and reassurance that new kinds of data can be gathered, the next task is to provide a way of thinking about interaction in terms of stimulus effects of both participants. First, some of the usual ways in which we think about stimuli, responses, and participants in social interactions are reviewed, then to these are added two special propositions which have proved useful in accounting for correlations between parent and child characteristics reported in studies of socialization (Bell, 1968). Two samples of mother-infant interaction are then interpreted with the concepts at hand to provide concrete illustrations in actual ongoing sequences. Following this, hypotheses developed from these concepts are tested against available quantitative data from interaction sequences. Sooner or later all theories of parent or child effects, all extrapolations from laboratory manipulations of child or parent behavior, must be tested against these minute-to-minute interactions occurring in their natural context—the kinds of interactions which Wright (1967) pointed out were the basis for less than 8 percent of 1409 empirical studies of childhood and adolescence.

Were it not for our history of slighting the child's contribution to parent-child interactions, it would seem superfluous to note, first of all, that we should consider these interactions as comprising a social system. Sears (1951) has described the properties of a simple dyadic social unit in which mutual expectancies make the behavior of the participants interdependent. Parsons and Bales (1955) elaborated the view of the family as a social system. Brim (1957) further spelled out the implications of this conceptual approach, including the need for investigators to make a number of distinctions called for by role theory, and to consider the reciprocal nature of parent and child roles. Glidewell (1961) followed the same line of thinking and pointed out the applicability of the basic paradigm in which a unit of behavior is both a response of one participant and a stimulus for the other. If we treat each response of the caretaker as a possible stimulus for the young, we should also look for all the ways in which stimuli from the young may affect the caretaker.

Marler and Hamilton (1966) have provided a classification of stimulus effects at a level of generality which should prove helpful in the initial stages of developing a theory of parent and child effects. Generally, stimuli are facilitating or inhibiting. More specifically, they orient, elicit, produce decrement or increment (sensitization, learning, habituation), and check or arrest behavior.

Next, it is necessary to select a level of complexity at which responses will be defined. Most investigators of parent-child interaction have set a minimum level which is appropriate to a social system, the discrete actions of each participant being at least of sufficient complexity to be recognized by the other and perceived as relevant. As Wright (1967) has pointed out in connection with the midwest studies of behavior ecology, crying might be recorded, but not expiration and setting the vocal cords.

Most investigators also attempt conceptual organization above the level of discrete, molar acts. In the present approach it is assumed that actions or responses may occur in subsets called repertoires. An example is fussing, crying, grimacing, and threshing shown by infants and comprising, in effect, an "alerting and proximity maintaining" repertoire. The several acts a mother uses in soothing also constitute a repertoire. Items in a repertoire are interdependent in the sense that given one response to a stimulus, others in the same set are more likely to occur than those in another set.

Further organization within the set may also exist. The various responses may be graded relative to which one is most likely, given various levels of stimulation which activate the repertoire. The several responses may be released in a certain order; that is, a sequentially ordered repertoire may exist.

It should be noted that although the repertoires have been labelled in terms of their likely effect on the other participant, they should be demonstrable from interaction data without reference to the stimulating effects of their component responses on the other. All that is required is that there be a temporal association of certain elements within the one participant's total set of responses occurring during the interaction.

It is expected that differences in repertoires will be found. For example, the young should have fewer, more sparse, and less well organized and differentiated repertoires than their caretakers.

An additional assumption is that each participant has upper and lower limits relative to the intensity, frequency, or situational appropriateness of behavior shown by the other. When the upper limit is reached, the reaction of the other is of a kind likely to redirect or reduce the excessive or inappropriate behavior (upper limit control reaction). When the lower limit is not reached, the reaction is to stimulate, prime, or in other ways increase the insufficient or nonexistent behavior (lower limit control reaction).

The limits and control reactions would vary with the participants, but by way of illustration, one could say that a characteristically high level of parent control reaction would be expected in response to excessive and sustained crying in infants or to impulsive, precocious, or overly assertive children. Parent lower-limit control behavior would be elicited by lethargy in infants, by low activity, overly inhibited behavior, shyness, and lack of competence in the young child. These widely different child behaviors are assumed to be similar only with respect to their effects on the appropriate parent control pattern.

The conceptual system is now applied to one sequence of mother-infant interaction which primarily involves caretaking, and to another consisting of more purely social interaction. Both were extracted from a continuous three-hour observation by Howard Moss of the National Institute of Mental Health, for a three-month-old male infant and its mother. The case selected was simply the first from a series of 14 for which an analysis had been completed for the contingencies of several maternal and infant behaviors. The sequence is presented while interweaving propositions from the conceptual system, using such terms as reinforcement descriptively, since to provide evidence for the interpretive use of such a term would require data from rather sustained units, or across several units.

CARETAKING SEQUENCE

The infant had been alone in his crib, awake and quiet for a thirteen-minute period in which there was no interaction. The interaction was initiated by a three and one-half minute period in which he changed to an awake fussing state. This oriented the mother to the infant but did not at that time disrupt other ongoing activities or elicit approach. Presumably, the level of fussing was below a level which activates her soothing repertoire. This period was followed by one and one-half minutes during which fussing progressed to full cyclic crying, and the latter did elicit the mother's approach. She looked and presumably saw grimacing and threshing—

further stimuli from the infant which had the effect of keeping the mother in the immediate vicinity. The mother stood over him, since he continued to thresh and cry. She then talked. The crying continued. The mother then picked him up and cradled him in her arms. This part of her repertoire was reinforced by the infant, who reduced motor movement but continued crying. After about eight seconds the mother again talked, but the crying continued, and another element was introduced from the maternal repertoire—she stressed his musculature by holding him so his weight was partially on his arms and legs. The crying was maintained, however, and the mother then showed another behavior: holding the infant up in the air in front of and above her. The crying continued. She then held him against her shoulder and relieved ingested air. This was followed by rather massive tactile stimulation, jiggling, rubbing, and patting, but, after a pause, the infant resumed crying. Continuation of the tactile stimulation by the mother was followed by a reduction of the crying to fussing. However, the infant started crying again. Then the infant opened his eyes and was quiet for several seconds. The mother talked again, and the infant provided reinforcement for this behavior by continuing to remain quietly awake for several seconds, then emitting a noncrying vocalization. This elicited responsive talking by the mother, who then placed her baby in an infant seat. He remained quiet and awake in his seat, smiled, and the mother left a few seconds later. The state of the infant apparently terminated the interaction sequence. The smile could have differentiated this unit into a reciprocal social interchange, but the mother at this time was apparently only set to quiet the infant. Eighteen minutes in which the infant remained quiet and awake elapsed before another unit of interaction.

To recapitulate, in the absence of a change in stimulation from the mother the infant showed a sequential repertoire in that he progressed from fussing through alternating fussing and crying to full crying. The fussing oriented the mother toward the infant, then the crying activated what is best described as a sequentially ordered quieting repertoire. The mother talked to the infant first, then, after trying different methods, including stressing the infant's musculature and relieving air, she finally reached the stage of jiggling, rubbing, and patting. She reverted to talking when the infant quieted, and his resumption of crying was followed by a recycling of her ordered repertoire, though in shortened form. She proceeded directly to points in the repertoire that were later in the order when first presented.

SOCIAL INTERACTION SEQUENCE

The next illustrative sequence is from the record of the same mother-infant pair one hour and fifty minutes later. The baby had been placed in a seat after he was fed, and ingested air had been relieved. Throughout this period the mother had talked to him, held him in different positions, and wiped his face. He either looked directly at her, smiled, or both, when being wiped. There followed an essentially social interaction sequence, lasting six minutes, in which infant vocalized and mother talked alternately eight times during the first three and one-half minutes. The infant then smiled for the first time in four minutes, and the mother's rate of talking increased from one utterance in the one-half-minute period before, to one every three seconds in the one-half minute afterward. The infant's rate of vocalization in turn increased from one per seventeen seconds to one every four seconds for the same periods. There followed an interval of two and one-half minutes in which the infant smiled no more and there were six alternations of infant vocalizing and mother talking. The infant maintained his rate of vocalization, though introducing no novel responses, while the mother's rate of talking declined to one every six seconds. This last rate of decline was superimposed on a more general response decrement for maternal talking over the entire six minutes. Finally, the mother abruptly picked up the infant, held him close upright, then at a distance, tickled him, and turned her attention away. This burst of maternal activity reduced the infant's vocalizations, but the mother's cessation of responding was followed by fussing, then crying. She repeated the physical contact, but the infant continued to cry, and the sequence was terminated by the mother leaving the immediate vicinity of the infant and remaining away for nearly three minutes.

Sequentially ordered repertoires are not evident in the social interaction period of this sequence as in the previous caretaking sequence. Rather, the interaction has the qualities of a well-practiced game, each participant alternating in providing a response which serves as a stimulus to the other's response. While the infant dependably provides his phase of the alternation, the lack of novelty in his response is a likely basis for the general maternal response decrement. This is evidenced both in the general reduction in the mother's rate of talking, but also in the fact that a brief introduction by the infant of novelty, the smile (novel relative to this segment of the interaction), is followed by an increase in the maternal rate.

The effect of the novelty of the smile within this sequence raises the question of infant stimulus effects operating to prevent maternal response decrement across episodes. It is possible that a decrement in maternal attachment would occur were it not for general changes in infant behavior, particularly eye-to-eye contact and smiles, that engender a feeling of reciprocal relations. It can be surmised that in the previous example the gradual emergence of "the game" was an exhilarating experience to the mother, especially in contrast with the drudgery of caretaking that primarily prevented the appearance of aversive stimuli. From normative data in other studies it is readily apparent that such a mother could hardly fail to note a procession of other changes, and these are of interest in themselves. She could have observed that the infant was more "choosey," ceasing to respond when her own behavior was repetitive (Lewis, Goldberg, and Rausch, 1967); that his visual attention had shown a steady increase, and that it was accompanied by qualitative changes— early motor quieting, then scanning and excitation, then particular attention to her face, with smiling and soft cooing rather than mouth and head movements (Carpenter and Stechler, 1967). She could have seen a progression from mere swiping at objects after the second month, through to ability to open the hand in anticipation of contact with the object (White, Castle, and Held, 1964).

The origin of the changes just listed is a very complex matter, but at least we can say that there is no simple and straightforward support from White's (1969) series of enrichment studies for the notion that the changes are simple creations of the mother's stimulation as such. Some enrichment disturbed and upset the infant, actually reduced the level of functioning; some functions were unaltered; others altered in very unpredictable fashion. It should be noted, however, that these enrichment studies did not use behavior contingencies provided by a caretaker.

Returning to the system of thinking about caretaker-young interactions and to the illustrative sequences, the latter exemplify content in which we can expect to find answers as to how infants develop. Sequences such as these are going on during these periods, and we know that infant and maternal behavior progress during the same time. The system of interaction analysis which was applied is a general model, but the infant's contribution was emphasized in the selection and description of the sequences because, in the past, the mother's contribution has been belabored at the expense of the infant. The purpose of the conceptual system is to open up for consideration, along with parental effects, a wide variety of ways in which stimuli coming from the young may control and guide parental behavior. These sources of control by the young are not now represented in customary ways of looking at parent-child interactions. If we can begin to think of interactions in this way, the necessary empirical work will follow.

QUANTITATIVE DATA FROM INTERACTION SEQUENCES

It has been possible to locate only two published reports from the period of infancy through early childhood which were based on analyses preserving natural sequences. In the material to follow, these will be supplemented with unpublished results from ongoing analysis of Moss' data already mentioned. More data are available in the files of various investigators, but informal communication indicates that analyses have been held down by the mass of such information resulting from only a few hours of observation on a single case, as well as the difficulties of hand, and even computer analysis. The absence of a method of conceptualizing sequences so that the contribution of both participants is identified lies at the heart of the problem of analysis.

The data which are available, however, can be brought to bear in a crude way on the question of whether it is necessary to consider the infant or young child's response as a stimulus preceding the caretaker's. Reasoning from the old model of training being carried out by an agent of culture, we would expect parents to start interactions, and any infant or child behavior which might be found occasionally preceding parent behavior would not be specific to that which followed. This expectation follows from the assumption that what the object of socialization is doing does not matter, only the fact that at a given moment the training agent elected to transfer a bit of culture. If this latter assumption appears to make a straw man of socialization theory, it is nonetheless justified in effect by the following: (a) the small number of studies concerned with effects of children on parents (Bell, 1968), (b) the equally small number which record parent and child behavior as mutually contingent (Hoffman, 1957), (c) in Brim's (1957) survey of research on parent-child relations, the number of studies primarily concerned with child behavior only approached those for parents in the case of role *prescriptions*—what children should do, while in the category of role *performance*, there were four times as many studies concentrated on the parent.

The assumption of the irrelevance of behavior shown by the young finds little support in the available data. Lawrence Harper of the author's laboratory analyzed discrete behaviors in the three-hour interaction sequence from which the first illustrative sequence was chosen. No interaction units were imposed. Of 29 instances of the mother looking at the infant, 15 were preceded by the crying of the infant. Of 13 instances of the mother holding the infant in the close-upright position, 9 were preceded by the infant crying or fussing. One other category accounted for all other preceding infant events (infant awake). The state of an infant is a very obvious kind of control over maternal behavior.

The previous data were from an intensive analysis of several categories for a single case.[1] Moss and Robson (1968) have reported maternal response contingencies for a single category of infant behavior in 54 mother-infant pairs studied during two six-hour home observations at one month, and one six hour observation at three months. The question was whether maternal contacts preceded or followed episodes of continuous crying or fussing. An episode was considered terminated if one minute or more elapsed without crying or fussing. The mean number of episodes preceding maternal contact was 21.3 at one month, 11.0 at three months, in contrast with means of 4.1 and 2.5 for episodes following maternal contacts for the same time periods. The mean differences at each point were not only significant ($p < .001$), there was only 1 mother-infant pair at one month, and 4 at three months which did not show the same direction of differences as the means. The potency of crying and fussing in eliciting maternal behavior has been the subject of much comment based on casual observation, but here is rather strong evidence from systematic direct observation.

Moving to a later period of development, Gewirtz and Gewirtz (1965) have reported caretaker-infant contingencies based on full-day home observation of two 32-week-old infants. Frequencies of four kinds of interaction sequences were sufficient to permit a statistical test by the present author from tabular data. For one infant, three-event sequences initiated and terminated by the infant, with one intervening response from the mother, were more frequent than the converse, those in which the mother initiated and terminated sequences with one intervening infant response (23 versus 10; $\chi^2 = 5.12, p < .05$).

For the same infant, differences in frequencies between two-event sequences were not significant, being 61 for those initiated by the infant and terminated by the mother, versus 67 for the converse. For the other infant, there was no significant difference in frequencies between the infant- versus mother-initiated three-event sequences (37 versus 48), but there were significantly more mother-initiated two-event sequences than the converse (148 versus 115; $\chi^2 = 4.14, p < .05$).

The data from the two infants show no preponderance of interactions initiated and terminated by the mother, as we would expect from the dominant role accorded the mother in most socialization studies. Instead, one infant appears to control one kind of interaction, one mother another kind, while mothers and infants play an equivalent role for the other two types of interaction.

Schoggen (1963) has reported a special analysis of 18 specimen records available from the Midwest study for three mother-toddler pairs, the shortest of the 18 covering over eleven hours. Interactions were coded into environmental force units. This unit is an action of the child or parent which is directed toward a recognizable end state and is identified as such by the other. These characterizations of the data were made years ago without reference to the problem of direction of effects. Because of the method of analysis, it is possible to answer the question of who starts interactions when the units are defined in this way. During this period of observation, for these particular pairs, from 49 to 61 percent of the units were initiated by behavior of the child.

This analysis was based on units of interaction; one can always argue with the units used. However, a different analysis into episodes resulted in very similar results for 11 children covering the age range from one year two months to ten years nine months. Children started approximately one-half of the episodes at all age periods (Wright, 1967). There was no significant change in age in this basic division of initiating exchanges, although many other interaction parameters showed considerable change with age.

Schoggen commented on the surprising fact that the most frequent judged goal of the mothers in all but 1 of the 18 records was that of getting the toddler to cease his demands on her (that is, to quit bothering her, not to question her further, to leave her alone, not to press a request, not to attack). And these are passive, impressionable children! The intensity of the toddler's behavior, and the nature of the mother's reactions, makes it possible to see the value of one feature in the conceptual scheme just applied to interaction sequences. The mothers may be described as showing upper limit control behavior, the toddlers, lower limit control behavior. Characterizing the toddlers as showing lower limit control behavior means that the mothers' behavior was perceived by the toddlers as generally insufficient in intensity, frequency, or variety.

To return to the primary question asked in this portion of the article, do the available data on interaction support the concept of the irrelevance of behavior of the young? The answer is negative. That is, if any inferences about dynamic action can be drawn from a knowledge of which participant's behavior came first and the specificity of antecedent-consequent behaviors, the young must be given considerable credit for impelling action.

CONCLUSION

The propriety of hereditary determination of position in society was discredited long ago, as was the notion of innate ideas and the view that children are just little adults who don't need education. More recently, but certainly long enough in the past that it is no longer necessary to pay scientific penance, we ceased looking for instincts in every human behavior pattern and for evidences of the doctrine of the survival of the fittest in national, ethnic, and individual differences.

While for the better part of this century social scientists have considered it necessary to deny these views, proclaiming the malleability of man, at this time some unfortunate consequences of the denial may be safely considered. It is now no longer necessary to assert the child's educability with such vehemence that we deny his contribution, whatever may be its origins. A way of restoring his contribution to our basic view of the process of parent-child interaction has been offered in this report. If we can come out from under the shadow of old ideological conflicts, the child can be recognized as a very lively educator himself. To quote Peter De Vries (1954), "The value of marriage is not that adults produce children but that children produce adults [p. 98]."

NOTE

[1] Though it appears that these data could be subjected to a chi-square test, no such test is offered since the expected value for any combination of categories is a function of the durations as well as the marginal frequencies of the categories involved. Pending development of an appropriate test, a program for generating random functions of duration and frequency is being developed. A chi-square test was appropriate for a test of interaction data reported later in this article.

REFERENCES

Becker, W. C., and R. S. Krug. "The parent attitude research instrument—a research review," *Child Development* 36: 329–365, 1965.

Bell, R. Q. "A reinterpretation of the direction of effects in studies of socialization," *Psychological Review* 75: 81–95, 1968.

Brim, O. G. "The parent-child relation as a social system: I. Parent and child roles," *Child Development* 28: 343–364, 1957.

Caldwell, B. "The effects of infant care." In: Hoffman, M. L., and L. W. Hoffman (eds.), *Review of child development research*. Vol. 1. New York: Russell Sage Foundation, 1964.

Carpenter, G. C., and G. Stechler. "Selective attention to mother's face from week 1 through week 8," *Proceedings of the 75th Annual Convention of the American Psychological Association* 2: 153–154, 1967 (Summary).

Clausen, J. A. "The organism and socialization," *Journal of Health and Social Behavior* 8: 243–252, 1967.

Etzel, B. D., and J. L. Gewirtz. "Experimental modification of caretaker-maintained high-rate operant crying in a 6- and a 20-week-old infant (Infans tyrannotearus): Extinction of crying with reinforcement of eye contact and smiling," *Journal of Experimental Child Psychology* 5: 303–317, 1967.

De Vries, P. *The tunnel of love.* Boston: Little, Brown, 1954.

Gewirtz, J. L. "A learning analysis of the effects of normal stimulation, privation, and deprivation on the acquisition of social motivation and attachment." In: Foss, B. M. (ed.), *Determinants of infant behavior*. New York: Wiley, 1961.

Gewirtz, J. L., and H. B. Gewirtz. "Stimulus conditions, infant behaviors, and social learning in four Israeli child-rearing environments: A preliminary report illustrating differences in environment and behavior between the 'only' and the 'youngest' child." In: Foss, B. M. (ed.), *Determinants of infant behavior*. Vol. 3. New York: Wiley, 1965.

Glidewell, J. C. "On the analysis of social intervention." In: Glidewell, J. C. (ed.), *Parental attitudes and child behavior*. Springfield, Ill.: Charles C Thomas, 1961.

Hilton, I. "Differences in the behavior of mother toward first- and later-born children," *Journal of Personality and Social Psychology* 7: 282–290, 1967.

Hinde, R. A., and Y. Spencer-Booth. "The study of mother-infant interaction in captive group-living rhesus monkeys," *Proceedings of the Royal Society (Ser. B)* 169: 177–201, 1968.

Hoffman, M. L. "An interview method for obtaining descriptions of parent-child interaction," *Merrill-Palmer Quarterly* 3: 76–83, 1957.

Kessen, W. "Research in the psychological development of infants: An overview," *Merrill-Palmer Quarterly* 9: 83–94, 1963.

Korner, A. F. "Mother-child interaction: One- or two-way street?" *Social Work* 10: 47–51, 1965.

Lewis, M., S. Goldberg, and M. Rausch. "Attention distribution as a function of novelty and familiarity," *Psychonomic Science* 7: 227–228, 1967.

Marler, P., and W. J. Hamilton. *Mechanisms of animal behavior*. New York: Wiley, 1966.

McCall, R. B., and J. Kagan. "Stimulus-schema discrepancy and attention in the infant," *Journal of Experimental Child Psychology* 5: 381–390, 1967.

Moss, H. A., and K. S. Robson. "The role of protest behavior in the development of the mother-infant attachment." In: Gewirtz, J. L. (Chm.), *Attachment behaviors in humans and animals*. Symposium presented at the meeting of the American Psychological Association, San Francisco, September 1968.

Nowlis, V. "The search for significant concepts in a study of parent-child relationships," *American Journal of Orthopsychiatry* 22: 286–299, 1952.

Orlansky, H. "Infant care and personality," *Psychological Bulletin* 46: 1–48, 1949.

Parsons, T., and R. F. Bales. *Family socialization and interaction process*. Glencoe, Ill.: Free Press, 1955.

Rheingold, H. L. "The social and socializing infant." In: Goslin, D. (ed.), *Handbook of socialization theory and research*. Chicago: Rand McNally, 1969.

Schoggen, P. "Environmental forces in the everyday lives of children." In: Barker, R. G. (ed.), *The stream of behavior: Explorations of its structure and content*. New York: Appleton-Century-Crofts, 1963.

Sears, R. R. "A theoretical framework for personality and social behavior," *American Psychologist* 6: 476–482, 1951.

Sears, R. R., E. E. Maccoby, and H. Levin. *Patterns of child rearing*. Evanston, Ill.: Row, Peterson, 1957.

Stott, D. H. *Studies of troublesome children*. New York: Humanities Press, 1966.

Sewell, W. H. "Some recent developments in socialization theory and research," *Annals of the American Academy of Political and Social Science* 349: 163–181, 1963.

Wenar, C., and S. C. Wenar. "The short term prospective model, the illusion of time, and the tabula rasa child," *Child Development* 34: 697–708, 1963.

White, B. L. "Child development research: An edifice without a foundation," *Merrill-Palmer Quarterly* 15: 49–79, 1969.

White, B. L., P. Castle, and R. M. Held. "Observations on the development of visually-directed reaching," *Child Development* 35: 349–364, 1964.

Wright, M. F. *Recording and analyzing child behavior.* New York: Harper & Row, 1967.

Yarrow, M. R., J. D. Campbell, and R. V. Burton. *Child rearing: An inquiry into research and methods.* San Francisco: Joscy-Bass, 1968.

Yee, A. H., and N. L. Gage. "Techniques for estimating the source and direction of casual influence in panel data," *Psychological Bulletin* 70: 115–126, 1968.

Twenty-seven

Continuities and Discontinuities
of Psychological Issues into Adult Life

Bernice L. Neugarten

I have chosen to play the devil's advocate today, for despite the title of the paper assigned to me, I am impressed more by the discontinuities than the continuities in the psychological issues that have thus far been preoccupying developmental psychologists who are concerned with the lifespan.

We shall not understand the psychological realities of adulthood by projecting forward the issues that are salient in childhood—neither those issues that concern children themselves, nor those that concern child psychologists as they study cognitive development and language development and the resolution of the Oedipal.

Many of those investigators who have been focusing upon infancy and childhood are dealing with issues that are not the salient issues to adults. To illustrate very briefly, and not to dwell upon the obvious:

In the adolescent we are accustomed to thinking that the major psychological task is the formation of identity. For the period that immediately follows, Kenneth Keniston has recently suggested the title "youth," distinguishing it from young adulthood, as the time when the major task for the ego is the confrontation of the society, the sorting out of values, and making a "fit" between the self and society.

In young adulthood, the issues are related to intimacy, to parenthood, and to meeting the expectations of the world of work, with the attendant demands for restructuring of roles, values, and sense of self—in particular, the investment of self into the lives of a few significant others to whom one will be bound for years to come.

In middle age, some of the issues are related to new family roles—the responsibilities of being the child of aging parents, and the reversal of authority which occurs as the child becomes the decision-maker for the parent . . . the awareness of the self as the bridge between the genera-

Neugarten, B. L. "Continuities and discontinuities of psychological issues into adult life." *Human Development*, 1969, 12, 121–130. Reprinted with permission of the author and publisher, S. Karger, Basel/New York.

tions . . . the confrontation of a son-in-law or daughter-in-law with the need to establish an intimate relation with a stranger under very short notice . . . the role of grandparenthood.

Some of the issues are related to the increased stock-taking, the heightened introspection and reflection that become characteristic of the mental life . . . the changing time-perspective, as time is restructured in terms of time-left-to-live rather than time-since-birth . . . the person-alization of death, bringing with it, for women, the rehearsal for widow-hood, and for men, the rehearsal of illness; and for both, the new atten-tion to body-monitoring.

Some of the issues relate to the creation of social heirs (in contrast to biological heirs) . . . the concomitant attention to relations with the young—the need to nurture, the care not to overstep the delicate bound-aries of authority relationships, the complicated issues over the use of one's power—in short, the awareness of being the social*izer* rather than the socialized.

And in old age, the issues are different again. Some are issues that relate to renunciation—adaptation to losses of work, friends, spouse, the yielding up of a sense of competency and authority . . . reconciliation with members of one's family, one's achievements, and one's failures . . . the resolution of grief over the death of others, but also over the ap-proaching death of self . . . the need to maintain a sense of integrity in terms of what one has been, rather than what one is . . . the concern with "legacy" . . . how to leave traces of one-self . . . the psychology of sur-vivorship. . . .

All these are psychological issues which are "new" at successive stages in the life cycle; and as developmental psychologists we come to their investigation ill-equipped, no matter how sophisticated our ap-proaches to child development.

The issues of life, and the content and preoccupation of the mental life, are different for adults than for children. Furthermore, as psycholo-gists, we deal in a sense with different organisms. Let me illustrate again, only briefly, that which we all know:

As the result of accumulative adaptations to both biological and social events, there is a continuously changing basis within the individual for perceiving and responding to new events in the outer world. People change, whether for good or for bad, as the result of the accumulation of experience. As events are registered in the organism, individuals inevi-tably abstract from the traces of those experiences and they create more encompassing as well as more refined categories for the interpretation of new events. The mental filing system not only grows larger, but it is reor-ganized over time, with infinitely more cross-references. This is merely one way of saying that not only do the middle-aged differ from the young because they were subject to different formative experiences, but because of the unavoidable effects of having lived longer and of having therefore a greater apperceptive mass or store of past experience.

Because of longer life histories, with their complicated patterns of personal and social commitments, adults are not only much more complex than children, but they are more different one from another, and increasingly different as they move from youth to extreme old age.

More important, the adult is a self-propelling individual who manipulates the environment to attain his goals. He creates his environment, more or less (and varying in degree, of course, by the color of his skin and the size of his own or his father's bank account). He invents his future self, just as he recreates or *reinvents* his past self. We cannot go far in understanding adult psychology, then, without giving a central position to purposive behavior, to what Charlotte Buhler calls intentionality, or to what Brewster Smith has called the self-required values, or to what Marjorie Lowenthal refers to as the reassessment of goals as itself the measure of adaptation.

These are not new ideas, but because they are such striking features of the adult as compared to the child, they create special problems for the student of the life cycle when he turns to problems of prediction, for we do not yet know how to capture the phenomena of decision-making.

Another factor is the adult's sense of time and timing. The adult, surely by middle age, with his highly refined powers of introspection and reflection, is continually busying himself in making a coherent story out of his life history. He reinterprets the past, selects and shapes his memories, and reassesses the significance of past events in his search for coherence. An event which, at the time of its occurrence, was "unexpected" or arbitrary or traumatic becomes rationalized and interwoven into a context of explanation in its retelling twenty years later.

The remembrance of things past is continually colored by the encounter with the present, of course; just as the present is interpreted in terms of the past. To deal with both the past and the present simultaneously is a unique characteristic of human personality. It is a set of mental processes which vary according to the sensitivity of the individual, probably with his educational level and his ability to verbalize, but a set of mental processes which probably also follow a distinguishable course with increasing age. In a study presently under way in Chicago, for instance, the data seem to show that middle-aged people utilize their memories in a somewhat different fashion than do old people. The middle-aged draw consciously upon past experience in the solution of present problems; the old seem to be busy putting their store of memories in order, as it were, dramatizing some, striving for consistency in others, perhaps as a way of preparing an ending for that life-story.

There is another way in which issues of time and timing are of central importance in the psychology of adulthood: namely, the ways in which the individual evaluates himself in relation to socially defined time. Every society is age-graded, and every society has a system of social expectations regarding age-appropriate behavior. The individual passes through a socially regulated cycle from birth to death as inexorably as he passes through the biological cycle; and there exists a socially prescribed timetable for the ordering of major life events: a time when he is expected to marry, a time to raise children, a time to retire. Although the norms vary somewhat from one socioeconomic, ethnic, or religious group to another, for any social group it can easily be demonstrated that norms and actual occurrences are closely related.

Age norms and age expectations operate as a system of social controls, as prods and brakes upon behavior, in some instances hastening an event, in others, delaying it. Men and women are aware not only of the social clocks that operate in various areas of their lives; but they are aware also of being "early," "late," or "on time" with regard to major life events.

Being on time or off time is not only a compelling basis for self-assessment with regard to family events, but also with regard to occupational careers, with both men and women comparing themselves with their friends or classmates or siblings in deciding whether or not they have made good.

Persons can describe ways in which being on-time or off-time has other psychological and social accompaniments. Thus, in a study of Army officers (the Army is a clearly age-graded occupation, where expectations with regard to age and grade are formally set forth in the official Handbook) the men who recognized themselves as being too long in grade—or late in career achievement—were also distinguishable on an array of social and psychological attitudes toward work, family, community participation, and personal adjustment.

When factors such as these are added to the inexorable biological changes, the individual develops a concept of the "normal, expectable life cycle"—a phrase which I have borrowed from Dr. Robert Butler and which owes much, of course, to Hartmann's "normal, expectable environment." Adults carry around in their heads, whether or not they can easily verbalize it, a set of anticipations of the normal, expectable life cycle. They internalize expectations of the consensually validated sequences of major life events—not only what those events should be, but when they should occur. They make plans and set goals along a time-line shaped by these expectations.

The individual is said to create a sense of self very early in life. Freud, for example, in describing the development of the ego; and George

Mead, in describing the differentiation between the "I" and the "me," placed the development of self very early in childhood. But it is perhaps not until adulthood that the individual creates a sense of the life cycle; that is, an anticipation and acceptance of the inevitable sequence of events that will occur as men grow up, grow old and die—in adulthood, that he understands that the course of his own life will be similar to the lives of others, and that the turning points are inescapable. This ability to interpret the past and foresee the future, and to create for oneself a sense of the predictable life cycle differentiates the healthy adult personality from the unhealthy, and it underlies the adult's self-assessment.

The self-concept of the adult has the elements of the past contained within it. The adult thinks of himself in the present in terms of where he has come from; what he has become; how content he is at fifty compared to the time when he was forty.

All this differentiates the adult as subject from the psychologist as observer. The adult has a built-in dimension of thought that is the present-relative-to-the-past—but the psychologist has not yet created dimensions of this type in capturing the psychological realities of the life cycle and in studying antecedent-consequent relations. In fact, it is the specific aim of most investigators to keep separate Time 1 from Time 2 observations and evaluations, on the premise that to do otherwise is to contaminate the data.

To put this differently, to the subject, the blending of past and present is psychological reality. To the investigator, validity (and therefore, reality) lies in keeping time segments independent of each other.

Thus, to repeat, some of the problems that face us in attempting to build a psychology of the life cycle stem from the facts that the salient issues of the mental life are different for adults than for children; the underlying relations of the individual to his social environment are different; the relations of the investigator to his subject are different; and the salient dimensions psychologists use to describe and measure mental and emotional life *should* be different.

I am suggesting, then, that our foremost problem in studying the lifespan is to create a frame of reference and sets of dimensions that are appropriate to the subject matter, and that are valid in the sense that they are fitting ways of capturing reality. To do this, it might be added, we need first a great wealth of descriptive studies, based on various methods that stem from naturalistic observational approaches.

Let me turn now more specifically to the studies which are emerging in which the attempt is made to relate findings on childhood and adulthood in the lives of the same individual—in short, to longitudinal studies which form the foundation of a psychology of the life cycle.

The longitudinal studies may be seen, in overly simplified and overly dichotomized terms, as being of two major types: first are those I shall refer to as "trait" or "dimension" oriented, studies addressed to questions of stability and change along given dimensions of ability and personality; second are those I shall call "life-outcomes" oriented, those which pose such questions as these: What kind of child becomes the achieving adult? The middle-aged failure? The successful ager? The psychiatric casualty? What constellation of events are predictive of outcomes?

In the first type of research, the investigators have been preoccupied with such problems as whether or not the individual who is aggressive at age three is aggressive at age thirty, or whether the high I.Q. child turns out to be the high I.Q. adult. There are also studies in which the ipsative approach is taken, and in which the stability of personality types is the question being pursued—the difference being, that is, that attention is focused upon the patterns of traits rather than individual traits, and the degree to which these patterns show stability or change.

In such studies, the investigator is plagued with questions of validity of his measures over time—is the concept of aggression or intelligence the same concept for three- and thirty-year-olds? Are we measuring the same phenomenon? These studies have proceeded without regard to the events of the life cycle, and the passage of calendar time is itself taken to be the sufficient variable. As in Kagan and Moss', studies from birth to maturity, or Nancy Bayley's studies of cognitive development, or Oden's latest follow-up of the Terman gifted group, the presumption seems to be that the same changes can be expected between age three and thirty whether or not marriage has intervened, or parenthood, or job failure or widowhood. "Time" is treated as independent from the biological and social events that give substance to "time," and independent from the events that might be regarded as the probable psychological markers of time.

In studies of life-outcomes, we need, of course, studies of traits and dimensions; but we are in particular need of studies aimed at determining which life events produce change and which do not—which ones leave measurable traces in the personality structure, and which ones call forth new patterns of adaptation.

Let me illustrate: parenthood might be presumed to be an event that has a transforming effect upon the personality, whether one reasons from psychoanalytic theory, or role theory, or learning theory; and whether one conceptualizes the event in terms of elaboration and differentiation of the ego, or in terms of adjustment to a major new set of social roles, or in terms of the development of new sets of responses to the demands of a new significant other. Yet we have no systematic studies of the effects of parenthood upon personality development; and no good evidence that parenthood is more significant, say, than college attendance or marriage or widowhood.

To take another example: some of my own work on middle-aged women has led me to conclude that the menopause is not the transforming event in personality development that puberty is; nor is the departure of children from the home of the same importance as parenthood.

We need to establish which life events are the important ones, but we need to study also when the event occurs, in terms of its social "appropriateness." (To marry at age thirty is a different psychological event than to marry at age sixteen; and to be widowed at age forty may be more significant than to be widowed at age sixty-five, for in either case, the event comes off-time and does not fit the anticipations of the normal, expectable life cycle.)

Among the longitudinal and long term follow-up studies presently available, investigators have taken both prospective and retrospective approaches. They begin with a group of infants or young children and follow them forward in time; or they begin with a group of adults and look backward in their life histories to identify the predictors or antecedents of present adult status. In both instances, what is most striking is the relative lack of predictability from childhood to adulthood with regard to life-outcomes. To mention only a few very recent studies: Hoyt's review of the literature indicates that we cannot predict from school success to vocational success . . . Robins' study of deviant children grown up shows that while anti-social behavior in childhood is predictive of sociopathic behavior in adults, the withdrawn personality characteristics of childhood are not associated with later adult pathology of any kind . . . Rogler and Hollingshead's study in Puerto Rico indicates that experiences in childhood and adolescence of schizophrenic adults do not differ noticeably from those of persons who are not afflicted with the illness . . . Baller's follow-up of mid-life attainment of the mentally retarded shows low predictability from childhood, with persons of below-70 I.Q. faring vastly better than anyone anticipated.

I recognize that, in some ultimate sense, we may never be able to make satisfactory predictions regarding life outcomes, no matter how well we chose our variables or how well we manage to identify the important and unimportant life events, for we shall probably never be able to predict the changes that will occur in an individual's social environment, nor the particular contingencies and accidents that will arise in an individual life, nor—equally important—the ways his life cycle is affected by those of the significant others with whom his life is intertwined. Furthermore, the psychologist, no matter how sophisticated his methods, will need the sociologist, the anthropologist, and the historian, to say nothing of the developmental biologist, to help him. Thus the study of lives will flourish only to the extent that a truly interdisciplinary behavioral science is created. Perhaps we shall have to leave the field to the creative writer, the philosopher, or the archivist for a long time still to come, and decide that the life cycle as a unit of study in the behavioral sciences is one with which we are not yet prepared to deal.

Yet we developmental psychologists are not likely to abandon the subject matter that intrigues us; and in the immediate future, as we work in our own areas, we can probably gain enormously in our ability to predict outcomes if we focus more of our attention upon the things that are of concern to the individuals we are studying—what the subject selects as important in his past and in his present; what he plans to do with his life; what he predicts will happen; and what strategies he elects—in short, if we make greater use of the subject himself as the reporting and predicting agent.

I am reminded in this connection that Jean McFarlane recently told me that after her intensive and intimate study of her subjects over a thirty-year period, she was continuously surprised to see how her people turned out. In going back over the data that she and Marjorie Honzik had painstakingly amassed, she found that much of what her subjects told her had been important to them when they were children or adolescents was not even to be found in her records—in other words, that which the investigators had regarded as important and had bothered to record was not the same as that which the subjects themselves had regarded as important at the time.

I suggest therefore that in future studies we pay more attention to gathering systematic and repeated self-reports and self-evaluations, and in doing so, to utilize what I shall call the "clinical" as well as the "observer's" approach. In the one case, the clinical psychologist tries to put himself into the frame of reference of his patient or client and to see the world through that person's eyes. In the other case, the "observer" psychologist brings his own frame of reference to the data and interprets according to his own theories.

We need to gather longitudinal data of both types (as by collecting autobiographies from our subjects at repeated intervals, and by creating a set of dimensions and measures that are appropriate to that data). We need, in other words, to use a double perspective: that of the observer and that of the person whose life it is.

In conclusion, I have been drawing attention to the discontinuities between a psychology of childhood and a psychology of adulthood, between the perspectives of the investigator and that of the subject himself, between the stances of the clinician and the psychometrician. If, as our chairmen had in mind in naming this symposium, we have presently available only a few elements of a lifespan developmental psychology, I am suggesting a few of the elements that are conspicuously missing.

Twenty-eight

Adult Personality: A Developmental View

Bernice L. Neugarten

In introducing this topic I should like to take cognizance of two facts: first, there is as yet no developmental psychology of adulthood in the sense that there is a psychology of childhood or adolescence; second, that at the same time, all of us are aware that changes in personality occur with the passage of time—not only as the child becomes the adolescent, and the adolescent, the adult, but also as the adult moves from youth to middle age to old age. Any sensitive adult, whether professional psychologist or amateur, will agree at once that there are observable changes as well as observable consistencies in adult personality; and that both stability and change are to be seen within the self as well as in the behavior of other adults. The problems lie not in obtaining agreement among psychologists that this is so. The problems are how to delineate those personality processes which are the most salient at successive periods in adulthood; how to describe those processes in terms which are appropriate to the phenomena; and how to isolate the changes that relate to age from those that relate, say, to illness on the one hand, or to social and cultural change, on the other.

The amateur psychologist may have difficulty in finding the appropriate words by which to describe the processes he recognizes. It has been said in this connection that our society is one in which there is increasing difficulty in expressing feelings and in giving articulation to one's personal experience, especially if one happens to be male. In the mass society, where the bases for decision-making are becoming increasingly statistical, increasingly rationalized (if *not* rational), and increasingly computerized, men and women both are attuned more to the outer than to the inner life. I am not sure that this is an accurate view, however, for as our own studies have proceeded, I have become increasingly aware of introspection as a salient characteristic of mental life in persons of all educational levels as they move through adulthood.

Neugarten, B. L. "Adult personality: A developmental view," *Human Development*, 1966, 9, 61–73. Reprinted with permission of the author and publisher, S. Karger, Basel/New York.

The difficulties of the amateur psychologist in finding words to express himself is unhappily matched by the difficulties of the professional psychologist who searches for appropriate concepts and appropriate terms by which to describe his observations, and by which he may proceed beyond description to more sophisticated methods of research—to the manipulation and control variables, to the prediction of behavior in the laboratory or in other controlled settings and finally to the more complex problems of prediction of normal behavior in the natural setting.

I am reminded, in using the word "normal," of the two most recent instances when my attention has been drawn to the term: one, when a noted personality theorist commented that there are so many more ways of being normal than of being abnormal that it is much easier for psychologists to study the abnormal. The second instance was when a well-known psychotherapist, in quite a different context, remarked that in training clinical psychologists he uses psychiatric patients in early phases of training, but that he must reserve "normal" subjects for those psychologists who are in advanced phases of their training.

In any case, as students of human behavior who are interested primarily in normal people and in the questions of stability and change in human behavior with the passage of time, most of us have been, understandably enough, beguiled by concepts of *child* development. The changes observable in behavior with the passage of short intervals of lifetime—a month, six months, a year—are dramatic when one regards a young biological organism. No less compelling are the overall regularities of biological changes, regularities that have come to be called "maturational."

It has been easy for students of behavior to draw parallels between biological phenomena and psychological, and then to establish a developmental psychology of childhood and adolescence. It also has been relatively easy first to assume, then to look for ways of describing, sequential and orderly progressions in behavior in infancy, young childhood and even in adolescence.

I am not unaware, of course, that in adults, as well as in children, some of the major problems that need elucidation are those that deal with the relationships between biological, psychological, and social factors. In adulthood, compared with childhood, the focus shifts from maturation to health, but this does not reduce the saliency of the biological; and one of the central research problems that needs further exploration is the disentangling of illness from normal aging. Once these two sets of biological processes are separated, the relationship of development to aging can then take new form as a fruitful area of inquiry.

In this same connection it might be pointed out that we do ourselves a disservice to assume that relationships found to hold between various classes of phenomena in the child or in the adolescent hold, also, for the adult. The very relevance of states of physiological reactivity to psychological states (or to social behavior) remains to be demonstrated, for the most part; but whatever such relationships are discovered in the child or adolescent, they do not necessarily hold true in the adult.

There are two general problems in adult personality that need investigation: (1) What are the salient psychological phenomena that characterize adulthood? (2) Which of these are, and which are not, developmental? I should like to comment first upon the second of these questions.

ARE PERSONALITY CHANGES DEVELOPMENTAL?

By the term "developmental," I am not referring to processes which are biologically inherent in the organism, or inevitable. I mean, instead, processes in which the organism, by interacting with the environment, is changed or transformed; so that, as the result of the life history with its accumulating record of adaptations to both biological and social events, there is a continually changing basis within the individual for perceiving and responding to new events in the outer world—processes which follow an *orderly progression* with the passage of time.

We assume some orderly connection between the early and later phases of life; yet there is no compelling logic, let alone any evidence, that many of the phenomena that we can succeed in measuring in adults—especially in the young and middle-aged—do, indeed, follow systematic changes with the passage of time. Some of us developmental psychologists who have moved into the area of adult personality—and there are few of us in total—have proceeded on the assumption that if only our measures are well enough focused, we will find the age-relationships that must exist. We have needed to be reminded that, no matter how good our measurements may become, the variances in our data may turn out to be related more to social and situational events than to developmental events.

I have already mentioned the dangers involved in drawing analogies between biological and psychological phenomena. Without forgetting that danger, the example nevertheless that comes to mind is the comparison between puberty and menopause in the female. Psychologists have been impressed with the changes in behavior that accompany puberty, and with the interpretation, therefore, that there are major developmental components in the personality changes that are measurable in adolescents. It has been reasoned by analogy that there must also be developmental components that underlie personality differences in middle-aged women. While there is a dearth of data on the menopause, there are some studies which would indicate that, in 85 of 100 females, there seems to be no major lasting effect of menopause upon physique or upon overt behavior. Certain

data of our own bear this out—data in which a large number of psychological and social variables were obtained on a sample of normal women, and
in which menopausal status (presence or absence of observed changes in
menstrual rhythm, or cessation of menses) was unrelated to any of the wide
array of measures. Furthermore, severity of somatic and psychosomatic
symptoms attributed to climacteric changes was also unrelated to most of
our variables. We could not even find relationships between a woman's
own positive or negative *attitudes* toward menopause and our other social-
psychological measures of behavior.

I mention this, not because I think the menopause is a meaningless
phenomenon in the lives of women, but because it indicates that we must
proceed cautiously in assuming developmental personality changes in adulthood; and that we need to gather a good deal of data by which to distinguish developmental from nondevelopmental changes.

THE SALIENT PSYCHOLOGICAL ISSUES

Now to return to the first question, that of delineating the salient
phenomena. The point is often made that in the absence of longitudinal
data over the lifespan there is a tendency for investigators to use different
methods and different concepts for studying different age levels; and sometimes to regard children, adolescents, adults and old people as members of
different species. The implication is that, if psychologists of childhood and
psychologists of adulthood had longitudinal data and had more conceptual
approaches in common, the lifespan could be seen in more continuous and
more meaningful terms, and antecedent-consequent relations could be
more readily investigated. While that point of view has merit, it is also
true that thus far not only longitudinal studies but many cross-sectional
studies of adult personality have a certain child-centered, or what one of
my colleagues has called a "childomorphic," orientation. In such studies,
the variables have been those that are particularly salient in childhood, or
they are variables that can be measured retrospectively from data that may
have been gathered when subjects were children. Either way, the investigator is confined to data, to variables and to concepts that may be less than
optimum or that may be of only secondary relevance when he turns
attention to adult behavior.

There are many studies in the child-development literature, for example, that deal with sex-role identity, dependence, aggression—dimensions
of behavior or issues that may not be major issues for adults.

Upon what, then, does the psychologist of adulthood focus his attention? At one level of generality, there are certain psychological issues in all
adult age groups: the individual's use of experience; his structuring of the

social world in which he lives; his perspectives of time; the ways in which he deals with the major life themes of work, love, time, and death; the changes in self-concept and changes in identity as individuals face the successive contingencies of marriage, parenthood, career advancement and decline, retirement, widowhood, and illness. At another level of generality, however, some issues are more salient in young adulthood, others in middle age, and still others in old age.

YOUNG ADULTHOOD

I take it for granted that members of this audience are familiar enough with adolescents and young adults, so that I need not elaborate upon the issues of young adulthood. Presumably as persons connected with educational institutions, whether as researchers or teachers, we deal already with a wide array of these issues: the ways certain of our "cool" young people move from uncommitted positions to committed, and the ways they adopt conventional or unconventional work, family and community roles; the ways in which young men and women try to establish emotional independence from parents without the accompanying financial independence; the initial job placements and launching of careers; the selection of appropriate adult sponsors, whether these be faculty members who provide apprenticeship relations in graduate school training or particular business corporations where young men and women will get what they regard as a proper start; the effects of legitimate as well as illegitimate pregnancies upon both male and female occupational plans; the use of contraceptives in creating what may turn out to be, for many young people, a changed motivational base for marriage and parenthood as compared to earlier generations; and the ways young men and women follow an internalized social clock that acts as a prod or a brake and that tells them they are on time, early, or late with regard to marriage, parenthood, and economic independence.

MIDDLE AGE

Let me illustrate at greater length from middle age. In an exploratory study of competent middle-aged men and women, we have found ourselves moving to more and more naturalistic-type data. As we learn to listen to what our subjects are saying—and, if I may say so, to listen with a third and sometimes with a fourth ear—and as we learn to give less attention to what we investigators had thought important to start with, we are discovering certain issues in personality change and in social behavior that seem age-related, yet do not appear at all or take quite different forms in younger and older people. There is, for instance, the forty-five-year-old's sensitivity to the self as the instrument by which he reaches his goals; the conscious and deliberate choices involved in creating more or less freedom for the self; what we might call an emphasis on self-utilization, rather than self-consciousness.

Professor Robert White, lecturing at The University of Chicago, commented in this connection that the personality is not properly to be regarded as a product of the forces that have acted upon the individual; and that the nervous system is *not* a switchboard without an operator. The individual (or the personality, in this sense) has a stake in the proceedings and in the transactions with the environment, so that there is selectivity, control, manipulation of experience, as well as the promotion of learning and the achievement of competence. In my own view, these processes are particularly true of middle age, where it is the preoccupation with past experience and the attempt to interpret experience that is the striking characteristic of the self.

Among other major preoccupations of mid-adulthood, as described by men and women who are presently in their forties and early fifties, are the ways that men rationalize their career achievements, and how, while some feel they have reached a plateau, others worry lest they have begun a downward slide, and still others look ahead to better things yet to come. There is the launching of one's children into the adult world and readjusting to changes in family relationships after the children are gone, coping with decrements in energy, physical health, and sexual potency, managing the changing relationships and responsibility for aging parents and adapting to the finiteness of lifetime as one faces the death of relatives and close friends.

It is apparent that middle-aged people look to their positions within different life contexts—body, career, family—rather than to chronological age for their primary cues in clocking themselves. Often there is a differential rhythm in the timing of events within these various contexts so that the cues used for placing oneself in a particular phase of the life cycle are not always synchronized. For example, one business executive regards himself as being on top in his occupation and assumes all the prerogatives and responsibilities that go with seniority in that context, yet, because he has children who are still young, he feels that he has a long way to go before completing his major goals within the family.

Women, more often than men, regard the middle years of life as a period of greater freedom for the self. It is not surprising, therefore, that many women tended to define middle age in terms of their present stage in the family cycle rather than by chronological age. Middle age is seen as beginning at the time the youngest child reaches high school. Energy and time which had previously been directed toward children and homemaking are now available for uses that can be self-determined—that can be inner-directed rather than other-directed in focus. Some women regard this period of life as an opportunity to expand their activities or develop previously latent or dormant talents. For these women, middle age is characterized not only by a marked change in activity, but by a major change in self-image, as well.

Men, on the other hand, tend to perceive and recognize their middle-age position in the life-line from cues received outside the family context. The deferential behavior accorded them by junior colleagues in various work, civic, or social settings; their sponsoring of younger persons for positions of responsibility; any disparity between career expectations and career achievements, that is, whether one is "on time" in reaching career goals and such, serve to trigger a heightened awareness of age.

A recurrent theme for some of the men we interviewed, for example, is the close relationship between lifetime and career movement. Career movement or change is viewed as feasible up to a certain age—generally not much later than the early fifties. Thus, a man in his early forties who had gone as far as he could in his firm said that he was giving "serious thought to a change now. If I'm ever going to make a satisfactory change, I must do it now." A forty-seven-year-old trust lawyer, who had moved at age forty-five from a large corporation to law firm, remarked, "I feel I got out at the last possible moment, because at forty-five it's very difficult to get another job. If you haven't made it by then, you better make it up fast, or you're stuck."

New and heightened awareness of age takes still other forms. A newspaper writer expressed some of the feelings of a sensitive middle-aged man in the following words:

". . . the realization suddenly struck me that I had become, perhaps not an old fogy but surely a middle-aged fogy. . . . For the train was filled with college boys returning from vacation. . . . They cruised up and down the aisles, pretending to be tipsy . . . they were boisterous, but not obnoxious; looking for fun, but not for trouble . . . Yet most of the adult passengers were annoyed with them, including myself. I sat there, feeling a little like Eliot's Prufrock, 'so meticulously composed, buttoned-up, bespectacled, mouth thinly set' . . . Squaresville." (Harris, 1965).

Many men regard their increased attention and concern with health problems and energy conservation to be a salient characteristic of middle age. Concern over health was seldom mentioned by the women we interviewed despite the obvious signs and manifestations of the climacterium during their late forties and early fifties. Both men and women recognized a shift in their general orientation to the body, and in body cathexis, with increased attention now being given to "body-monitoring"—a term we used to describe the large variety of protective strategies used by middle-aged persons for maintaining the body at a given level of performance or in preparing for future decrements in function. Closer attention was focused upon diet, rest, and sleep than had been true at earlier periods; there was a shift in emphasis from a "youth-vigor" value system to one of "health-comfort-grooming."

Thus, for example, a fifty-two-year-old business executive, recognizing that regular physical examinations had now become a routinized part of his life, remarked, "I began to go in for semi-annual check-ups about ten years ago. When you reach this age there are various changes in blood chemistry and so on that take place without your noticing them." Or, consider the forty-five-year-old attorney, who said, "I think the physical changes occur first. Mentally you still feel young, but you begin to notice that your legs ache if you run up the stairs. You remember when they didn't. You get winded more quickly when you do physical activity, and those things all add up."

There were other indications of a changing time-perspective, in the ways individuals orient themselves to time and personalize the phenomenon of death as they move from young adulthood to middle adulthood. The restructuring of life in terms of time-left-to-live rather than time-since-birth, the provision for social as well as biological heirs, the "rehearsal for widowhood" that was common in women, but not in men; the awareness that time is finite—these take on a saliency in mid-adulthood that is not so evident at earlier stages in the life cycle. Thus a building contractor, aged forty-eight, remarked, "You hear so much about deaths that seem to be premature. That's one of the changes that comes over you over the years, whereas young fellows never give it a thought."

The recognition that there is "only so much time left" was a frequent theme in the interviews. In referring to the death of a contemporary, one man, aged forty-eight, stated, "There is now the realization that death is very real. Those things don't quite penetrate when you're in your twenties and you think that life is all ahead of you. Now you realize that those years are gone and with each passing year you are getting closer to the end of your life."

For some men and women, the death of the last surviving parent introduces a feeling of personal vulnerability. A forty-seven-year-old author said in describing her reactions, "Both of my parents died within the last year . . . and all of a sudden I have the sense of being vulnerable myself, a feeling that I've never had before."

OLD AGE

I shall shortcut the discussion of the issues for aged persons, but not because the aged are any less important a group to study than younger adults, if we are to establish a psychology of adulthood. In fact, some of us now recognize that studies of aged persons are contributing to some of the basic theoretical and substantive issues in psychology in much the same way as studies of children. Because many of the issues have already been implied in my comments about the middle-aged, and because of the attention old people have been attracting in recent years, the issues of old age are likely to be better known to this audience.

It is perhaps evident enough, by now, that the social-psychological phenomena which I have been describing are ones to which the investigator comes unprepared, as it were, from his studies of children and adolescents.

Perhaps it is also evident why I have dealt at such length with what I call the "issues" of adulthood. It is not that I am confusing the issues or problems people face with the *processes* of personality, or that I am unaware that personality as such is not synonymous with the content of life. It is, instead, that you may see why, to psychologists of adult personality, the terms and the theories that exist around us seem to offer something less than a good fit for the phenomena with which we deal. Neither psychoanalytic theory nor learning theory nor social-role theory provides us with adequate concepts or with suitable variables. Where, except perhaps to certain egopsychologists who use terms such as "competence," "self," or "effectance," shall we turn if we wish to describe the incredible complexity of psychological phenomena shown, for instance, in the behavior of a business executive, age forty-five, who deals with the thousand and one questions and decisions that arise in the course of a single day? What terms shall we use to describe the strategies with which such a person manages his time, deals with troublesome as well as helpful other people, buffers himself from certain stimuli, while selecting other stimuli to which to pay full attention, engages in elaborate time schedules for attending to his physical and his mental (to say nothing of his emotional) preoccupations, sheds some of his cognitive "load" and some of his emotional "load," delegating some tasks to people around him over whom he has certain forms of control, while accepting other tasks as being singularly appropriate to his own competencies and skills? It is the incongruity between existing psychological concepts and theories, on the one hand, and the phenomena of real-life adult behavior, on the other, to which I draw your attention.

RESEARCH ON INTRAPSYCHIC PROCESSES

Lest I give only one side of the picture, however, let me say that some research is beginning to appear that is relevant to developmental lines of inquiry. Some of our own research has attempted to deal with adult phenomena, and at the same time to be concerned with personality *processes*. We have carried out, for instance, several related studies of middle and old age, all based upon large samples of normal people and drawn by probability techniques from the community at large. Although these were cross-sectional studies, they were designed specifically to clarify differences between age groups with regard to ego processes. In one of this group of studies, analysis of TAT data indicated that forty-year-olds saw themselves as possessing energy congruent with the opportunities perceived in the outer world; the environment was seen as rewarding boldness and risk-taking. Sixty-year-olds, however, perceived the world as complex and dangerous, no longer to be reformed in line with one's own wishes; and the individual was perceived as conforming and accommodating to outer-world demands. The protagonist, to the sixty-year-old, was no longer a forceful manipulator of the object world, but a relatively passive object manipulated by the environment.

The same study suggested important differences between men and women as they age. For instance, men seem to become more receptive to their affiliative, nurturant, and sensual promptings; women seem to become more responsive toward and less guilty about their aggressive and egocentric impulses.

Another of the studies confirmed these trends, and indicated that with increasing age, ego functions are turned inward, as it were. With the change from active to passive modes of mastering the environment, there was also a movement of energy away from an outer-world to an inner-world orientation.

Whether or not such age differences in personality have inherent as well as reactive qualities cannot yet be established from the findings presently available. We have chosen to interpret certain of these changes as developmental, however, because they seem to occur well before the social or biological "losses" of aging can be said to begin, and before there was evidenced change on any of our gross measures of social personality or social role performance. In other words, the appearance of such personality changes by the mid-forties in a group of well-functioning adults seems congruent with a developmental view of adult personality.

I have already implied that the middle years of life—probably the fifties for most persons—represent an important turning point in personality organization with the re-evaluation of time and the formulation of new perceptions of time and death. It is at this point in the life-line that introspection seems to increase noticeably and contemplation of one's inner thoughts seems to become a characteristic form of mental life. The implication is not that the introspection of middle age is the same as the reminiscence of old age; but it seems to be its forerunner—perhaps a preparatory step in a restructuring of the ego that Erikson has called the attainment of integrity, the symbolic putting of one's house in order which is described as characteristic of old age.

Without attempting to summarize the evidence that exists (Neugarten, 1964; Neugarten et al., 1964), the view I should like to set before you is that there are sets of personality processes, primarily intrapsychic in nature, which show developmental changes throughout the lifespan. As the individual moves from childhood and adolescence into adulthood, ego processes become increasingly salient in personality dynamics. In the broadest terms, the development of the ego is, for the first two-thirds of the lifespan, outward toward the environment; for the last part of the lifespan, inward toward the self. In these impressionistic terms, it is as if the ego, in childhood, is focused upon the development of physical, mental, and emotional tools to manipulate both the inner and the outer worlds. In young adulthood, the thrust is toward the outer world, toward mastery of the environment. In middle age there comes a realignment and restructuring of ego processes, and, to the extent that these processes become conscious, a reexamination of the self. In old age, there is a turning inward, a withdrawal of investment from the outer world, and a new preoccupation with the inner world.

Although our evidence is still very inadequate, there are data to support the position that changes occur in intrapsychic processes, then, as well as in more readily observable behavior; that such changes are probably orderly and developmental in nature. The realignment and redirection of ego processes seems to begin in middle rather than in old age. There is a general direction of change from active to passive modes of relating to the environment; and there is a general movement of energy from an outer-world to an inner-world orientation.

REFERENCES

Harris, S. "Strictly personal," *Chicago Daily News*, May 11, 1965.
Neugarten, B. L. "A developmental view of adult personality." In: Birren, J. E. (ed.), *Relations of development and aging*. Springfield, Ill.: Charles C Thomas, 1964.
Neugarten, B. L., et al. *Personality in middle and late life*. New York: Atherton, 1964.

Twenty-nine

The "Sexless Older Years"—
A Socially Harmful Stereotype

Isadore Rubin

ABSTRACT: The stereotype of the "sexless older years," which has placed its stamp upon our entire culture and which, in many cases, acts as a "self-fulfilling prophecy," has done considerable damage to our aging population. Although no studies of sexual behavior and attitudes of the aging have been done on a sufficiently representative sample to provide us with norms, a growing body of research makes clear that there is no automatic cutoff to sexuality at any age and that sex interests, needs, and abilities continue to play an important role in the later years. This is true not only for the married, but also for the single and widowed. Unless our entire culture recognizes the normality of sex expression in the older years, it will be impossible for older persons to express their sexuality freely and without guilt.

It has been suggested that our culture has programmed marriage only until the child-raising period has been completed.[1] If this is true of marital roles in general, it is especially true of sexual roles in the later years. Society has not given genuine recognition to the validity of sexual activity after the child-bearing years, creating a dangerous stereotype about the "sexless older years" and defining as deviant behavior sex interest and activity which may continue vigorously into these older years. Thus, for example, the opprobrious term "lecher" is never coupled with any age group but the old; the young are "lusty" or "virile."

Rubin, I., "The 'sexless older years'—a socially harmful stereotype," *Annals of the American Academy of Political and Social Science*, 1968, 376, 86–95. Reprinted with permission of the author and publisher.

Dr. Rubin was editor of *Sexology*, Chairman of the Publications Committee of SIECUS, Fellow of the Society for the Scientific Study of Sex, and an affiliate of the American Association of Marriage and Family Counselors. He was author of *Sexual Life after Sixty* (1965) and *Sexual Life in the Later Years* (SIECUS Study Guide No. 12, 1970); editor of *Sexual Freedom in Marriage* (1969); and co-editor of *Sex in the Adolescent Years* (1968) and *Sex in the Childhood Years* (1970).

A SELF-FULFILLING PROPHECY

This stereotype has until recently placed its unchallenged stamp upon our culture. In the late 1950s, undergraduates at Brandeis University were asked to take a test to assess their attitudes toward old people.[2] Those taking the test were requested to complete this sentence: "Sex for most old people. . . ." Their answers were quite revealing. Almost all of these young men and women, ranging in age from seventeen to twenty-three, considered sex for most old people to be "negligible," "unimportant," or "past." Since sex behavior is not only a function of one's individual attitudes and interactions with a partner, but also a reflection of cultural expectations, the widespread belief about the older person being sexless becomes for many a "self-fulfilling prophecy." Our society stands indicted, says psychiatrist Karl M. Bowman, of grave neglect of the emotional needs of aging persons:

Men and women often refrain from continuing their sexual relations or from seeking remarriage after loss of a spouse, because even they themselves have come to regard sex as a little ridiculous, so much have our social attitudes equated sex with youth. They feel uncertain about their capacities and very self-conscious about their power to please. They shrink from having their pride hurt. They feel lonely, isolated, deprived, unwanted, insecure. Thoughts of euthanasia and suicide bother them. To prevent these feelings, they need to have as active a sex life as possible and to enjoy it without fear.[3]

Most of our attitudes toward sex today still constitute—despite the great changes that have taken place in the openness with which sex is treated publicly—what a famous British jurist has called "a legacy of the ascetic ideal, persisting in the modern world after the ideal itself has deceased."[4] Obviously, the ascetic attitude—essentially a philosophy of sex-denial—would have far-reaching effects upon our attitude toward the sexual activity of those persons in our society who have passed the reproductive years. Even so scientific a writer as Robert S. de Ropp, in his usually excellent *Man against Aging*, betrays the unfortunate effects of our ascetic tradition when he says:

For sexual activity, enjoyable as it may seem in itself, still has as its natural aim the propagation of the species, and this activity belongs to the second not the third act of life's drama.[5]

In addition to our tradition of asceticism, there are many other factors which undoubtedly operate to keep alive a strong resistance to the acceptance of sexuality in older people. These include our general tradition of equating sex, love, and romance solely with youth; the psychological difficulty which children have of accepting the fact of parental intercourse; the tendency to think of aging as a disease rather than a normal process; the focusing of studies upon hospitalized or institutionalized older people rather than upon a more typical sample of persons less beset by health, emotional, or economic problems; and the unfortunate fact that—by and large—physicians have shared the ignorance and prejudices equally with the rest of society.[6]

It is significant, however, that centuries of derogation and taboo have not been successful in masking completely the basic reality that sex interest and activity do not disappear in the older years. Elaine Cumming and William E. Henry point out that our jokes at the expense of older people have revealed considerable ambivalence in the view that all old people are asexual.[7] The contradictory attitude which people possess about sexuality in the later years is also well illustrated by the history of the famous poem "John Anderson, My Jo," written by Robert Burns almost two centuries ago. In the version known today, the poem is a sentimental tribute to an old couple's calm and resigned old age. The original folk version—too bawdy to find its way into textbooks—was an old wife's grievance about her husband's waning sex interest and ability which makes very clear that she has no intention of tottering down life's hill in a passionless and sexless old age.[8] It is also interesting to note that sexuality in older women was an important part of one of Aristophanes' comedies. In his play *Ecclesiazusae* ("Women in Parliament"), Aristophanes described how the women seized power and established a social utopia.[9] One of their first acts was to place sexual relations on a new basis in order to assure all of them ample satisfaction at all times. They decreed that, if any young man was attracted to a girl, he could not possess her until he had satisfied an old woman first. The old women were authorized to seize any youth who refused and to insist upon their sexual rights also.

THE HARMFUL INFLUENCE OF THE MYTH

A British expert in the study of aging has suggested that the myth of sexlessness in the older years does have some social utility for some older women in our society who may no longer have access to a sexual partner.[10] However, the widespread denial of sexuality in older persons has a harmful influence which goes far beyond its effect upon an individual's sexual life.[11] It makes difficult, and sometimes impossible, correct diagnoses of medical and psychological problems, complicates and distorts interpersonal relations in marriage, disrupts relationships between children and parents thinking of remarriage, perverts the administration of justice to older persons accused of sex offenses, and weakens the whole self-image of the older man or woman.

A corollary of the failure to accept sexuality as a normal aspect of aging has been the tendency to exaggerate the prevalence of psychological deviation in the sexual behavior of older men and to see in most old men potential molesters of young children. Seen through the lenses of prejudice, innocent displays of affection have often loomed ominously as overtures to lascivious fondling or molestation. It is common, too, to think of the exhibitionist as being, typically, a deviation of old age.

Actually, the facts indicate the falsity of both of these stereotypes. As research by Johann W. Mohr and his associates at the Forensic Clinic of the Toronto Psychiatric Hospital showed, "contrary to common assumption the old age group is the relatively smallest one" involved in child-molesting.[12] The major age groups from whose ranks child-molesters come are adolescence, the middle to late thirties, and the late fifties. The peak of acting out of exhibitionism occurs in the mid-twenties; and, in its true form, exhibitionism is rarely seen after the age of forty.

In relatively simple and static societies, everyone knows pretty much where he stands at each stage of life, particularly the older members of the group. "But in complex and fluid social systems," notes Leo W. Simmons, "with rapid change and recurrent confusion over status and role, no one's position is so well fixed—least of all that of the aging."[13] For many aging persons, there is a crisis of identity in the very sensing of themselves as old, particularly in a culture which places so great a premium upon youth. David P. Ausubel notes that, just as in adolescence, the transition to aging is a period where the individual is in the marginal position of having lost an established and accustomed status without having acquired a new one and hence is a period productive of considerable stress.[14] Under such conditions of role confusion, aging persons tend to adopt the stereotype which society has molded for them, in sex behavior as in other forms of behavior. But they do so only at a very high psychic cost.

For many older people, continued sexual relations are important not so much for the pleasurable release from sexual tension as for the highly important source of psychological reinforcement which they may provide. Lawrence K. Frank has said:

> Sex relations can provide a much needed and highly effective resource in the later years of life when so often men face the loss of their customary prestige and self-confidence and begin to feel old, sometimes long before they have begun to age significantly. The premature cessation of sexual functioning may accelerate physiological and psychological aging since disuse of any function usually leads to concom-

itant changes in other capacities. After menopause, women may find that continuation of sexual relations provides a much needed psychological reinforcement, a feeling of being needed and of being capable of receiving love and affection and renewing the intimacy they earlier found desirable and reassuring.[15]

THE GROWING BODY OF RESEARCH DATA

Gathering data about the sexual behavior and attitudes of the aging has not been an easy task. To the generalized taboos about sex research have been added the special resistance and taboos that center around sexuality in older persons. For example, when the New England Age Center decided to administer an inventory to its members, they included only nine questions about sex among the 103 items.[16] The nine questions were made deliberately vague, were confined largely to past sexual activities, and were given only to married members. Leaders of the Center felt that if they had asked more direct questions or put them to their unmarried members, these people would not have returned to the Center. In California, a study of the attitudes of a sample of persons over sixty years old in San Francisco during the early 1960s included just one general open-ended question about sexual attitudes, apparently because of the resistance which many of the researchers had about questioning subjects in the area of sex.[17] Psychiatrists reporting on this research before the Gerontological Society noted that the people involved in research in gerontology are being hamstrung by their own attitudes toward sex with regard to the elderly in much the same way in which the rest of society is hamstrung with regard to their attitudes toward the elderly in such matters as jobs, roles, and those things which go into determining where a person fits into the social structure.

Fortunately, although no sample has yet been studied that was sufficiently broad or typical to present us with a body of norms, a sufficient amount of data now exists which leaves no doubt of the reality of sex interests and needs in the latter years. While it is true that there are many men and women who look forward to the ending of sexual relations, particularly those to whom sex has always been a distasteful chore or those who "unconsciously welcome the excuse of advancing years to abandon a function that has frightened them since childhood,"[18] sexual activity, interest, and desire are not the exception for couples in their later years. Though the capacity for sexual response does slow down gradually, along with all the other physical capacities, it is usually not until actual senility that there is a marked loss of sexual capacity.

With the research conducted by William H. Masters and Virginia E. Johnson, who observed the anatomy and physiology of sexual response in the laboratory, confirmation has now been obtained that sexual capacity can continue into advanced old age.[19] Among the subjects whose orgasmic cycles were studied by these two investigators were 61 menopausal and post menopausal women (ranging from forty to seventy-eight) and 39 older men (ranging from fifty-one to eighty-nine). Among the women, Masters and Johnson found that the intensity of physiologic reaction and the rapidity of response to sexual stimulation were both reduced with advancing years. But they emphasized that they found "significant sexual capacity and effective sexual performance" in these older women, concluding:

> The aging human female is fully capable of sexual performance at orgasmic response levels, particularly if she is exposed to regularity of effective sexual stimulation. . . . There seem to be no physiologic reasons why the frequency of sexual expression found satisfactory for the younger woman should not be carried over into the postmenopausal years. . . . In short, there is no time limit drawn by the advancing years to female sexuality.

When it comes to males, Masters and Johnson found that there was no question but that sexual responsiveness weakens as the male ages, particularly after the age of sixty. They added, however:

> There is every reason to believe that maintained regularity of sexual expression coupled with adequate physical well-being and healthy mental orientation to the aging process will combine to provide a sexually stimulative climate within a marriage. This climate will, in turn, improve sexual tension and provide a capacity for sexual performance that frequently may extend to and beyond the eighty-year age level.

These general findings have been supported by various types of studies which have been made over the course of the years. These studies include the investigation by Raymond Pearl in 1925 into the frequency of marital intercourse of men who had undergone prostatic surgery, all over the age of fifty-five;[20] Robert L. Dickinson and Lura E. Beam's studies of marriages and of single women, including a number of older single women and widows;[21] the Kinsey studies of the male and the female;[22] older men studied at outpatient clinics by urologists at the University of California School of Medicine at San Francisco;[23] extended study by Duke University psychiatrists of Negroes and whites living in the Piedmont area of North Carolina;[24] Joseph T. Freeman's study of older men in Philadelphia;[25] a study of patients attached to a geriatric clinic in New York;[26] a survey of veterans applying for pensions;[27] a questionnaire survey by *Sexology* magazine of men over sixty-five who were listed in

Who's Who in America;[28] and a study of sex attitudes in the elderly at the Langley Porter Neuropsychiatric Institute in San Francisco.[29]

NO AUTOMATIC CUTOFF DATE

All of these studies indicate the continuation of sex needs, interests, and abilities into the later years despite the gradual weakening that may take place. The Kinsey group, quite contrary to general conceptions of the aging process in sex, found that the rate at which males slow up sexually in the last decades of life does not exceed the rate at which they have been slowing up and dropping out of sexual activity in the previous age groups.[30] For most males, they found no point at which old age suddenly enters the picture. As far as females were concerned, the Kinsey investigators—like Masters and Johnson later—found little evidence of any aging in their capacities for sexual response.[31] "Over the years," they reported, "most females become less inhibited and develop an interest in sexual relations which they then maintain until they are in their fifties or even sixties." In contrast to the average wife, the responses of the husband dropped with age. Thus, many of the younger females reported that they did not desire intercourse as often as their husbands. In the later years of marriage, however, many of the wives expressed the desire for coitus more often than their husbands were then desiring it.

The Duke University survey—reported by Gustave Newman and Claude R. Nichols—found that only those persons who were seventy-five or older showed a significantly lower level of sexual activity.[32] This study found that Negro subjects were sexually more active than white subjects; men were more active than women; and persons lower in the social and economic scale were more active than those in the upper-income group. A possible explanation of the greater activity reported by males lies in the fact that men and women of the same age were reporting on different age groups. The wives, on the average, would be reporting on sex activity with a husband who was perhaps four years older.

Despite the fact that masturbation has been usually considered an activity that ends with maturity, for many older persons, this practice apparently continues to serve as a satisfactory form of release from sexual tensions when a partner, is for one reason or another, not available.[33]

Several of the studies suggest a correlation between early sex activity and a continuation into the late years. The Kinsey group found that, at age fifty, all of the males who had been sexually active in early adolescence were still sexually active, with a frequency about 20 percent higher than the frequency of the later-maturing males.[34] They report:

Nearly forty years maximum activity have not yet worn them out physically, physiologically, or psychologically. On the other hand, some of the males (not many) who were late adolescent and who have had five years less of sexual activity, are beginning to drop completely out of the picture; and the rates of this group are definitely lower in these older age periods.

They conclude:

The ready assumption which is made in some of the medical literature that impotence is the product of sexual excess, is not justified by such data as are now available.

Freeman[35] found that the sex urge of persons in advanced years correlated strongly with their comparative sex urge when young, and a similar finding was reported by the Duke University survey.[36]

Masters and Johnson report the same finding, with additional emphasis upon regularity of sexual expression as the essential factor in maintaining sexual capacity and effective performance for both males and females:[37]

When the male is stimulated to high sexual output during his formative years and a similar tenor of activity is established for the 31–40 year range, his middle-aged and involutional years usually are marked by constantly recurring physiologic evidence of maintained sexuality. Certainly it is true for the male geriatric sample that those men currently interested in relatively high levels of sexual expression report similar activity levels from their formative years. It does not appear to matter what manner of sexual expression has been employed, as long as high levels of activity were maintained.

FACTORS RESPONSIBLE FOR DECLINING SEX ACTIVITY

On the basis of present data, it is not possible to sort out the emotional element from the purely physiologic factors in the decline in sexual activity of the older male. Some animal experiments have shown that changes in the external environment can result in changes in sexual drive. When aging rats had the opportunity for sex activity with a number of partners, for example, the number of copulations increased considerably.[38] However, as soon as male rats reached a certain age, they failed to respond to females.[39]

Many men also find that, with a new partner, a new stimulus is given to their virility.[40] However, often these men return to their old level within comparatively short periods of time.[41] Present data lead us to conclude, with the Kinsey investigators:

> The decline in sexual activity of the older male is partly, and perhaps primarily, the result of a general decline in physiologic capacity. It is undoubtedly affected also by psychologic fatigue, a loss of interest in repetition of the same sort of experience, an exhaustion of the possibilities for exploring new techniques, new types of contacts, new situations.[42]

Masters and Johnson, on the basis of their clinical work with older males, describe six general groups of factors which they believe to be responsible for much of the loss of sexual responsiveness in the later years: (1) monotony of a repetitive sexual relationship (usually translated into boredom with the partner); (2) preoccupation with career or economic pursuits; (3) mental or physical fatigue; (4) overindulgence in food or drink; (5) physical and mental infirmities of either the individual or his spouse; and (6) fear of performance associated with or resulting from any of the former categories.

The most constant factor in the loss of an aging male's interest is the problem of monotony, described by the Kinsey group as "psychologic fatigue." According to Masters and Johnson, many factors may produce this: failure of the sexual relationship to develop beyond a certain stage; overfamiliarity; lack of sexual interest on the female's part; aging and loss of personal attractiveness of the female.

A major deterrent for many men is preoccupation with the outside world and their careers. Overindulgence in food and drink, particularly the latter, takes a high toll. According to Masters and Johnson, secondary impotence developing in the late forties or early fifties has a higher incidence of direct association with excessive alcohol consumption than with any other single factor.

As each partner ages, the onset of physical or mental infirmities is an ever-increasing factor in reducing sexual capacities. The harmful effect of this is sometimes multiplied by the negative or discouraging attitude of the physician. Once a failure in performance has occurred because of any of the factors, the fear of failure becomes an additional factor in bringing about withdrawal from sexual activity. "Once impotent under any circumstances," remark Masters and Johnson, "many males withdraw voluntarily from any coital activity rather than face the ego-shattering experience of repeated episodes of sexual inadequacy."

The very scanty data concerning the sexual attitudes of older persons suggest a more positive attitude toward sex among men than among women, with women being more "culture-bound" and still showing strong evidences of the effects of the Victorian age in which they acquired their attitudes toward sex.[43] A study of dreams of residents of a home for the aged and infirm, on the other hand, indicates a contrasting difference in emotional tone of the sexual content of the dreams of men and women: "Whereas in men sexual dreams revealed anxiety, failure, and lack of mastery, in women they usually depicted passive, pleasurable gratification of dependent needs."[44]

THE UNMARRIED HAVE SEX NEEDS TOO

It is not only the married who have sexual needs. Aging widows, widowers, and single persons, who make up an increasingly large segment of our population, face even greater problems in respect to sex than do the married. In the survey by Newman and Nichols, only seven of the 101 single, divorced, or widowed subjects reported any sexual activity with partners.[45] Apparently, the strength of the sexual drive of most elderly persons is usually not great enough to cause them to seek a sexual partner outside of marriage in the face of social disapproval and the difficulties of such an endeavor. Interestingly, however, thousands of older couples were reportedly living "in sin—or what they think is sin" because marriage would mean loss of social security payments.[46]

Dickinson and Beam reported that in their study of widows ranging from sixty to eighty years of age there was evidence of masturbation.[47] They reported that when these women underwent pelvic examinations they showed such marked sexual reactions that they found that "it is desirable to relieve the patient's embarrassment by hurting her, lest she have orgasm." Since many older women are quite troubled by their practice of masturbation, marriage counselors have stressed the importance of helping older persons to accept this practice as a valid outlet when they feel the need for it.[48]

THE GREAT NEED FOR INFORMATION

Persons who have worked with "senior citizens" and "golden age" clubs have reported the great need for knowledge, the confusion, and the eager hunger for information about sex shown by persons in these clubs.[49] The many perplexing problems that they raise indicate the extent to which such information is needed to help people solve broader questions of remarriage and interpersonal relationships during their later years. The growing incidence of disease states in these years—each of which may require a difficult readjustment in sexual and other relationships—makes it essential that older people be provided with this information openly and consistently.[50]

It should be clear, however, that unless our entire culture recognizes the normality of sex expression in the older years, it will be impossible for older persons to express their sexuality freely and without guilt. Physicians are particularly crucial in this respect; unless they are convinced of the psychological importance of sexual functioning in the later years, they can do irreparable harm to their patients' sexuality.[51] Fortunately, at long last, medical schools and medical publications have begun to take steps to correct the glaring lacks in the education of medical students, which have in the past resulted in the creation of a body of medical practitioners who, by and large, shared the general prejudices of our society concerning sexuality in older persons.

NOTES

[1] E. Cumming and W. E. Henry, *Growing Old*. New York: Basic Books, 1961, p. 155.

[2] P. Golde and N. Kogan, "A Sentence Completion Procedure for Assessing Attitudes Toward Old People," *Journal of Gerontology*, Vol. 14, July 1959, pp. 355–363.

[3] K. M. Bowman, "The Sex Life of the Aging Individual," in M. F. DeMartino (ed.), *Sexual Behavior and Personality Characteristics*. New York: Citadel, 1963, pp. 372–375.

[4] G. Williams, *The Sanctity of Life and the Criminal Law*. New York: Alfred A. Knopf, 1957, p. 51.

[5] R. S. de Ropp, *Man against Aging*. New York: Grove Press, 1962, p. 252.

[6] H. I. Lief, "Sex Education of Medical Students and Doctors," *Pacific Medicine and Surgery*, Vol. 73, February 1965, pp. 52–58.

[7] Cumming and Henry, *op. cit.*, footnote, p. 21.

[8] R. Burns, *The Merry Muses of Caledonia*, J. Barke and S. G. Smith (eds.). New York: Putnam, 1964, pp. 147–148.

[9] H. Einbinder, *The Myth of the Brittanica*. New York: Grove Press, 1964, p. 94.

[10] A. Comfort, "Review of *Sexual Life after Sixty*, by Isadore Rubin," *British Medical Journal*, II, March 25, 1967, p. 750.

[11] Isadore Rubin, *Sexual Life after Sixty*. New York: Basic Books, 1965, Chap. i.

[12] J. W. Mohr, R. E. Turner, and M. B. Jerry, *Pedophilia and Exhibitionism*. Toronto: University of Toronto Press, 1964.

[13] L. W. Simmons, "Social Participation of the Aged in Different Cultures," in M. B. Sussman (ed.), *Sourcebook in Marriage and the Family*, 2d ed. Boston: Houghton Mifflin, 1963.

[14] D. P. Ausubel, *Theory and Problems of Adolescent Development*. New York: Grune and Stratton, 1954, pp. 53 ff.

[15] L. K. Frank, *The Conduct of Sex*. New York: Morrow, 1961, pp. 177–178.

[16] E. B. Armstrong, "The Possibility of Sexual Happiness in Old Age," in H. G. Beigel (ed.), *Advances in Sex Research*. New York: Hoeber-Harper, 1963, pp. 131–137.

[17] E. H. Feigenbaum, M. J. Lowenthal and M. L. Trier, "Sexual Attitudes in the Elderly." Unpublished paper given before the Gerontological Society, New York, November 1966.

[18] W. R. Stokes, *Married Love in Today's World*. New York: Citadel, 1962, p. 100.

[19] W. H. Masters and V. E. Johnson, *Human Sexual Response*. Boston: Little, Brown, 1966, sec. on "Geriatric Sexual Response," pp. 223–270.

[20] R. Pearl, *The Biology of Population Growth*. New York: Alfred A. Knopf, 1925, pp. 178–207.

[21] R. L. Dickinson and L. E. Beam, *A Thousand Marriages*. Baltimore: Williams & Wilkins, 1931, pp. 278–279, 446; and R. L. Dickinson and L. E. Beam, *The Single Woman*. Baltimore: Williams & Wilkins, 1934, p. 445.

[22] A. C. Kinsey, W. B. Pomeroy, and C. E. Martin, *Sexual Behavior in the Human Male*. Philadelphia: W. B. Saunders, 1948; and A. C. Kinsey, W. B. Pomeroy, C. E. Martin, and P. H. Gebhard, *Sexual Behavior in the Human Female*. Philadelphia: W. B. Saunders, 1953.

[23] A. L. Finkle et al., "Sexual Function in Aging Males: Frequency of Coitus among Clinic Patients," *Journal of the American Medical Association*, Vol. 170, July 18, 1959, pp. 1391–1393.

[24] G. Newman and C. R. Nichols, "Sexual Activities and Attitudes in Older Persons," *Journal of the American Medical Association*, Vol. 173, May 7, 1960, pp. 33–35.

[25] J. T. Freeman, "Sexual Capacities in the Aging Male," *Geriatrics*, Vol. 16, January 1961, pp. 37–43.

[26] L. Friedfeld, "Geriatrics, Medicine, and Rehabilitation," *Journal of the American Medical Association*, Vol. 175, February 18, 1961, pp. 595–598; and L. Friedfeld et al., "A Geriatric Clinic in a General Hospital," *Journal of the American Geriatrics Society*, Vol. 7, October 1959, pp. 769–781.

[27] L. M. Bowers, R. R. Cross, Jr., and F. A. Lloyd, "Sexual Function and Urologic Disease in the Elderly Male," *Journal of the American Geriatrics Society*, Vol. 11, July 1963, pp. 647–652.

[28] I. Rubin, "Sex over Sixty-five," in H. G. Beigel (ed.), *Advances in Sex Research*. New York: Hoeber-Harper, 1963.

[29] Feigenbaum et al., *op. cit.*

[30] Kinsey et al., *Sexual Behavior in the Human Male*, pp. 235–237.

[31] Kinsey et al., *Sexual Behavior in the Human Female*, pp. 353–354.

[32] Newman and Nichols, *op. cit.*

[33] Rubin, "Sex over Sixty-five"; and Dickinson and Beam, *A Thousand Marriages*.

[34] Kinsey et al., *Sexual Behavior in the Human Male*, pp. 319–325.

[35] Freeman, *op. cit.*

[36] Newman and Nichols, *op. cit.*

[37] Masters and Johnson, *op. cit.*

[38] J. Botwinick, "Drives, Expectancies, and Emotions," in J. E. Birren (ed.), *Handbook of Aging and the Individual*. Chicago: University of Chicago Press, 1959, pp. 739–768.

[39] L. F. Jakubczak, Report to the American Psychological Association, August 31, 1962.

[40] J. Bernard, *Remarriage*. New York: Dryden, 1956, p. 188.

[41] Kinsey et al., *Sexual Behavior in the Human Male*, pp. 227–229; and A. W. Spence, "Sexual Adjustment at the Climacteric," *Practitioner*, Vol. 172, April 1954, pp. 427–430.

[42] Kinsey et al., *Sexual Behavior in the Human Male*, pp. 226–235.

[43] Feigenbaum et al., *op. cit.*

[44] M. Barad, K. Z. Altshuler, and A. I. Goldfarb, "A Survey of Dreams in Aged Persons," *Archives of General Psychiatry*, Vol. 4, April 1961, pp. 419–424.

[45] Newman and Nichols, *op. cit.*

[46] *New York Times*, January 12, 1965.

[47] Dickinson and Beam, *A Thousand Marriages*.

[48] L. Dearborn, "Autoerotism," in A. Ellis and A. Abarbanel (eds.), *The Encyclopedia of Sexual Behavior*. New York: Hawthorn, 1961, pp. 204–215; and L. Hutton, *The Single Woman*. London: Barrie & Rockcliff, 1960, p. 58.

[49] Feigenbaum et al., *op. cit.*

[50] Rubin, *Sexual Life after Sixty*, Chaps. xi–xiii.

[51] J. S. Golden, "Management of Sexual Problems by the Physician," *Obstetrics and Gynecology*, Vol. 23, March 1964, pp. 471–474; and A. L. Finkle and D. V. Prian, "Sexual Potency in Elderly Men before and after Prostatectomy," *Journal of the American Medical Association*, Vol. 196, April 11, 1966, pp. 139–143.

Thirty

Women's Attitudes toward the Menopause

Bernice L. Neugarten, Vivian Wood, Ruth J. Kraines, and Barbara Loomis

The menopause, like puberty or pregnancy, is generally regarded as a significant event in a woman's life; one that is known to reflect profound endocrine and somatic changes; and one that presumably involves psychological and social concomitants as well. Although there is a large medical and biological literature regarding the climacterium,[1] there are few psychological studies available, except those reporting symptomology or those based on observations of women who were receiving medical or psychiatric treatment. Even the theories regarding the psychological effects of the climacterium are based largely upon observations of clinicians—psychoanalytic case studies, psychiatric investigations of climacteric psychoses, and observations of their middle-aged patients made by gynecologists and other physicians (August, 1956; Barnacle, 1949; Deutsch, 1945; Fessler, 1950; Hoskins, 1944; Ross, 1951; Sicher, 1949).

Unlike the case with puberty or pregnancy, developmental psychologists have not yet turned their attention to the menopause, and to the possible relationships, whether antecedent or consequent, between biological, psychological, and social variables. [A paper by the psychoanalyst Therese Benedek (1952) is a notable exception.] Neither is there a body of anthropological or sociological literature that describes the prevailing cultural or social attitudes to be found in America or other Western societies regarding the menopause.

As preliminary to a larger study of adjustment patterns in middle age, a number of exploratory interviews were gathered in which each woman was asked to assess her own menopausal status (whether she regarded herself as pre-, "in," or postclimacteric). She was then asked the basis of this assessment; what, if any, symptoms she had experienced; what her anticipations of menopause had been, and why; what she regarded as the worst and

Neugarten, B. L., V. Wood, R. J. Kraines, and B. Loomis. "Women's attitudes toward the menopause," *Vita Humana*, 1963, 6, 140–151. Reprinted with permission from the authors and publisher, S. Karger, Basel/New York. This research was facilitated by USPHS Grant No. M-3972, Bernice L. Neugarten, Principal Investigator.

what the best aspects; and what, if any, changes in her life she attributed to the menopause.

It was soon apparent that women varied greatly in their attitudes and experiences. Some, particularly at upper-middle-class levels, vehemently assured the interviewer that the menopause was without any social or psychological import; that, indeed, the enlightened woman does not fear, nor—even if she suffers considerable physical discomfort—does she complain about the menopause.

"Why make any fuss about it?"

"I just made up my mind I'd walk right through it, and I did. . . ."

"I saw women complaining, and I thought I would never be so ridiculous. I would just sit there and perspire, if I had to. At times you do feel terribly warm. I would sit and feel the water on my head, and wonder how red I looked. But I wouldn't worry about it, because it is a natural thing, and why get worried about it? I remember one time, in the kitchen, I had a terrific hot flush. . . . I went to look at myself in the mirror. I didn't even look red, so I thought, 'All right . . . the next time I'll just sit there, and who will notice? And if someone notices, I won't even care. . . .' "

Others confessed to considerable fear:

"I would think of my mother and the trouble she went through; and I wondered if I would come through it whole or in pieces. . . ."

"I knew two women who had nervous breakdowns, and I worried about losing my mind. . . ."

"I thought menopause would be the beginning of the end . . . a gradual senility closing over, like the darkness. . . ."

"I was afraid we couldn't have sexual relations after the menopause— and my husband thought so, too. . . ."

"When I think of how I used to worry! You wish someone would tell you—but you're too embarrassed to ask anyone. . . ."

Other women seemed to be repeating the advice found in women's magazines and in newspaper columns:

"I just think if a woman looks for trouble, she'll find it. . . ."

"If you fill your thinking and your day with constructive things—like trying to help other people—then it seems to me nothing can enter a mind already filled. . . ."

"If you keep busy, you won't think about it, and you'll be all right. . . ."

Underlying this variety of attitudes were two common phenomena: first, whether they made much or made little of its importance, middle-aged women were willing, even eager, to talk about the menopause. Many volunteered the comment that they seldom talked about it with other women; that they wished for more information and more communication. Second, although many professed not to believe what they termed "old wives' tales," most women had nevertheless heard many such tales of the dangers of menopause, and could recite them easily: that menopause often results in mental breakdown; that it marks the end of a woman's sexual attractiveness as well as her sexual desires; and so on. Many women said, in this connection, that while they themselves had had neither fears nor major discomforts, they indeed knew other women who held many such irrational fears or who had suffered greatly. (The investigators interpreted such responses as indicative, at least in part, of the psychological mechanism of projection.)

THE INSTRUMENT

Following a round of preliminary interviews, a more systematic measurement of attitudes toward the menopause was undertaken. A checklist was drawn up containing statements culled from the exploratory interviews and from the literature on the subject. For example, the statement, "Women generally feel better after the menopause than they have for years," appears in a pamphlet about menopause for sale by the U.S. Government Printing Office (*U.S. Public Health Service*, 1959). "Women who have trouble with the menopause are usually those who have nothing to do with their time," is a statement made by a number of interviewees. "A woman in menopause is likely to do crazy things she herself does not understand" is a statement made by a woman describing her own behavior. Respondents were asked to check, for each statement, (1) agree strongly; (2) agree to some extent; (3) disagree somewhat; or (4) disagree strongly. Because of the projective phenomenon already mentioned, the statements were worded in terms of "other women," or "women in general" rather than "self" (see Table 30.2).

The checklist was then pre-tested on a sample of 50 women aged forty to fifty. Following the analysis of those responses, the instrument was revised and the number of items reduced to 35. Certain statements were eliminated because they drew stereotyped responses; others, because of overlap.

THE SAMPLES

The revised Attitudes-toward-Menopause Checklist, hereafter referred to at the ATM, was administered as part of a lengthy interview to a sample of 100 women aged forty-five to fifty-five on whom a variety of other data were being gathered. These 100 women, referred to here as the C, or Criterion, group had been drawn from lists of mothers of graduates from two public high schools in the Chicago metropolitan area. None of these women had had surgical or artificial menopause, and all were in relatively good health.

Once the data on the ATM had been analyzed for the Criterion group, the question arose, how do women of different ages view the menopause? Accordingly, the instrument was administered to other groups of women contacted through business firms and women's clubs. Directions for filling out the ATM were usually given in group situations, and the respondents were asked to mail back the forms to the investigators along with certain identifying information (age, level of education, marital status, number and ages of children, and health status). The proportion responding varied from group to group, with an average of about 75 percent responses. From this larger pool, Groups A, B, and D were drawn.

The composition of the four samples, by age and level of education, is shown in Table 30.1. All the women in all four groups were married, all were mothers of one or more children and, with the exception of a few in Groups B and D, all were living with their husbands. None of these women reported major physical illness or disability.

TABLE 30.1   The Samples: By Age and Level of Education

Group	Number	Age	High school graduation or less	One or more years of college	No information
			Percents		
A	50	21–30	8	90	2
B	52	31–44	33	50	17
C	100	45–55	65	35	0
D	65	56–65	54	46	0

These groups of women, although by no means constituting representative samples, are biased in only one known direction: Compared with the general population of American women, they are higher on level of education, for they include higher proportions of the college-educated. This is especially true of Group A and Group D.

FINDINGS

LEVEL OF EDUCATION

When responses to the ATM were analyzed for differences between the women in each age group who had and those who had not attended college, only a few scattered differences appeared, a number attributable to chance.

As already indicated, however, the four samples of women represent relatively advantaged groups with regard to educational level. It is likely that in more heterogeneous samples educational level would emerge as a significant variable in women's attitudes toward menopause.

AGE DIFFERENCES

As shown in Table 30.2, consistent age differences were found. The statements are grouped in the table according to the pattern that emerged from a factor analysis carried out on the responses of Group C.[2] That analysis, although serving a purpose extraneous to the present report, provided groupings of the statements that are meaningful also for studying age differences.

Overall inspection of Table 30.2 shows first that, as anticipated, young women's patterns of attitudes toward the menopause are different from those of middle-aged women. When each group is compared with the Criterion group, the largest number of significant differences are found between Groups A and C; then between B and C.

At the same time, it appears that it is not age alone, but age and experience-with-menopause that are probably operating together. There are very few differences between Groups C and D; and relatively few between A and B. The major differences lie between the first two and the last groups—in other words, between women who have and those who have not yet experienced the changes of the climacterium. Although there is not a one-to-one correlation between chronological age and age-at-menopause, approximately 75 percent of the women in Group C, as well as all those in D, reported that they were presently experiencing or had already completed the "change of life." Only a few of Group B had yet entered "the period of the change."

It can be seen from Table 30.2, also, that age differences follow a particular pattern from one cluster of statements to the next. Thus, on the first cluster, "negative affect," there are no significant differences between age groups nor between statements. In each instance, about half the women agree that the menopause is a disagreeable, depressing, troublesome, unpleasant, disturbing event; and about half the women disagree.

On the second cluster of statements, however, there are sharp age differences, and in general, all in the same direction: middle-aged women recognize a "recovery," even some marked gains occurring once the menopause is past. The postmenopausal woman is seen as feeling better, more confident, calmer, freer than before. The majority of younger women, by contrast, are in disagreement with this view.

On the third and fourth clusters age differences, while not so sharp as on the second cluster, are numerous and are again consistent in direction: namely, middle-aged women take what may be interpreted as the more positive view, with higher proportions agreeing that the menopause creates no major discontinuity in life, and agreeing that, except for the underlying biological changes, women have a relative degree of control over their symptoms and need not inevitably have difficulties. This is essentially the view that one woman expressed by saying, "If women look for trouble, of course they find it."

Of the remainder of the statements, those that form the fifth, sixth, and seventh clusters as well as those that fit none of the clusters, age differences are scattered and inconsistent in direction, depending evidently upon the particular content and wording of each statement. It is interesting to note, for instance, that on No. 18, "Women worry about losing their minds," it is the Criterion group, Group C, which shows the highest proportion who agree.

It is also of interest that on Nos. 3 and 19, it is the youngest group who disagree most with the view that menopausal women may experience an upsurge of sexual impulse. In this connection, the interviews with Group C women, many of whom had not completed the change of life, showed a wide range of ideas about a woman's interest in sex relations after the menopause. The comments ranged from, "I would expect her to be less interested in sex, because that is something that belongs more or less to the child-bearing period," to, "She might become more interested because the fear of pregnancy is gone." Many women expressed considerable uncertainty about the effects of the menopause on sexuality.

DISCUSSION

That there should be generally different views of menopause in younger and in middle-aged women is hardly a surprise. Any event is likely to have quite different significance for persons who are at different points in the life-line.

One reason why fewer middle-aged as compared to younger women in this study viewed menopause as a significant event may be that loss of reproductive capacity is not an important concern of middle-aged women at either a conscious or unconscious level. In the psychological and psychiatric literature it is often stated that the end of the reproductive period— the "closing of the gates," as it has been described (Deutsch, 1945, p. 457) —evokes in most women a desire for another child. If so, women might be

Table 30.2 **Attitudes toward Menopause: By Age**

Items subgrouped	Percent who agree[a] Age groups			
	A 21–30 (N=50)	B 31–44 (N=52)	C 45–55 (N=100)	D 56–65 (N=65)
I. "Negative Affect"				
28. Menopause is an unpleasant experience for a woman	56	44	58	55
32. It's not surprising that most women get disagreeable during the menopause	58	51	57	43
34. Women should expect some trouble during the menopause	60	46	59	58
30. Menopause is a disturbing thing which most women naturally dread	38*	46	57	53
20. It's no wonder women feel "down in the dumps" at the time of the menopause	54	46	49	49
33. In truth, just about every women is depressed about the change of life	48	29	40	28
II. "Post-Menopausal Recovery"				
24. Women generally feel better after the menopause than they have for years	32*	20*	68	67
26. A woman has a broader outlook on life after the change of life	22*	25*	53	33*
27. A woman gets more confidence in herself after the change of life	12*	21*	52	42
16. Women are generally calmer and happier after the change of life than before	30*	46*	75	80
23. Life is more interesting for a woman after the menopause	2*	13*	45	35
17. After the change of life, a woman feels freer to do things for herself	16*	24*	74	65
31. After the change of life, a woman has a better relationship with her husband	20*	33*	62	44*
35. Many women think menopause is the best thing that ever happened to them	14*	31	46	40
21. After the change of life, a woman gets more interested in community affairs than before	24*	31*	53	60
III. "Extent of Continuity"				
14. A woman's body may change in menopause, but otherwise she doesn't change much	48*	71*	85	83
15. The only difference between a woman who has not been through the menopause and one who has, is that one menstruates and the other doesn't	34*	52	67	77
12. Going through the menopause really does not change a woman in any important way	58*	55*	74	83

IV. *"Control of Symptoms"*

4. Women who have trouble with the menopause are usually those who have nothing to do with their time	58	50*	71	70
7. Women who have trouble in the menopause are those who are expecting it	48*	56*	76	63
8. The thing that causes women all their trouble at menopause is something they can't control—changes inside their bodies	42*	56*	78	65

V. *"Psychological Losses"*

29. Women often get self-centered at the time of the menopause	78	63	67	48*
25. After the change of life, women often don't consider themeslves "real women" anymore	16	13	15	3*
1. Women often use the change of life as an excuse for getting attention	60*	69	80	68
18. Women worry about losing their minds during the menopause	28*	35	51	24*
11. A woman is concerned about how her husband will feel toward her after the menopause	58*	44	41	21*

VI. *"Unpredictability"*

6. A woman in menopause is apt to do crazy things she herself does not understand	40	56	53	40
10. Menopause is a mysterious thing which most women don't understand	46	46	59	46

VII. *"Sexuality"*

3. If the truth were really known, most women would like to have themselves a fling at this time in their lives	8*	33	32	24
19. After the menopause, a woman is more interested in sex than she was before	14*	27	35	21

Ungrouped Items[b]

2. Unmarried women have a harder time than married women do at the time of the menopause	42	37	30	33
5. A woman should see a doctor during the menopause	100	100	95	94
9. A good thing about the menopause is that a woman can quit worrying about getting pregnant	64	38*	78	63*
13. Menopause is one of the biggest changes that happens in a woman's life	68*	42	50	55
22. Women think of menopause as the beginning of the end	26	13*	26	9*

[a] Those subjects who checked "agree strongly" or "agree to some extent."

[b] These statements did not show large loadings on any of the seven factors represented by the groupings in the table.

* The difference between this percentage and the percentage of Group C is significant at the .05 level or above.

expected to view menopause as most significant at that time in life when the loss of reproductive capacity is imminent. Yet this was not the case in these data.

There is additional evidence on the same point from our interview data. Of the 100 women in the Criterion group, only 4, in responding to a multiple-choice question, chose, "Not being able to have more children" as the worst thing in general about the menopause. (At the same time, 26 said the worst thing was, "Not knowing what to expect"; 19 said, "The discomfort and pain"; 18 said, "It's a sign you are getting old"; 4 said, "Loss of enjoyment in sex relations"; 22 said, "None of these things"; and 7 could not answer the question.) It is true, of course, that all these women had borne children; but the same was true of all the younger women in Groups A and B. Many Group C women said, in interview, that they had raised their children and were now happy to have done, not only with menstruation and its attendant annoyances, but also with the mothering of small children.

The fact that middle-aged as frequently as younger women view the menopause as unpleasant and disturbing is not irreconcilable with their view of the menopause as an unimportant event. As one woman put it, "Yes, the change of life is an unpleasant time. No one enjoys the hot flushes, the headaches, or the nervous tension. Sometimes it's even a little frightening. But I've gone through changes before, and I can weather another one. Besides, it's only a temporary condition."

Another woman joked, "It's not the pause that refreshes, it's true; but it's just a pause that depresses."

The middle-aged woman's view of the postmenopausal period as a time when she will be happier and healthier underscores her belief in the temporary nature of the unpleasant period, a belief that is reinforced perhaps by hearing postmenopausal women say, as two said to our interviewer:

"My experience has been that I've been healthier and in much better spirits since the change of life. I've been relieved of a lot of aches and pains."

"Since I have had my menopause, I have felt like a teen-ager again. I can remember my mother saying that after her menopause she really got her vigor, and I can say the same thing about myself. I'm just never tired now."

The fact that most younger women have generally more negative views is perhaps because the menopause is not only relatively far removed, and therefore relatively vague; but because, being vague, it becomes blended into the whole process of growing old, a process that is both dim and unpleasant. Perhaps it is only the middle-aged or older woman who can take a differentiated view of the menopause; and who, on the basis of experience, can, as one woman said, "separate the old wives' tales from that which is true of old wives."

SUMMARY

An instrument for measuring attitudes toward the menopause was developed, consisting of 35 statements on which women were asked if they agreed or disagreed. The instrument was administered to 267 women of four age groups: 21–30; 31–44; 45–55; and 56–65. Differences were most marked between the first two and the last two groups, with the younger women holding the more negative and more undifferentiated attitudes.

NOTes

[1] *Menopause and climacterium* (as well as the more popular term, *the change of life*) are often used interchangeably in the literature. In more accurate terms, *menopause* refers to the cessation of the menses; and *climacterium* to the involution of the ovary and the various processes, including menopause, associated with this involution.

[2] Responses were scored from 1 to 4; and on the matrix of intercorrelations, the principal component method of factor extraction by Jacoby and the Varimax program for rotation were used. Seven factors, accounting for 85 percent of the variance, emerged from that analysis, factors which have been named "negative affect," "post-menopausal recovery," and so on, as indicated in Table 30.2. Within each group of statements, the order is that of their loadings on the respective factor. It should be kept in mind that a somewhat different factor pattern might have emerged from the responses of Groups A, B, or D.

REFERENCES

August, H. E. "Psychological aspects of personal adjustment." In: Gross, I. F. (ed.), *Potentialities of women in the middle years*. East Lansing: Michigan State University Press, 1956.

Barnacle, C. H. "Psychiatric implications of the climacteric," *American Practitioner 4*: 154–157, 1949.

Benedek, T. *Psychosexual functions in women*. New York: Ronald, 1952.

Deutsch, H. *The psychology of women. Vol. II. Motherhood*. New York: Grune & Stratton, 1945.

Fessler, L. "Psychopathology of climacteric depression," *Psychoanalytic Quarterly 19*. 28–42, 1950.

Hoskins, R. G. "Psychological treatment of the menopause," *Journal of Clinical Endocrinology* 4: 605–610, 1944.

Ross, M. "Psychosomatic approach to the climacterium," *California Medicine* 74: 240–242, 1951.

Sicher, L. "Change of life: A psychosomatic problem," *American Journal of Psychotherapy* 3: 399–409, 1949.

U. S. Public Health Service. *Menopause. Health Information Series Publication No. 179.* Washington, D.C.: U. S. Government Printing Office, 1959.

Thirty-one

On the Meaning of Time in Later Life

Robert Kastenbaum

A. INTRODUCTION

In the time that it takes to read this paragraph, one might have conducted a complete data-gathering operation involving the use of a tachistoscope, stopwatch, or electrical timing device. The measurement of human behavior with respect to short temporal durations has long been a familiar technique in psychological research. Time is kept by the psychologist in his modest stopwatch or elaborate timing system. The subject is permitted to use a certain small quantity of the timekeeper's commodity: for example, for reacting to a stimulus or attempting to replicate a specified short temporal duration.

When the study has been completed, the subject is "on his own time" again. It is one's "own time" that will be considered in this paper, especially the time one organizes and experiences when one is in the later years of one's life. In recent years an increasing number of psychologists have moved away from the role of a timekeeper to that of a participant-observer in the individual's own organization and use of time. In this general approach, time is "kept" (or "created") by the subject, who permits the psychologist to enter partially into his temporal experiences and constructions. This approach is fraught with difficulties, for the investigator has relinquished much of his external control in order to learn how the subject goes about the process of controlling, organizing, and planning when he is left pretty much to his own devices. It is more difficult, if not impossible, to achieve the same precision when, let us say, one is attempting to learn how a centenarian is weaving together strands of past, present, and future, than when one is limiting one's interest to how well he could reproduce a twenty-second time interval. So our research involves a constant struggle to do justice both to the richness and complexity of human temporal

Kastenbaum, R. "On the meaning of time in later life," *Journal of Genetic Psychology*, 1966, *109*, 9–25. Reprinted with the permission of the author and publisher. A preliminary version of this paper was presented at the September 30, 1965, meeting of the Boston Society for Gerontological Psychiatry.

experience, and to scientific requirements for operational definitions, replicable procedures, and adequate, unbiased analysis of the findings. The methods, concepts, and conclusions that will be sketched rather briefly in this paper should be regarded as quite preliminary, tentative, and selective. Studies of personal time are considered to be particularly important in developing conceptual frameworks for understanding personality in later life (for example, Kastenbaum, 1965a, 1966).

To keep this paper within reasonable bounds, nothing will be ventured about the biological correlates of temporal experience, and very little about psychoanalytic conceptions of time or differences associated with socioeconomic variables. These omissions are necessary because the biological, psychoanalytic, and sociological approaches to the understanding of time deserve extensive consideration in their own right.[1]

B. TEMPORAL EXPERIENCE
BEFORE ADVANCED AGE

What are the meanings of time in later life? For perspective on this question let us take an earlier point on the lifespan, and see what shape has been given to temporal experience before advanced age. Adolescence is a logical choice, both for its intrinsic interest and for the happy circumstance that we know or think we know a little more about temporal experience in adolescence than at earlier and later periods. The discussion here is limited to adolescents in our society in our own times.

The typical adolescent has already developed many resources for coming to terms with the temporal dimensions of experience. He is able to distinguish between his realm of personal time and the realm of public time that moves unidirectionally, continuously, and in fixed units regardless of his own state of being. Within his realm of personal time he recognizes a predominantly biological periodicity, such as in the activity-and-rest cycle, but also an arena of fantasy where he can engage in free play with time. In his fantasy the adolescent can suspend time altogether, or project himself selectively back and forth in time without regard for the steady unidirectional pulse of public or astronomical time.

Of particular importance is the fact that the adolescent has come into the possession of most if not all of the specific intellectual operations that characterize the adult human being (for example, Piaget, 1958). He is capable of thinking about thought, of taking future contingencies into account, of liberating himself from the demands of the present situation by utilizing that which is *not* here-and-now. Obviously, a person can manage to detach himself so much from the here-and-now situation that his use of futurity takes on a predominantly escapist quality. But in general, as

Hartmann (1958), Piaget, Werner (1957), and many others have observed, the ability to scan the future is a milestone in the early development of the normal individual.

There are at least two lines of research that indicate that the future has become more "real" for the adolescent than it was for the child. Some investigations have indicated that the adolescent projects himself further into the future than does the child, so one might say that the future is more "real" in that there is more of it, and it seems to have increased influence upon the individual's present behavior (see Wallace and Rabin, 1960; and Lhamon, Goldstone, and Boardman, 1957). The other, and perhaps even more relevant aspect, might be illustrated by results of an unpublished study by the present writer. In this study the author selected "normal" children and adolescents in groupings that ranged from age ten to age nineteen (total sample equalled 75). Three brief and fictitious, but plausible, case histories were devised. Each described a fifteen-year-old boy, in one instance a creative and admired leader, in another instance a boy who had been in trouble with authorities for a series of minor delinquencies, and in the third instance a fairly typical adolescent with no salient creative or delinquent characteristics. Ss were asked to make several judgments about each of these young people, judgments pertaining mostly to what they would be like when they grew up. But the author was most interested in one particular judgment: namely, how certain the subject was that his prediction would come true. Here it was found that with increasing age from late childhood through adolescence there were increasing expressions of *un*certainty. The child was pretty sure that he knew what the future held; but the adolescent felt that the future could go in a number of different ways: one could judge and plan, but the future remained a zone of contingency.

This writer suggests that there is an intimate relationship between a sense of genuine futurity and a sense of contingency. The adolescent, better than the child, appreciates the qualitative difference between time that is past and time that is yet to come and might bring forth new events. This sense of the future as a qualitatively different kind of time also stands in contrast to cyclical views that have prevailed in some cultures; that is to say, time is regarded as a wheel that provides a sort of stationary motion— movement that, in the short or long run, does not go any place. Another way of putting it is that the adolescent is aware of the directionality of time.

The sense of forward motion is linked with the adolescent's idea that he is on the verge of coming into "his own time." At the moment he is still partially controlled by the parental generation, but soon he will be enfranchised to use his time as he pleases (or, so he thinks!), and to fill this time with the really significant events and experiences in life.

Yet research findings have indicated that the sense of forward motion in time is not necessarily associated with planning and organization of future experiences, particularly in the remote future. Study after study discloses that it is chiefly the near future, the next few years, that occupies the thoughts and emotions of the so-called typical adolescent. Furthermore, the keen sense of forward motion also seems to exclude for many young people an appreciation of the past, including their own.

One investigation of 260 high school juniors involved a battery of time perspective procedures (Kastenbaum, 1959). It was found that the next few years of life were crammed with expectations, but that the later years of life—here defined as about age twenty-five—were considered much less interesting than the next few years; and the very advanced years—about age thirty-five—were almost devoid of content. We seemed to be dealing with a poverty of thought concerning the later years of life, and also with a definite resistance to dwelling upon what was considered to be an unpleasant zone of the lifespan that contained no significant values. In the same study it appeared that many adolescents also were quite uncomfortable with their own past, and were attempting to put it out of mind. There was the implication that both the past and the remote future were seen as vague, confusing zones in which one could not be sure of one's personal identity, or could not be sure that one wanted the identity that these zones seemed to confer.

A second investigation confirmed some of these findings, and added the conclusion that an adolescent's sense of forward motion, as gleaned from his choice of static or dynamic time metaphors (see Knapp and Garbutt, 1958), has no necessary relationship with his proclivity for extending his thoughts to the future, or imposing an organization upon the anticipated contents (Kastenbaum, 1961).

The results up to this point implied that many adolescents have a sense of hurtling rapidly toward a future that stands but a short distance away. What happens next? Will the adolescent reinterpret his life so that a new future looms before him, or will he eventually pass through what once was "The Future" and from then on have to look back over his shoulder? In other words, we might be dealing chiefly with a once-only phenomenon, the establishment of a personal future that is anchored to a particular point in time, or we might be dealing with the first of several futures that will be established by the individual as he continues to experience life. This question will be considered later from the vantage point of old age. For now, it is relevant to add the finding from another set of investigations, that adolescents and young adults who come from a variety of backgrounds and are engaged in a variety of career preparations tend to limit their scanning of their own lifespan to the range of years approximately between age twelve and age forty-six; actually, age forty-six is our highest figure, with the more typical upper limit being around age thirty-five (Kastenbaum and Durkee, 1964a).

Another word is in order about the use of the past in adolescence. In a separate investigation, the author asked 104 high school seniors to construct a story for each of 12 story-beginnings that were provided to them (Kastenbaum, 1965a). The story-beginnings were divided conceptually into three sets according to the affective tone: a neutral set, a pleasant set, and an unpleasant set. Each set contained four stories. Here is a sample of matched story-beginnings in each set: "Wally and Carol were at the dance together . . ." (neutral); "Pete and Sally were at the dance together, having a miserable time . . ." (unpleasant affect); "Bill and Louise were at the dance together, having a marvelous time . . ." (pleasant affect).

Most Ss took all of these story-beginnings into the future. But there was a good deal of variability concerning whether or not the subject would also make use of the past. In the examples given, use of the past would mean that the story might include some information about what had been going on before Wally and Carol went to the dance. The emotional element, either positive or negative, produced more attention to the past than did the stories that were framed with neutral roots. Furthermore, the brightest students in this sample, as categorized by intelligence test scores, were much more likely to introduce pastness into their stories than were the least intelligent students.

What does all this mean? The line of interpretation offered here might take us closer to discerning possible differences in the meaning of time between adolescence and later life. It will be easiest to quote from the discussion section of this investigation:

> The data suggest that the "easiest" way to develop a unified production or completed meaning from the introductory fragment (story-beginning) is to project into the future. Even the most minimal stories involved some projection into the future. This implication is supported not only by the statistical information reported above, but also by inspection of the stories. It is clearly evident in the type of story the writer has come to classify as "So-He-Went-Out-And." In this terse and simple form of meeting task requirements, the S merely appends to the meaning fragment another phrase or two which serves the purpose of closure. Typically, such stories are routine and unimaginative. They appear to reflect the operation of a powerful drive toward the "principle of least effort" in dealing with this somewhat unfamiliar and perhaps uncongenial task. Here are two illustrations: "*Hal woke up and didn't have nothing to do. So he went out for a drive.*" "*Curt and Don met each other in the center of town early one morning,* and so they went some place and hung around." Rudimentary productions of this type seem to move instantly into the future—without developing that

future or elaborating upon the situation described in the root. By contrast, it is rare to find a rudimentary story that deals exclusively with the past.

The easy-way interpretation of futurity receives further support when the differences related to intelligence are considered. The brightest Ss conceptualized futurity no less frequently than did the dullest Ss, but gave a good deal more attention to the past. Thus, it would appear that with a higher grade of intellectual effort comes an increasing concern with antecedent conditions.

Affective set might be regarded as a factor that complicates the S's task: The affective tinge increases the stimulus intensity with which the subject must cope. Somehow, the task has become more significant or challenging. One way to accommodate the stimulus increment is to contain the affect within a psychological structure, such as a "time perspective." This alternative appears most likely to be taken by those who have relatively good cognitive skills, although doubtlessly other variables are important. Another alternative would be to discharge the stimulus increment in a manner that is irrelevant to the task: for example, by motoric restlessness (foot tapping, wriggling around in the chair, and such). The present study does not provide data regarding the latter alternative, although it would now be hypothesized that those who produced only simple bursts into the future when presented with the affective roots would be especially apt to discharge their tensions motorically.

If we take seriously the "perspective" in "time perspective," then it is apparent that there must always be at least two reference points and a relationship between the points. A person who thinks exclusively of the future, for example, does not have a strong future time perspective—he has no time perspective at all. It is the person who manages to keep past, present, and future in mind that has the opportunity of developing a genuine perspective.

From these considerations, then, it seems that "pastness" is a variable element in the construction of a time perspective. While it would appear that routine psychological development includes the activation of a future-scanning function, the past-scanning function requires additional cognitive or motivational factors on the part of the individual or special cues in the environment. To learn why some people develop a temporal perspective on life and others do not we would wish to give particular attention to the conditions upon which the appreciation of the past depends.

If a time perspective can serve to accommodate affect and, therefore, afford the individual an alternative to quick response and impulsive discharge of tension, then the person who is disinclined to consider the past will tend to be deprived of this coping procedure. A previous study by the writer (Kastenbaum, 1959) suggested that most adolescents in a normal sample had an aversive, blocking-out reaction toward their personal past. The implication might be that one of the developmental tasks that still lies ahead for many adolescents is the ability to take "pastness" into account in elaborating a cohesive view of life. By contrast, the developmental task of the aged person might be to find a way of maintaining the future-scanning function. Neither the adolescent who ignores the past nor the aged person who ignores the future could develop or maintain a genuine time perspective, according to the present line of reasoning (Kastenbaum, 1965a, pp. 198-199).

Two further comments will be offered to round off this sketch of personal time in adolescence. The tendency to ignore the later years of one's own life seems to be related, as we might expect, to fears both of aging and death. We have the impression, not yet adequately supported by controlled observation, that adolescents and young adults more readily express death fears than do elderly adults. If time is experienced by adolescents primarily in its directional aspect—an arrow pointing to the new events and the new self in the future—then death becomes a massive threat, the termination of a sequence that is still in process of development.

Along with the adolescent's heightened concern for futurity, he develops the ability to delay gratification. The work of Singer and his colleagues suggests that increase in the ability to delay gratification is associated with various other measures of ego development, including relatively mature Rorschach perceptions and behavioral indices of impulse control (for example, Singer, 1955). The juvenile delinquent, the culturally disadvantaged, the mentally subnormal are among those adolescents who tend not to develop adequate capacity to delay gratifications.

C. DELAY OF GRATIFICATION IN LATER LIFE

Wolk, Rustin, and Scotti recently devoted a paper to the so-called "geriatric delinquent" who acts out his conflicts in criminal or socially undesirable forms (Wolk, Rustin, Scotti, 1963). According to these observers, the geriatric delinquent is "dominated by the pleasure principle—orientation toward immediate gratification—frequently at the expense of long-range plans . . ." (Wolk, Rustin, Scotti, 1963, p. 658). Such a person is thought to have suffered a breakdown in his ego control. Thus we have here the suggestion that delay of gratification and future planning diminish in at least one subgroup of the aging population, and that this constitutes an undesirable regression.

In a recent experiment, the author attempted to set up a simple analogue to study delay of gratification comparatively in young adults and in aged adults who are normal, self-sufficient members of the community (Pollock and Kastenbaum, 1964). Six hypothetical processes were raised as possibilities in advance of the investigation:

It might be ventured that delay of gratification remains strong in later life for several reasons:
1. Because delay of gratification has been practiced for so many

years, it has become a well-established habit and, therefore, should be resistant to change and ensure its own perpetuation.

2. The normal individual will have utilized delay of gratification with success. He knows from favorable experience that patience and planning frequently will bring him what he wants.

3. Many instinctual drives which impel one toward impulsive behavior early in life have passed the peak of their urgency. Furthermore, these drives will have become more or less integrated into the total personality structure. Therefore, in later life one should find it easier to delay gratification because there is reduced pressure for urgent relief.

One might also predict, however, that delay of gratification will weaken in later life for such reasons as the following:

1. There may be a general decline with age of the sensory and integrative capacities of the individual. Advanced senility would be an extreme case of this decline, but even more moderate impairment might significantly reduce ability to delay gratification.

2. Instinctual drives previously under the control of developmentally higher structures may "break the leash," so to speak, and emerge as urgent goads to direct gratification of needs.

3. The aging individual may perceive that time is running out on him. With this thought in mind (consciously or unconsciously) he might decide that there is no point in continuing to delay gratification. It is important to note that if this factor is the dominant one, then the abandonment of the "waiting game" need not be viewed as a pathological regression. Rather, it would represent a realistic adaptation to a new situation in life (Pollock and Kastenbaum, 1964, pp. 282–283).

Results of the present study most strongly supported this last statement, that there is a decline in delay of gratification, but that the decline is related in some measure to the individual's new interpretation of his situation in life. As many elderly persons told us in the course of this study, there is no point in wasting time, in delaying, when so little time is left. It should be added that the elderly Ss in this investigation were all men in good mental and physical health who were living in the community and active in church affairs. So we have here the possibility that many elderly persons are less inclined than formerly to delay gratification, but that this change does not necessarily signify a pathological regression; instead, it may signify a new adaptation to a new life situation.

In this connection we might call attention to another subpopulation among the aged: The responsible citizen who established a lifelong pattern of denying himself luxuries and pleasures while working hard to raise a family and prepare for the future—only to arrive at last in that future with a sense of having been cheated of his long-delayed gratifications. He ends up embittered against life, possessing neither the pot of gold nor memories of golden days. Here, perhaps, is future-orientation and delay of gratification gone wrong, valuable components of normal development that have turned on their host.

What is lacking here? Perhaps it is the failure to subordinate the dimension of futurity within a larger perspective that includes a place for engrossment in the present and utilization of the past (Kastenbaum, 1965b). Furthermore, this man's sense of futurity was forged relatively early in his life and was not allowed to modify itself by subsequent experience. He is growing old in the pursuit of a young man's future.

D. "OLD AGE" AS THE UNEXPECTED

Consider now the fate in later life of another tendency that shows up in adolescence. We have seen that most adolescents and young adults do not project themselves into their own remote futures—although it is possible that recent national attention to problems of aging might alter this situation. We now can add the finding that most patients at Cushing Hospital, according to their self-report on what we call the Past Futures Technique, also did not think ahead to advanced age when they were young, nor when they were middle-aged (Kastenbaum and Durkee, 1964b). And we can add the finding that in a limited sampling of middle-aged people there are relatively few who are inclined to project into the years that lie ahead of them (Kastenbaum and Durkee, 1964a).

This disinclination to take one's later years into account no doubt is related in part to negative social attitudes toward aging, which have been amply documented. But it may also be that a young person simply finds it very difficult to understand that some day he might *really* be old, or that his grandfather (or father) once *really* was young. The immediate situation and one's own limited experience provide strong evidence that "I am young" ("I have never been old"), and that "Grandfather is old" ("I have never seen him young"). By implication, then: "I will *always* be young, and "He was *always* old."

For some people, then, "old age" seems to be encountered as if it were unexpected. Clinical experience and some limited research at Cushing Hospital suggest that there is a subpopulation of elderly people who are thrown into an identity crisis when confronted with their situation (Kastenbaum, 1964a). These patients did not think about later life when they were young adults, and still do not regard themselves as "old," when they are ill and in their eighth decade. These people seem to have followed a life plan that relies upon the achievements, resources, and affiliations of their early years, with little provision for developing new goals or techniques of coping. Thus, they neither had a place for "growing old" in their early interpretations of futurity, nor the flexibility to add the dimension of aging to their time perspective as the years went by. Physical symptoms seem to provide these people with their explanation for what has happened to them. Since "old age" had no place in their personal time framework, its occurrence, if acknowledged, would strike like a mysterious, devastating thunderbolt from afar. Thus, instead of becoming "old," the patient "had a bad leg" that required him to curtail his activities. The patient in this way rescued himself from the brink of catastrophic anxiety by pinpointing a symptom that he could consider to be apart from his true and ageless self. The "true self" could now be regarded as having been battered by the world, but not mysteriously enveloped by "old age."

As one might expect, when in some instances the symptom was removed by effective medical and nursing care, the patient was thrown into a state of confusion. Unable to find any explanation that would permit him to retain a sense of self-consistency, he withdrew to a markedly decreased level of functioning. So it appears that a person who has never prepared himself psychologically for the later years of life—as, for example, by failing to include "old age" within his time perspective—may in advanced age be left without a firm sense of his own identity, having no ready alternative but some form of regression.

E. TWO MEANINGS OF FUTURITY

The last few points have had a somewhat dysphoric quality; so let us shift to an aspect of time that reveals the older person in a more radiant light. It seems to be the case that many older people are living in a sort of surplus time that extends far beyond the futures they envisioned when young. But does this mean that they are not able to deal adequately with the future dimension? In one investigation a distinction was made between personal and cognitive futurity (Kastenbaum, 1963). *Personal futurity* is bound to the individual's own lifespan. From the vantage of his present moment of existence, the individual can look back toward a past that is his own personal history, and ahead toward a future that holds his own personal "destiny." Time has the quality of an intimate personal possession. *Cognitive futurity* is the orientation toward utilizing time as an abstract cognitive category for organizing and interpreting experience in general. Time is a tool of the intellect. We attempted to learn if elderly people tend to suffer a loss in the use of both personal and cognitive futurity.

Personal futurity was studied by the Important Events Technique. The subject is asked to report the most recent important event or experience in his life, and the temporal distance between that event and the present time is ascertained. He is then asked for the most recent event prior to that, and, again, for the most recent event prior to that. These three steps into the past are followed by three steps into the future. S is requested to give his expectations for the next most important event in his life, and so on. S is free to give his own interpretation of what constitutes an important event. Cognitive futurity was studied with a story-completion technique similar to the one described earlier. The Ss were 24 alert Cushing Hospital patients, and an equal number of college students who were used not as a strict control group, but as a point for comparison.

There was clear evidence for a restriction of personal futurity, no matter how the data were analyzed. However, there was no corresponding decline in cognitive futurity. In fact, there was a tendency for the elderly people to construct more stories that were well integrated in comparison with the younger Ss. There was the hint that older people were likely to construct better integrated stories because they took the dimension of pastness into account more frequently than did the younger people. This result brings to mind the earlier investigation that implied that pastness is still a variable element in the adolescent's cognitive use of time (Kastenbaum, 1965a).

This study, then, suggested that a limited or bleak outlook on personal futurity in later life does not necessarily inhibit the older person from making good use of futurity as a dimension in the organization of experience and that, in fact, he might possibly have improved his ability to put events into perspective because he has retained his facility with the past, a skill that perhaps develops most rapidly after adolescence.

F. LIVING IN THE PAST

Let us remain with this fascinating concept of "pastness" for a moment. When it is said that an old person lives in the past, often this statement is issued in a tone of irritation or derogation. Yet we have seen that the ability to use pastness is one that may have to be acquired through years of experience, and one that may serve significant functions. Robert Butler has emphasized a process of life review in which the normal aged person reintegrates his personality and prepares himself for death (Butler, 1963). One would like to see increased experimental evidence for the existence of frequency of this process, but the suggestion at least is in the direction of positively evaluating a certain kind of past-centeredness in later life.

Perhaps what irritates us sometimes about an elderly person's past-orientation is our own reaction of having been snubbed. The present moment is consensual, public; we all share in it. We also have potentiality for sharing the future together; so the person who dwells on the present or the future has company. But when an aged person dwells on his past he is moving in a realm that is not directly accessible to us; we were not a part of it. Living in the past, then, tends to isolate the aged person from his younger contemporaries, and the effects may be the same as in any other form of isolation. When two or three centenarians get together to talk about old times that they shared, the past is brought alive into the present, and no longer has that solitary, private quality that we Americans tend to suspect and resent.

Furthermore, what we dislike in an old person's use of the past may be a particularly defensive or dulled quality that is linked not so much with past-orientation per se but with a withdrawal from the present. Some evidence in this direction was provided in a study by Fink, who found that elderly people who lived in an institution were more concerned with the past and less concerned with the future than were older persons who lived at home (Fink, 1957). Perhaps, then, we are reacting to some of the negative qualities associated with institutionalization rather than with normal changes in temporal orientation with age.

G. TIME AND DEATH IN LATER LIFE

Jung (1959) and Cumming and Henry (1961) are among those who have emphasized that preparation for death is a major developmental task for the aging person—indeed, for the second half of life. This significant concept is difficult to subject to experimental analysis. However, we might well expect that ideas of death would be closely related to ideas of time. One possibility is that time is felt to become of decreased importance as one ages either because one can do less with his time, or because the proximity of death overrules any realistic devotion to future prospects. This question is far from being settled. One investigation has found that no less than 50 percent of a population of centenarians residing in the community retain personally significant future ambitions, and that at least half of this subgroup take these ambitions as serious goals that might be achieved with appropriate effort on their part (Costa, Kastenbaum, and McKenna, 1965). Yet clinical experience in a geriatric hospital suggests that time frequently decreases in importance in the lives of aged persons —an impression that is supported to some extent by the observations of Henry (1965) based upon aged persons residing in the community.

The observations made within the framework of clinical services to people who are not only aged but also ill and institutionalized further suggest that "one pathway of individual experience . . . is gradually to constrict and shallow out the personality so that when death does approach it is a smaller event, because it is terminating a life that has already become more modest in scope and released its hold: image of a dry leaf blown away by a gentle breeze, as contrasted with the burgeoning young tree uprooted by a violent storm. In slightly less fanciful terms, it is the difference between the directional flow of time (as in adolescence) with its prospective new events and experiences coming to an abrupt termination, and the cyclical repetition of routine and diminishing events simply not repeating itself anymore. . . . (For some aged persons, at least) time is less fascinating and precious; death less formidable and devastating" (Kastenbaum, 1966).

But still another viewpoint has been advanced recently that regards the possible reduction of interest in time as a predominantly defensive maneuver. Back argues that "consciousness of the future implies consciousness of death, and when transcending time the thinker also tries to transcend death. It would seem significant that exalted states, maximum experiences, mystical states, or drug-induced concepts typically are described as involving some feeling of timelessness, which is the real attraction of these states. . . . When so much effort is devoted to denying the reality of the importance of time as a concept, it should not be surprising that consideration of time as an important variable has been delayed and neglected" (Back, 1965).

Back then studied a group of elderly people living in rural western areas and asked them, among other things, "If you knew you were going to die within 30 days, what would you do?" and "If you knew that everybody were going to die within 30 days, what would you do?" These men were relatively less likely than younger men to say that their activities would change at all under either of the hypothetical conditions. They were also relatively unconcerned with the fate of other people. They tended to "set the end at their own death—which they accepted as relatively soon—and did not care anything beyond this point. This, then, is the meaning of short time perspective" (Back, 1965).

One might wonder if the apparent lack of concern for other persons and even for one's own future is necessarily associated with short time perspective per se. For some individuals the notion of a limited time perspective might increase the intensity and value of the remaining time because it has become that much more precious. Perhaps Back's elderly subjects were dealing with private, isolated, nonmodulated futures somewhat in the way that some aged people dwell in worlds of pastness that have little communication with present or future. The implication might be that in later life one needs more than past, present, or future orientation taken separately; one needs also a skill in interweaving these realms and thus modulating each through its contact with the others. An example that comes to mind is the case of a 106-year-old magistrate who actively practices his complex and influential profession. This man apparently has a great reverence for the past, both of his profession and of local history, and a keen eye both to the present and the future. In hearing present testimony and evaluating it with historical precedents in mind, he is simultaneously making decisions for the future—here, then, is a natural interpenetration of temporal dimensions. The situation is quite different from that of the person of any age who, when he looks into his future, sees a time that is completely disconnected from past and present, and images only his own death (an image to be followed, perhaps, by a steadfast refusal to think of the future or of time in general).

H. THE INHERITANCE AND THE CHALLENGE

The adolescent bequeaths both riches and debits to that namesake who will succeed him on his traversal of the lifespan. Impetus toward the future, a sense of moving into new and valuable time, and the ability to plan and imagine are among the gifts that the adolescent brings to his place on the lifespan. But he leaves much unfinished business. Too often, it seems, the adolescent constructs a world-view that ignores the later stages of development. It is also likely to emphasize futurity beyond the point where this is adaptive for the mature person, and correspondingly to de-emphasize the past. The person who accepts uncritically what has been created by the adolescent within him thus finds himself rushing forward toward an imagined future that was not made for durability. If he cannot disembark from "The Future Express," he may eventually become the embittered old man who delayed or prepared too long. If his basic identity has been absorbed by the notion of perpetual youth moving toward or existing in a perpetual future, then he may become the anxious, panicky old man who can make no sense at all of a life situation that never was anticipated.

But now we must say a word for a person who has been pretty much ignored in this discussion as he has been elsewhere—surprising neglect since he exists in such large numbers. Reference is made to the person who never developed much in the way of a time orientation, who takes each day as it comes, and is neither delighted nor depressed by the rumble of distant drums. He seems to make his way along the lifespan with relatively few discontinuities. Taking life one moment or one hour at a time, he may not find things profoundly different at age eighty than they were at age twenty or age forty. He is the person we neglect when, for example, in the name of Disengagement Theory we postulate a universal and intrinsic process of aging. Many people do not gain a new appreciation of the finitude of time nor adjust their social style accordingly. They simply go on as best they can from day to day. Furthermore, there is that subgroup of "Young Prometheans" who early in their lives have surveyed what might lie before them. These people enter old age and the portals of death not completely as strangers, nor do they need to experience radical discontinuities as they move along the lifeline.

For the moment, let us concentrate upon the person who is in the tradition with which we have been chiefly concerned—the person who has always dealt with time, but never quite mastered it. His challenge is how to develop and then appropriately modify his self-identity over his entire lifespan, how to make both realistic and creative use of past, present, and future. The challenge can be phrased in many ways: how to be engrossed in the present moment—the only moment we ever have—and yet retain a perspective on what has been and what will be; or how to be committed to a core of identity and values and yet be adaptive to changing times, both in society and our intrapsychic milieu.

Engrossment and perspective, dedication and flexibility—how do some people manage to accomplish this kind of integration? Perhaps when we learn enough from those who age as good wine ages we will have a place in the technical jargons of psychology for the concept of wisdom.

NOTE

[1] For illustrative biological approaches, see Gooddy (1958), Davis (1956), and Campbell (1954). Among the many psychoanalytic contributions, one might note those of Dooley (1941) and Bonaparte (1940). LeShan (1952) and Graves (1962) are among those who have investigated social class correlates of temporal experience. Reviews of time perspective in general have been offered by Wallace and Rabin (1964), Fraisse (1963), Wohlford (1964), Kastenbaum (1964b), and Craik (1965).

REFERENCES

Back, K. W. *Time perspective and social attitudes.* Paper presented at the annual meeting of the American Psychological Association, Symposium on Human Time Structure, Chicago, Illinois, September 7, 1965.

Bonaparte, M. "Time and the unconscious," *Internat. J. Psychoanal.* 21: 427–468, 1940.

Butler, R. N. "The life review: An interpretation of reminiscence in the aged," *Psychiatry* 76: 65–76, 1963.

Campbell, J. "Functional organization of the central nervous system with respect to orientation in time," *Neurology* 4: 295–300, 1954.

Costa, P. T., R. Kastenbaum, and W. McKenna. *Application of developmental-field principles to the time perspectives of centenarians, geriatric patients, and young adults.* Paper presented at the annual meetings of the Gerontological Society, Los Angeles, California, November 11, 1965.

Craik, K. H. *Of time and personality.* Paper presented at the annual meetings of the American Psychological Association, Symposium on Human Time Structure, Chicago, Illinois, September 7, 1965.

Cumming, E. M., and W. E. Henry. *Growing Old.* New York: Basic Books, 1961.

Davis, H. "Space and time in the central nervous system," *Electroenceph. & Clin. Neurophysiol.* 8: 185–191, 1956.

Dooley, L. "The concept of time in defence of ego integrity," *Psychiatry* 4: 13–23, 1941.

Fink, H. H. "The relationship of time perspective to age, institutionalization, and activity," *J. Geront.* 12: 414–417, 1957.

Fraisse, P. *The Psychology of Time.* New York: Harper & Row, 1963.

Gooddy, W. "Time and the nervous system," *Lancet,* 1958, Whole No. 7031, 1139–1214.

Graves, T. *Time perspective and the delayed gratification pattern in a tri-ethnic community.* Doctoral dissertation, University of Pennsylvania, Philadelphia, Pennsylvania, 1962.

Hartmann, H. *Ego Psychology and the Problem of Adaptation* (trans. by David Rapaport). New York: Internatl. University Press, 1958.

Henry, W. E. "Engagement and disengagement: Toward a theory of adult development." In: Kastenbaum, R. (ed.), *Contributions to the Psychobiology of Aging.* New York: Springer, 1965.

Jung, C. J. "The soul and death." In: Feifel, H. (ed.), *The Meaning of Death.* New York: McGraw-Hill, 1959, pp. 3–15.

Kastenbaum, R. "Time and death in adolescence." In: Feifel, H. (ed.), *The Meaning of Death.* New York: McGraw-Hill, 1959, pp. 99–113.

Kastenbaum, R. "The dimensions of future time perspective, an experimental analysis," *J. Gen. Psychol.* 65: 203–218, 1961.

Kastenbaum, R. "Cognitive and personal futurity in later life," *J. Individ. Psychol.* 19: 216–222, 1963.

Kastenbaum, R. "The crisis of explanation." In: Kastenbaum, R. (ed.), *New Thoughts on Old Age.* New York: Springer, pp. 316–323, 1964. (a)

Kastenbaum, R. "The structure and function of time perspective," *J. Psychol. Res.* (India) 8: 1–11, 1964. (b)

Kastenbaum, R. "The direction of time perspective: I. The influence of affective set," *Gen. Psychol.* 73: 189–201, 1965. (a)

Kastenbaum, R. "Engrossment and perspective in later life: A developmental-field approach." In: Kastenbaum, R. (ed.), *Contributions to the Psychobiology of Aging.* New York: Springer, pp. 3–18, 1965. (b)

Kastenbaum, R. "Theories of human aging: The search for a conceptual framework," *J. Soc. Issues* 21: 13–36, 1965. (c)

Kastenbaum, R. "As the clock runs out," *Ment. Hyg.* 50: 332–336, 1966. (a)

Kastenbaum, R. "Developmental-field theory and the aged person's inner experience," *Gerontologist* 6: 10–13, 1966. (b)

Kastenbaum, R., and N. Durkee. "Young people view old age." In: Kastenbaum, R. (ed.), *New Thoughts on Old Age.* New York: Springer, pp. 237–249, 1964. (a)

Kastenbaum, R., and N. Durkee. "Elderly people view old age." In: Kastenbaum, R. (ed.), *New Thoughts on Old Age.* New York: Springer, pp. 250–264, 1964. (b)

Knapp, R. H., and J. T. Garbutt. "Time imagery and the achievement motive," *J. Person.* 26: 426–434, 1958.

LeShan, L. L. "Time orientation and social class," *J. Abn. & Soc. Psychol.* 47: 589–592, 1952.

Lhamon, W. T., S. Goldstone, and W. K. Boardman. *The time sense in the normal and psychopathological states.* Progress Report, May 1, 1957, M-1121. College of Medicine, Baylor University, Waco, Texas.

Piaget, J. *The Growth of Logical Thinking from Childhood to Adolescence.* New York: Basic Books, 1958.

Pollock, K., and R. Kastenbaum. "Delay of gratification in later life: An experimental analog." In: Kastenbaum, R. (ed.), *New Thoughts on Old Age.* New York: Springer, pp. 281–290, 1964.

Singer, J. L. "Delayed gratification and ego development: Implications for clinical and experimental research," *J. Consult. Psychol.* 19: 259–266, 1955.

Wallace, M., and A. I. Rabin. "Temporal experience," *Psychol. Bull.* 57: 213–236, 1960.

Werner, H. *Comparative Psychology of Mental Development* (rev. ed.). New York: Internatl. University Press, 1957.

Wohlford, P. *An investigation into some determinants of extension of personal time.* Doctoral dissertation, Duke University, Durham, North Carolina, 1964.

Wolk, R. L., S. L. Rustin, and J. Scotti. "The geriatric delinquent," *J. Amer. Geriat. Soc.* 11: 653–659, 1963.

Thirty-two

Some Aspects of Memories
and Ambitions in Centenarians

Paul Costa and Robert Kastenbaum

Well, this is a time for introspection, to take inventory of himself, to evaluate the mistakes made and opportunities lost, to recognize and appreciate the many virtues of others that had been overlooked. It is also a time for retrospection, and here begins the real second childhood. To many this is a comforting and rewarding period. He now begins to relive his life backward. He is astonished to see so much beauty in nature, and also in people, he has by-passed. He finds it more difficult to recall quite recent incidents than those which occurred further back, that the farther he goes the better his memory becomes until he reaches early childhood when everything becomes clearly bright.

The unedited passage above is from the essay "Second Childhood" written by Mr. Albert Davis, age 100, in August, 1959. He died in March, 1963, at the age of 104.

A. BACKGROUND

Relatively little psychological research has been done with centenarians, yet the long period of time during which the centenarian has been developing and functioning offers a truly unique opportunity to study the processes by which a person organizes his total life experiences.

At present there are few theoretical positions available to the interested practitioner or researcher in gerontology. What are the theoretical and applied implications of very advanced age? How does the centenarian integrate his extensive past experiences with his present moment of existence? What does the present mean to him, and how does he conceive

Costa, P., and Kastenbaum, R. "Some aspects of memories and ambitions in centenarians," *Journal of Genetic Psychology*, 1967, *110*, 3–16. Reprinted with the permission of the authors and publisher. The authors are grateful to the Social Security Administration for making its centenarian data available for their analysis. A preliminary version of this paper was presented at the 1965 meetings of the Gerontological Society, Los Angeles, California.

of futurity? Does the centenarian plan ahead or does he organize his life on a day-by-day basis? This study attempts to explore some of these issues, a few of which were expressed in the "Second Childhood" essay by Mr. Davis. The present exploratory approach is at the level of verbal self-reports, with complete dependence upon interview data. Analysis will focus on the responses to three memory items, and one item concerning future outlook.

This paper reports an analysis of centenarian interviews that were devised and conducted by the Social Security Administration for people 100 years old or over who were on the benefit rolls as of 1963. The S.S.A. dispatched field representatives from district offices throughout the country to gather material for the four-volume publication: *America's Centenarians: Reports of interviews with Social Security beneficiaries who have lived to 100*. The interviews followed a standard 22-question outline, and the material gathered by the field representatives was reproduced in *Amercia's Centenarians*, in most instances in the centenarians' own words. The above publication consists entirely of raw interview data, without analysis or interpretation. Because all information in S.S.A. records is confidential, only the stories of those who consented to the publication of this information have been included. Some centenarians, or their families, did not wish publicity because of delicate health or other personal reasons.

The interview items were age at time of interview, sex, place of birth, marital status, number of children, number of grandchildren, number of great-grandchildren, participation in armed services, earliest memory, most salient historical event remembered, most exciting event remembered, presidents remembers having voted for, nature of first job, nature of the work done most of life, thoughts about the Social Security program (in general), part of income played by Social Security checks, thoughts about the Social Security program (personal), current health, reason for longevity, church participation, current use of time, and future ambitions.[1]

There are two major limitations within the data. One limitation is the possible bias of the published centenarian interviews. It appears that the protocols that comprise the sample favor the healthier and more intact centenarians. The other limitation is the fact that the theoretical framework (developmental-field) was applied after the interviewing had been conducted. An interview schedule designed with a definite theoretical perspective would have yielded additional data of scientific interest. One important specific would have been a measure of the transaction between the interviewer and S, which is an essential level of analysis in developmental-field theory. Despite the methodological limitations of the data, the authors are grateful to the S.S.A. for permission to utilize their information.

B. CHARACTERISTICS OF THE SAMPLE

The dates of the interviews encompass the 1958–1962 period, with the majority of the interviews conducted during 1959–1960. More than half of the sample were alive at the time these reports were published. The following facts were true of the sample as of 1963. The sample reported here is comprised of 276 centenarians, 90 percent of whom were between 100 and 103 years of age when interviewed. Men predominate (70 percent). Of the Ss born in the United States, most come from the Midwest, 43 percent; 24 percent from the Northeast; 23 percent from the South and Southwest; and 10 percent from the Far West. An additional 34 or 12 percent of the total sample were born in Europe. Four-fifths of the sample were interviewed in a place different from their birth place.

With respect to marital status, 75 percent of the total sample were widowed, most having been married only once; some 20 percent of the widowed were multiple widows or widowers. Fourteen percent were married with their first or second spouse still alive. The divorce rate was very low, less than 1 percent. The number of single people was less than 3 percent. Only 14 percent had no children. Approximately three-fourths of the Ss had grandchildren and great-grandchildren. The overwhelming majority of the centenarians (90 percent) were able to dress and feed themselves, and maintain personal hygiene. Most of the centenarians (91 percent) reported some degree of current social contact. Eighteen percent remained active churchgoers.

C. RESULTS

1. TEMPORAL LOCATION OF MEMORIES

To explore the psychological structure of centenarians, the authors focused upon limited aspects of their time perspective. The guidelines of the present theoretical orientation are those of Kastenbaum's developmental-field approach (Kastenbaum, 1965b, 1966), which places strong emphasis on time perspective research.

One possible measure of the centenarian's perspective is the time period in life in which he locates his memories. The location in time of the available three recall items may not only delineate the degree of extension into the past, but also indicate which periods of life are most salient to the person, now at age 100.

The three memory recall items were (a) the Earliest Memory (EM), (b) the Most Salient Historical Event (MSH), and (c) the Most Exciting Event (ME). The categories used in locating the time periods for the three recall items were Early Childhood, ages 0 to 5; Childhood, ages 6 to 12; Adolescence, ages 13 to 19; Young Adult, 20 to 39; Middle Age, 40 to 70; and Old Age, over 70.

First, it is interesting to note that the percentage of centenarians responding varies considerably among the three recall items. The highest percentage of the sample responding is for the "Earliest Thing Remembered" (88 percent); 56 percent respond for the "Most Exciting Event," and 48 percent for the "Most Salient Historical Event" (see Table 32.1). The earliest events were recalled more often than the Most Exciting or the Most Salient Historical Event.

The analysis of the distribution of time periods for the three recall items excludes those Ss who either did not answer or did not localize the event in time. Thus, the percentages given for each of the time periods are based on differing sample sizes for the three recall events.

The distribution of time periods for the earliest memory (EM) is markedly skewed toward the early years of life. Early Childhood is most frequently cited for the time period of the EM (58 percent of sample, $N = 187$). Ninety-two percent of the EM fall within the first twelve years of life.

For the MSH event the distribution of time periods is multimodal with the greater percentage of historical events (49 percent, $N = 123$) falling within the first two time periods, 0 to 5 years (27 percent) and 6 to 12 years (22 percent). The second mode includes the time periods 20 to 39 years (23 percent) and 40 to 70 years (19 percent). The distribu-

TABLE 32.1   Centenarian Distribution of Time Periods for the Three Recall Items

Variable	Earliest Thing Remembered[a]			Most Salient Historical Event[b]			Most Exciting Event[c]		
	F	Per-cent	Per-cent*	F	Per-cent	Per-cent*	F	Per-cent	Per-cent*
. Not given	62	22		140	52		121	44	
2. Answered but not localized in time	27	10		10	4		37	13	
3. Old Age (over 70)	0	0	(0)	1	1	(1)	15	5	(13)
4. Middle Age (40–70)	1	0	(0)	24	9	(19)	21	8	(18)
5. Young Adult (20–39)	8	3	(4)	29	11	(23)	40	15	(34)
6. Adolescence (13–19)	8	3	(4)	7	3	(6)	16	6	(14)
7. Childhood (6–12)	62	22	(34)	28	12	(22)	16	6	(14)
8. Early Childhood (0–5)	108	40	(58)	34	12	(27)	10	4	(7)
Total	276	100	(100)	276	100	(100)	276	100	(100)

* Recomputed percentages based on categories 3 through 5; those memories not given or not localized in time are not included in this table.
[a] $N = 187$.
[b] $N = 123$.
[c] $N = 118$.

tion of time periods for the ME event is skewed toward the higher ages or time periods. The time period most frequently cited is Young Adult with 34 percent of the sample ($N = 118$). Thus, Early Childhood and Childhood times are most salient in terms of the centenarians' EM. Early Childhood, Young Adult, and Middle Age are most often cited in terms of the MSH event. For the ME event, the period of Young Adulthood is most salient.

As might have been expected, the EM was almost always located within the first few years of life (ages 0 to 12), with just a few cases in the Young Adult range and none in Middle Age. It is also understandable that the MSH was localized over a larger range, and with approximately the same frequency from time period to period. However, there was relatively little *a priori* basis for predicting the distribution of ME recollections. While the Young Adult period was most salient, as already noted, it is also worth bearing in mind that approximately one-sixth of these centenarians reported their ME as occurring in Old Age (beyond 70).

It is also interesting to note the percentage of Ss citing Adolescence as a salient period in their past experience among all the recall items. Only 4 percent of the Ss cited Adolescence as the time of the EM; 6 percent cite Adolescence as the time of the MSH; while 14 percent cite Adolescence as the time in which the ME occurred.

2. S's Relating to His Memories

In an attempt to measure the processes by which the centenarian locates himself within his own temporal experience, a set of coding categories was devised. These categories attempt to delineate the person's *relation* to the event recalled as distinct from the *content* of the event recalled. The relation to the event was dichotomized into (a) the degree of involvement the centenarians exhibit during the retelling of this memory and (b) the situational context of the memory.

The *engrossment-perspective* dimension is employed in coding the person's involvement in telling the event to the interviewer. If the person retells the event with vividness and great detail, then his relation to the event recalled is classified as engrossment. If the person seems to be retelling the event as an objective "thing" that he now sees as part of his past life, then S's relation to the event is classified as perspective. Engrossment in more general terms has been defined ". . . as complete psychological involvement in one unitary situation, while perspective may be regarded as simultaneous involvement at two or more points in the life-span. The quality of engrossment vanishes when we compare, judge, plan, or seek to explain" (Kastenbaum, 1965a, p. 7).

In classifying the *situational* context of the memory, the authors employed two dimensions. The *observer-participant* dimension considers the S's relation to the event when it took place—was he an active agent or passive recipient? The private-public dimension concerned the event itself. Was the event shared with others, or was it of a solitary nature?

The distribution of the centenarians' relations to the recall items reveals that the categories of perspective, participant, and public (as contrasted with engrossed, passive, and private) are most frequently encountered in the total sample. The engrossment relation is approximately equal for ME (32 percent, $N = 118$), and for EM (30 percent, $N = 187$), but only 18 percent ($N = 123$) for the MSH.

For EM, 45 percent of the Ss report a private event, 26 percent are private for the ME, and 4 percent are private for the MSH. Ss have greatest participation in the ME—69 percent, as compared with 53 percent for the EM and 12 percent for the MSH.

Next, let us consider the interaction between the S's relation to the recall items and the distribution of the time periods cited for each of the recall items. For both EM and MSH there is a greater proportion of engrossments for the earliest time periods cited for the recalled events. In other words, where a comparison is made between the dimension of engrossment versus perspective for the time period of the recollection, we find that the earlier the time period cited the higher the proportion of engrossments, and the later the time period in life cited for the event, the greater the proportion of perspectives (see Table 32.2).

Interestingly enough, Mr. Davis' notions about early childhood memories becoming clear, and memory being better for earlier events, are in harmony with the statistical findings. He appears to be correct also in believing that the earlier the age at which the episode is cited, the greater the degree of engrossment as contrasted with perspective.

3. UNREALIZED AMBITIONS

Future time perspective is a complex topic to investigate at any age level, requiring a broad range of carefully selected assessment procedures (Butler, 1964; Wohlford, 1964). With only one futurity item available for analysis, the authors can do no more than gain a few hints as to the nature and meaning of future perspective for centenarians. The item was: "Do you have any unrealized ambitions?" (the final item in the schedule). The present analysis is concerned not only with the response to this item per se, but also with its relationship to the set of recall items.

Approximately two-thirds of the Ss answered the future ambition inquiry. These 192 respondents were almost equally divided with respect to presence or absence of a future ambition. Each of the subgroups can be subdivided again according to their attitudes toward future ambitions.

Those who said they had no ambitions were about equally divided between centenarians who simply reported "no ambitions" and those who pointed out that they had satisfied their ambitions earlier in life and, thus, there was nothing further to be accomplished. Among those who reported future ambitions, again there was an approximately even split—52 percent expressed ambitions that they considered unrealizable in their lifetime, while 48 percent specified ambitions that they hoped to fulfill.

Those who seek improvements in material welfare for themselves, their family, or their immediate environment were more optimistic about their ambitions being fulfilled than were those who referred either to restorative or self-development ambitions. The former were seen as capable of fulfillment within the centenarian's lifetime twice as frequently as were those ambitions that concern restoration or self-development (by restoration is meant the repossession of some lost value, either in terms of health or ways of functioning; self-development includes betterment in the intellectual, religious, and ethical spheres). The future ambition that was seen most optimistically was that of reaching a certain age. Twelve of the 13 Ss whose future ambition was solely to reach a certain age were optimistic or hopeful of fulfillment.

TABLE 32.2 Interaction between Relation of Engrossment and Perspective and Distribution of Time Periods Cited (Early versus Late) for Three Recall Items

Earliest Memory[a]		Historical Memory[b]		Exciting Memory[c]	
Later years (13–70)	Earlier years (0–12)	Later years (13–70)	Earlier years (0–12)	Later years (13–70)	Earlier years (0–12)
Engrossment					
17	40	6	17	27	11
Perspective					
61	69	52	48	65	15

[a] $N = 187$; $\chi^2 = 3.98$, $df = 1$, $p < .05$.
[b] $N = 123$; $\chi^2 = 6.45$, $df = 1$, $p < .02$.
[c] $N = 118$; $\chi^2 = 1.53$, $df = 1$, $p < .05$.

4. RELATIONSHIP BETWEEN RECALL AND FUTURE AMBITIONS

The following analysis is an attempt to apply theoretical principles to the relationship between the past and future outlooks of the centenarian Ss, although the limited nature of the data makes this a rudimentary effort.

An important characteristic of the structure of any system is degree of *differentiation*. "In broad psychological terms differentiation refers to the complexity of a system's structure. A less differentiated system is in a relatively homogeneous structural state; a more differentiated system is in a relatively heterogeneous state. . . . Among the major characteristics of the functioning of a highly differentiated system is specialization. When used to describe an individual's psychological system, specialization means a degree of separation of psychological areas, as feeling from perceiving, thinking from acting. It means as well, specificity in manner of functioning within an area" (Witkin et al., 1962, p. 11). Thus, developmentally one can conceptualize the organization of centenarians' memories as a subsystem of a more general psychological system. Theoretically, the memory subsystem may vary from a minimal to an optimal degree of differentiation. One implication is that highly differentiated organizations imply a finer articulation, or separation, among past experiences. Parts of the "field" (of memories) are experienced as discrete and unique. A low level of differentiation implies a global or fused state of organization and functioning.

At the outset of this paper, the authors raised questions, such as "How does the centenarian integrate his extensive past experiences with his present moment of existence?" "How does he conceive of futurity?" and "Does the degree of differentiation and level of organization of the S's past experiences bear any relationship to his differentiation of future aspirations?" To assess the degree of differentiation of memories in the present sample, a scale was constructed based upon the number of significant memories recalled by each S. This corollary scale was dimensionalized by using as criterion recall of the three significant memories: Earliest Memory (EM), Most Salient Historical Event (MSH), and Most Exciting Event (ME). For each S, recall of all three memories was Level 4; the recall of any two memories Level 3, and so on. Level 1 represents the state of least differentiation as indicated by no recall. Inspection of Table 32.3 reveals that 34.4 percent of 95 Ss recalled two of the three memories (Level 3). Level 4 accounts for 30.8 percent ($N = 82$) of the sample, recalling all three memories. The third most frequent level of differentiation is Level 2, 22.1 percent ($N = 55$) of the sample. Only 12.7 percent ($N = 35$) of the sample recalled none of the three memories (Level 1).

From a developmental viewpoint one would hypothesize that the relative state of differentiation and integration of the S's memory subsystem organization of past experiences will bear a direct relationship to his conception of futurity. More specifically, the authors' expectation is that those Ss who recall all three memories will have a greater frequency of future ambitions than will those Ss who recall fewer than three memories.

In the reduced sample of 268 Ss, future ambitions were elicited from

TABLE 32.3   Corollary Scale of Differentiation of Memories
(N = 268)

Level	Number and kinds of items recalled	Total	Percent of sample	N
4	Recall all three	$f = 82$	30.8	82
3	Recall two events	$f = 96$	34.4	96
	Early-Exciting	$f = 52\ (54.7\%)$		
	Early-Historical	$f = 34\ (35.4\%)$		
	Exciting-Historical	$f = 10\ \ (9.9\%)$		
2	Recall one event	$f = 55$	22.1	55
	Early	$f = 41\ (77\%)$		
	Exciting	$f = 11\ (18\%)$		
	Historical	$f = 3\ \ \ (5\%)$		
1	No recall	$f = 35$	12.7	35

90 centenarians (33.5 percent), while 101 (38.3 percent) declared they had no ambitions and 77 (28.2 percent) did not respond to the question.

Next, Table 32.4 represents the frequency breakdown of the future ambition responses for each (corollary) level of differentiation of the recalled memories. In column one we notice that there is a linear increase in the percentage of Ss not answering (or unable to answer) the future ambition item as we proceed from Level 4 (greatest differentiation of memories) to Level 1 (least differentiation or no recall). Only 15.6 percent of the Ss at Level 4 did not respond; percentages for the remaining levels were, 23.4 percent Level 3, 28.6 percent Level 2, and 32.4 percent Level 1. Chi square (4 × 3) analysis of all cell frequencies in Table 32.4 yields a χ^2 value significant beyond the .01 confidence level.

As further analysis will deal with the differentiation of responses in Table 32.4, the entries in column one (no responses to the future ambition item, N = 77) and the entries in row four (Level 1 of memory differentiation scale, N = 10) have been excluded, reducing the total analyzable responses to N = 181. By excluding "no" responses which are

TABLE 32.4   Frequency of Future Ambition Responses across Levels of Differentiation

Level of differentiation	No response	No ambition	Yes ambition	N
4	12	26	44	82
3	18	45	33	96
2	22	22	11	55
1	25	8	2	35
Total	77	101	90	268

TABLE 32.5 χ^2 Table (3 × 2) of Future Ambition Responses for
Three Levels of Differentiation

Level of differentiation	No ambition	Yes ambition	N
4	26	44	70
3	45	33	78
2	22	11	33
Total	93	88	181

qualitatively different from differentiated responses (either in terms of number of memories or future ambitions), the authors restrict the heterogeneity of variance to differentiated responses only.

Thus, Table 32.5 lists the frequency of future ambition responses for three levels of memory differentiation. Chi square analysis (3 × 2) for the total table yields a significant χ^2 value of 9.88, $df = 2$, < .01. The relationship between differentiation of the past and the presence or absence of future ambitions ("yes" responses) obtains for the total table.

Within Table 32.5 comparisons were made between the following rows or levels; Level 4 and Level 3, $\chi^2 = 6.27$, $df = 1$, $p < .02$; Level 3 and Level 2, $\chi^2 = 0.79$ ($df = 1$); $p > .05$ (n.s.); and Level 4 and Level 2, $\chi^2 = 7.80$, $df = 1$, $p < .01$. Level 4 responses are distributed significantly different from Levels 3 and 2 with respect to the presence or absence of future ambitions. Level 3 is not significantly different from Level 2.

The above findings indicate that the corollary scale of differentiation of memories is reduced to a dichotomy. Successful recall or differentiation of all three memories significantly relates to or predicts the presence of future ambitions. We would expect the present results to obtain greater relationships of significance were one to sample a greater number of memories.

Methodologically, it is important to sample enough behavior—in this case, a minimum of three qualitative memory items. Partial sampling of the memory field may not discriminate among Ss who have varying degrees of differentiation of recall.

D. DISCUSSION

The corollary scale of differentiation (memories) is only a simple *index* of the process of differentiation. There is no intention to equate the number of recalled memories with the psychological process of differentiation. More systematic analyses can be achieved by examination of differences within the memory responses themselves, which is beyond the scope of this paper.

The present research is guided by a model of development that enables one to select out from a mass of data a set of relationships that pertain to time perspective. Analysis of time perspective requires a determination of where the individual locates himself in his own "temporal life-space" or "life-gestalt," and how he is using other points of reference to organize a perspective.

One might construct, out of limited data, a picture of a certain group of centenarians who have a rich reservoir of recollections dated very early in life (5–12 years). These people are vividly engrossed in these experiences in a manner that suggests a well maintained and preserved "pastness" dimension. They appear motivated to scan their remote past experiences. These people also have a rich set of recollections throughout their total lifespan. For those centenarians whose early memories are bleached out or who do not enjoy reaching into their past, it is doubtful that they possess a full sense of time perspective. Such persons perhaps are more affected by the pressures of the immediate environment.

An excellent example of using earliest memories, and the reconstruction of past life as research data on current and covert states of the aged, is that of Tobin (1965). Considering early reminiscence as the overt reflection of ongoing adaptations to current stress, Tobin tested the hypothesis that institutionalization would evoke a restructuring of the earliest memory. "That is, the implicit hypothesis is that a *current* stress would evoke a *change* in the content and elaboration of the earliest memory" (Tobin, 1965, p. 2). It was found that, after institutionalization (two months), those Ss showed significantly more shifts toward themes of loss in their memories than did a matched community sample.

In an attempt to explain increased reminiscence in the aged, Butler postulates the "life review." Butler suggests that the basis of the life review is the sense of impending death and its function is the reintegration of personality and preparedness for death. The engrossment in early time periods found for the present sample of centenarians is compatible with observations Butler has reported: "Older people report the revival of the sounds, taste and smells of early life. ('I can feel a spring afternoon of my boyhood')" (Butler, 1963, p. 525). The finding that the earliest memory (EM) is recalled most frequently by the centenarians in this study supports Butler's comment that ". . . older people may be preoccupied at various times by particular periods of their life and not the whole of it" (Butler, 1964, p 88)

Other investigators have theorized about the nature and function of time perspective among the aged as a means of maintaining self-continuity throughout the lifespan. Kastenbaum (1965b) has suggested several functions of a time perspective for the individual:

> If a time perspective can serve to accommodate affect and, therefore, afford the individual an alternative to quick response and impulsive discharge of tension, then the person who is disinclined to consider the past will tend to be deprived of this coping procedure.
> A previous study suggested that most adolescents in a normal sample had an aversive, blocking-out reaction toward their personal past. The implication might be that one of the developmental tasks which still lies ahead for many adolescents is the ability to take pastness into account in elaborating a cohesive view of life. By contrast, the developmental task of the aged person might be to find a way of maintaining the future-scanning function. Neither the adolescent who ignores the past, nor the aged who ignores the future could develop or maintain a genuine time perspective, according to the present line of reasoning (Kastenbaum, 1965b, p. 199).

It might be suggested here that one of the developmental tasks of the centenarian is to find a way of maintaining the past-scanning function and linking it with his other modalities. The vivid early childhood memories and rich recollections of his total past offer the centenarian a means for creating a perspective on his present and future. The reservoir of memories helps sustain his present moment of existence and also aids him in creating a perspective in preparation for his eventual death.

E. SUMMARY

An inquiry was made into some aspects of the past and future outlooks of 276 centenarians, based upon interview protocols provided by the Social Security Administration. Developmental-field concepts were applied to the analysis of items pertaining to Earliest Memory, Most Exciting Event, and Most Salient Historical Event, as well as Future Ambitions. Earliest Memory was the recall item most frequently answered. The centenarians tended to put their memories into perspective rather than show engrossment in their recollections. There was relatively more engrossment in the telling of memories that were dated in the remote as compared with the recent past. Those centenarians who were able to offer responses for all the memory items more frequently stated future ambitions than did their peers who had less command over the past. Implications of the study are discussed with particular reference to the functions of time perspective in the later years of life.

NOTE

[1] Some of the items were analyzed, but not included in the present report. In several cases the same general item was analyzed in terms of multiple coding categories. Copies of the detailed coding categories and information about responses not reported here may be obtained from the authors.

REFERENCES

Butler, R. N. "Recall in retrospection," *J. Amer. Geriat. Soc.* 2: 523–529, 1963.

Butler, R. N. "The life review: An interpretation of reminiscence in the aged." In: Kastenbaum, R. (ed.), *New Thoughts on Old Age.* New York: Springer, pp. 265–280, 1964.

Kastenbaum, R. "Engrossment and perspective in later life: A developmental-field approach." In: Kastenbaum, R. (ed.), *Contributions to the Psychobiology of Aging.* New York: Springer, pp. 3–18, 1965. (a)

Kastenbaum, R. "The direction of time perspective. I: The influence of affective set," *J. Gen. Psychol.* 73: 189–201, 1965. (b)

Kastenbaum, R. "Developmental-field theory and the aged person's inner experience," *Gerontologist* 6: 10–13, 1966.

Tobin, S. S. *The effect of institutionalizations on the earliest memory of the very old.* Paper read at Eighteenth Annual Meeting of the Gerontological Society, Los Angeles, California, 1965.

Witkin, H. A., R. B. Dyk, H. F. Faterson, D. R. Goodenough, and S. A. Karp. *Psychological Differentiation.* New York: Wiley, 1962.

Wohlford, P. *An investigation into some determinants of personal time.* Unpublished dissertation, Duke University, Durham, North Carolina, 1964.

Concluding Note

Not many of us will reach the age of 100, as have the persons discussed in the last article in this book. Nevertheless, the memories and reflections of these very aged people seem a fitting closure to this book on lifespan developmental psychology. The papers presented here are, of course, only samples of the currently available body of literature demonstrating concern for psychological aspects of the human lifespan. The editors hope that these articles will stimulate the reader to broaden his view of human development to its fullest extent. Perhaps at this point in time we may be ready to adopt better research tactics and new perspectives that will enable us to improve our understanding of the patterns of continuity and change that are revealed as the person moves through the course of his life.

Author Index

Abarbarel, A., 379
Aberle, D. F., 241, 318, 331
Adorno, T. W., 235, 242, 321, 331
Alexander, F. G., 65, 70
Alfieri, 104
Allen, G., 118, 120
Allen, L., 11, 14, 17
Allport, G., 309
Altshuler, K. Z., 379
Amatruda, C. S., 169
Ames, L. B., 162, 168, 169
Ames, R., 24, 25, 27
Anderson, J. E., 3, 11, 16, 30–41, 174, 188
Angell, J. R., 66, 70
Ankus, M. W., 136, 147–157
Appenzeller, O., 111, 112
Archer, E. J., 219, 221
Archimedes, 51
Aristophanes, 369
Armstrong, E. B., 378
Aronfreed, J., 54, 59
Ashby, W. R., 36, 41
August, H. E., 380, 389
Ausubel, D. P., 370, 377
Axelrod, S., 147, 156

Babkin, B. P., 192, 295
Bach, J. S., 165
Back, K. W., 403, 405
Bacon, M. K., 241, 320, 331
Baer, D. M., 73, 78, 84
Bajema, C. J., 215, 221
Baker, C. T., 8, 17
Baldwin, A. L., 44, 59, 81
Bales, R. F., 336–337, 346
Baller, W. R., 205, 255–272, 354
Baltes, P. B., 206
Balzac, 101
Bancroft, G., 291, 292, 295

Bandura, A., 306, 310
Barad, M., 379
Barke, J., 377
Barker, A., 201
Barker, R. G., 10, 16, 184, 188, 201, 346
Barnacle, C. H., 380, 389
Baroff, G. S., 118, 120
Barry, H., 320, 331
Basowitz, H., 147, 156
Bateson, P. P. G., 53, 59
Bauer, R., 166, 241
Bayley, N., 3, 5–17, 18, 19, 21, 22, 23, 27,
    29, 44, 60, 274, 281, 353
Beam, L. E., 372, 376, 378
Bearison, D. J., 144, 146
Becker, W. C., 44, 59, 334, 345
Beier, H., 240
Beigel, H. G., 378
Bell, E. T., 285, 295
Bell, R. Q., 298, 332–347
Bender, I. E., 277, 281
Benedek, T., 380, 389
Bennet, E., 113
Bernard, J., 379
Bernhardt, S., 104
Bernstein, B., 241
Bijou, S. W., 4, 71–85, 309, 310
Binet, A., 203, 303
Birren, J. E., 28, 110, 113, 156, 177, 181,
    183, 189, 201, 366, 379
Bleibtreu, H. K., 112
Bliss, E. L., 133
Block, J., 22, 26, 27, 28
Bloom, 234
Boardman, W. K., 393, 407
Bonaparte, M., 405
Bondy, E., 17
Boren, H. G., 113
Boring, E. G., 70
Botwinick, J., 181, 188, 379

423

Subject Index